THE LIFE OF BIRDS
VOLUME I

The Life of Birds

Volume I

JEAN DORST

Translated by

I. C. J. Galbraith

COLUMBIA UNIVERSITY PRESS
New York 1974

Reprinted 1982

Published in Great Britain in 1974 by Weidenfeld and Nicolson
Printed in Great Britain

Library of Congress Cataloging in Publication Data

Dorst, Jean, 1924-
The life of birds

French translation has title: La vie des oiseaux.
Includes bibliographies.
1. Birds 2. Birds- -Behavior I. Title
QL673.D58 1974 598.2'5 74-8212
ISBN 0-231-03909-3

To Eliane

Contents

Acknowledgements

I would like to express my deep gratitude to all who helped me in writing a book dealing with such a large number of subjects. Several sets of rough drafts were circulated for comments and I am particularly grateful to Dr Christian Jouanin and Dr Jean Prévost for their useful suggestions. Others very generously provided me with unpublished information. Miss Odile Jachiet coped with typing and retyping the manuscript. I would also like to express my warmest appreciation of I. C. J. Galbraith, who, in translating this book, rendered very accurately the original text and made useful comments, and to Polly Curds who then edited the text. Finally, I would like to evoke the memory of my late friend Richard Carrington, who suggested this book and gave me the benefit of his advice and wide experience. It is very sad that he cannot witness the publication of a book inspired by his enthusiasm and friendship.

List of Plates

The author and publisher would like to thank the following for providing photographs for this volume: Jacana, Paris: pl. 1, Milwaukee; pl. 2, Varin 1; pl. 3, 5, 6, Visage; pl. 4, Fritz; pl. 7, Brosset; pl. 10, Renard; pl. 11a, 11b, Tollu; pl. 15, Robert; pl. 16, A. Fatras; pl. 17, Bel and Vienne.
L. Ricciarini, Milan: pl. 9, R. Longo; pl. 8, A. P. Rossi; pl. 12, S. Prato.
A. Fatras, Paris; pl. 13, 14.

Chapter 1

Introduction

OF all animals birds have attracted man's keenest interest, because many of them are so easily seen in the wild – in contrast to most mammals, many of which are nocturnal and very secretive. The colours and lively varied behaviour of birds have stimulated interest and curiosity from the earliest times, and their scientific study also dates from far back in the past. Buffon and other famous eighteenth-century naturalists wrote classic works on birds. During the next century many expeditions, especially to tropical regions, provided the opportunity to establish collections and compile detailed avifaunal lists. Since the beginning of the present century, the systematics of birds has reached a much more advanced state than that of any other group of animals.

Though sometimes still considered as an amateur science, ornithology is also an especially fruitful branch of biology, since birds have played an important part in many fundamental scientific discoveries. They are preferred material for certain specialized research, while many disciplines have benefited widely from conclusions arising out of their study. Embryology, endocrinology, ecology, ethology and the study of migration and of speciation have drawn largely upon work on birds. These results have then been successfully applied to other zoological groups, and have been accepted as general biological rules.

The number of works on birds has grown at an enormously increasing rate during the last few decades, and at present about 2,500 articles, reports and books on them are published every year. This one can deal only with certain aspects of bird life, and deliberately concentrates on the adaptations which birds have made to the various environments they have colonized across the globe. It is thus, most importantly, an essay on the ecology of birds, written not so much for the specialist as for the well-informed public.

If we consult the definitions of birds given in classic manuals, we find that they are vertebrates whose fore-limbs have been transformed into wings, which enable them to fly from place to place; whose bodies are covered in feathers; whose toothless jaws are cased in horny bills; whose internal temperature is constant and high; and which reproduce by oviparity. These are thus highly evolved animals whose principal adaptation, and the one which chiefly differentiates them from other vertebrates, is the conquest of aerial space. It is true that the bats among mammals have also succeeded in flying, and that during evolution various groups of reptiles have made attempts without lasting success. However, in their organization these animals show none of the related adaptations, necessary for a vertebrate to fulfil the very diverse requirements for complete evolutionary success.

We shall first consider the essential characteristics of birds as a whole. Their morphological, anatomical and biological peculiarities are very closely linked with the power of flight, whose demands have stamped them with common characters so that the class is remarkably homogeneous. We shall go on to see how birds are adapted to the most diverse natural environments to be found on earth, exploiting resources which no other animal can, reinforced by well-regulated homoiothermy which protects them from extremes of environmental temperature, great dietary range, and sense organs and a mode of reproduction which are well adapted to their ways of life. Of all vertebrates, birds best resist extreme conditions, and every habitat supports an avian community with precise and well defined ecological characteristics. Thus birds are extremely adaptable, despite their structural uniformity. In choosing among the variety of habitats available for study, we shall pay special attention to adaptations which underline ecological flexibility within the fixed framework of the class.

Treatment here is necessarily selective, and many facts of equal importance are excluded for want of space. This is especially regrettable because these ecological adaptations deserve to be set out precisely and in detail. As always where almost every statement is true only for certain species, generalizations are bound to introduce distortions.

Despite the great number of recent studies on birds, huge areas of knowledge are still very incomplete. The habits of many species, especially in the tropics, are unknown. Quantitative ecology is in its infancy, and this prevents us from determining accurately the position of birds in their communities. Although recent studies have thrown a new light on migration and its causes, the ways in which birds orient themselves during their tremendous journeys remain hypothetical. Even the evolution of the class as a whole deserves to be precisely determined by more detailed studies.

The last chapters of the book consider the connections between birds and man. We have successfully taken advantage of many species, domesticating some and exploiting wild populations in various ways, putting to use mainly empirical knowledge of the ecological and ethological characteristics discussed here. Furthermore, birds play an important part in habitats which have been transformed by man. This presents a final problem, which from various points of view is the most important: the place of birds in the modern world which we have so profoundly modified. Over a long time, and especially during the last 300 years or so, hunting, real vandalism and the destruction of habitats have reduced many species to rarity. Some which were once abundant have gone for ever, while others survive only as small populations, narrowly dependent on their habitats and at the mercy of the smallest ecological disturbance. Extermination and the reduction of stocks have been accelerating for several centuries, and man's relations with nature as a whole are now at a critical stage. Birds are among the worst threatened, or at any rate the most spectacularly threatened, species. There can be no doubt of our moral duty to preserve them all, as part of the world we have inherited. Furthermore, conservation is in our material interest, since birds with all other living things contribute to the balance of the biosphere, interacting under natural laws which we cannot circumvent. Thus there is ample reason for us to respect nature, and to protect not only birds but the whole life of our planet.

Chapter 2

The Conquest of Aerial Space

THE 8,600 or so existing bird species vary greatly in their gait, outline, form of the bill and feet, coloration, habits and size. The ostrich, the largest surviving bird, measures eight feet in height and weighs almost three hundred and fifty pounds, while certain hummingbirds are as little as two inches long and weigh only a tenth of an ounce. Nevertheless, this range of variation is small compared with that of the mammals. The blue whale weighs up to nearly a hundred and ninety tons, and the elephant, the largest land animal, seven tons, whereas the pygmy shrew weighs only a tenth of an ounce.

Birds' adaptations to different ways of life are equally varied. Penguins and ratites are incapable of flight, but the penguins are well adapted to swimming and the ratites to running. Among flying birds, the ability to fly fast and far, and different styles of movement through the air, are very unevenly distributed. However, birds do not show the extreme evolutionary divergence in locomotion seen in mammals. Adaptation to marine life is less fully developed among birds than among mammals. Even the most pelagic birds, such as the petrels and albatrosses, have to come to land for nesting, whereas whales live permanently in the water which it would be death for them to leave. Nor has any bird become truly subterranean like a vole or a mole, though some do shelter in tunnels which they dig themselves.

Despite extensive variations birds present a remarkable structural unity, all being built on the same general plan. This unity in diversity is a result of their common evolution as flying animals. Their most important features are adaptations to the conquest of the air. Their present success, following an explosive radiation in the Tertiary, can be explained in terms of this fundamental adaptation. By annual migrations they have been able to exploit important resources, despite great fluctuations in physical conditions and the availability of food. Every aspect of their biology, and especially their ways of obtaining food and raising their young, is profoundly affected by their ability to move rapidly in all three planes of space.

The acquisition of such highly-specialized powers of locomotion has demanded modifications to the general vertebrate plan. All other attempts by vertebrates to master aerial space, such as by the flying reptiles of the Secondary and Tertiary epochs, have miscarried and left no survivors.

The first essential property is lightness. Every anatomical and physiological feature tending towards useless weight has been suppressed in birds – for example a chewing structure in the head; the intake of low-energy foods which take a long time to be digested; the accumulation of food reserves; and viviparous reproduction. Every part, even the bones, is lightened to the limit. Thus the specific gravity of a duck, a 'heavy' bird, is only 0·6, whereas that of most vertebrates is about 1·0.

The second property is streamlining. Birds must be able to slip through the air offering the least possible resistance. Every projection, is suppressed. The legs can be retracted into the plumage, like the undercarriage of a modern aeroplane, or stretched out horizontally. The head, neck and body are shaped to part the air and ensure a smooth airflow. The muscular and visceral masses are concentrated near the centre of gravity of the whole animal, while the propulsive organs are lightened so as to reduce the energy wasted on inertia when they move.

The third property is strength. A bird travels through a fluid which provides no effective buoyancy. A fish needs to use its muscles only for propulsion, since its weight is balanced by that of the water in which it floats without effort. A bird on the other hand must work, not only to move forward but to keep itself airborne. It must therefore have strong and efficient muscles attached to a rigid skeleton. The rigidity of the skeleton, the strength of the musculature, and the lightness of the flying surfaces made up of feathers, combine to satisfy this demand.

Most quadrupeds, whether reptiles or mammals, move on all four limbs, so that they are supported in the same way during movement as when at rest. In contrast a bird is supported in flight by its wings, and hence by its thoracic cage, whereas when resting support comes from the legs to the pelvic girdle. The distribution of weight is different in the two configurations, and the layout of the skeletal and muscular support must be able to meet both situations.

The last property is the possession of an 'engine' capable of powering the mechanism. This must work at high efficiency in order to convert energy without introducing useless weight. The circulatory, respiratory, and digestive systems of birds are remarkably efficient, with a high performance, and a bird's metabolic rate is always very high.

A bird's complex of mechanisms is adapted to its power of flight, and the conquest of aerial space has been bought at the price of very far-reaching modifications.

The skeleton

The avian skeleton, though it shows undeniable reptilian features, is strongly marked by its adaptations to flight.

The axial skeleton consists of a variable number of vertebrae, from thirty-nine in passerines to sixty-three in swans, and this variability arises mainly from very different lengths of the neck. The number is more or less proportional to the length and mobility of the neck, varying between eleven in certain parrots to twenty-five in swans, with fourteen or fifteen as the commonest values.

There are great differences between the various parts of the spinal column. The neck is built up of complex vertebrae which move freely on one another, thanks to articular surfaces which allow very large movements both in the sagittal plane and in rotation. Many birds can thus easily turn their heads to the back and round in a complete circle. This arrangement compensates for the lack of mobility of their eyes, which are more or less fixed in their orbits, and is also adapted to snatching food as in herons, cormorants and especially darters.

In contrast, the dorsal or thoracic vertebrae show a marked tendency to lose their mobility and form a rigid system, by the development of strong intervertebral ligaments which become more or less ossified. Sometimes, as in grebes, currasows, cranes, sand-grouse and pigeons, these vertebrae even fuse to form a single dorsal bone separated from the pelvic bone by from one to three free vertebrae. They thus form a strong dorsal axis for the attachment of the ribs.

The lumbar region is still more rigid, since the ten to twenty-three sacral vertebrae are completely fused to one another at their vertebral bodies, neural spines and transverse processes to form a highly-developed synsacrum, which occasionally incorporates up to six rib-bearing thoracic vertebrae. This sacrum is firmly joined to the pelvic girdle.

Behind this are the caudal vertebrae, of which the first four to nine are free while up to ten fuse to form the pygostyle which supports the rectrices. Modern birds differ in this respect from their reptilian ancestors and from the Jurassic birds (*Archaeopteryx*,) which still had long jointed tails. Only the ratites and tinamous retain the primitive feature of separate caudal vertebrae.

The skull, which is of great interest in the study of the evolution and phylogeny of birds, is profoundly modified, partly by adaptations towards flight. Fully ossified and made up of bones fused together it forms a cranial

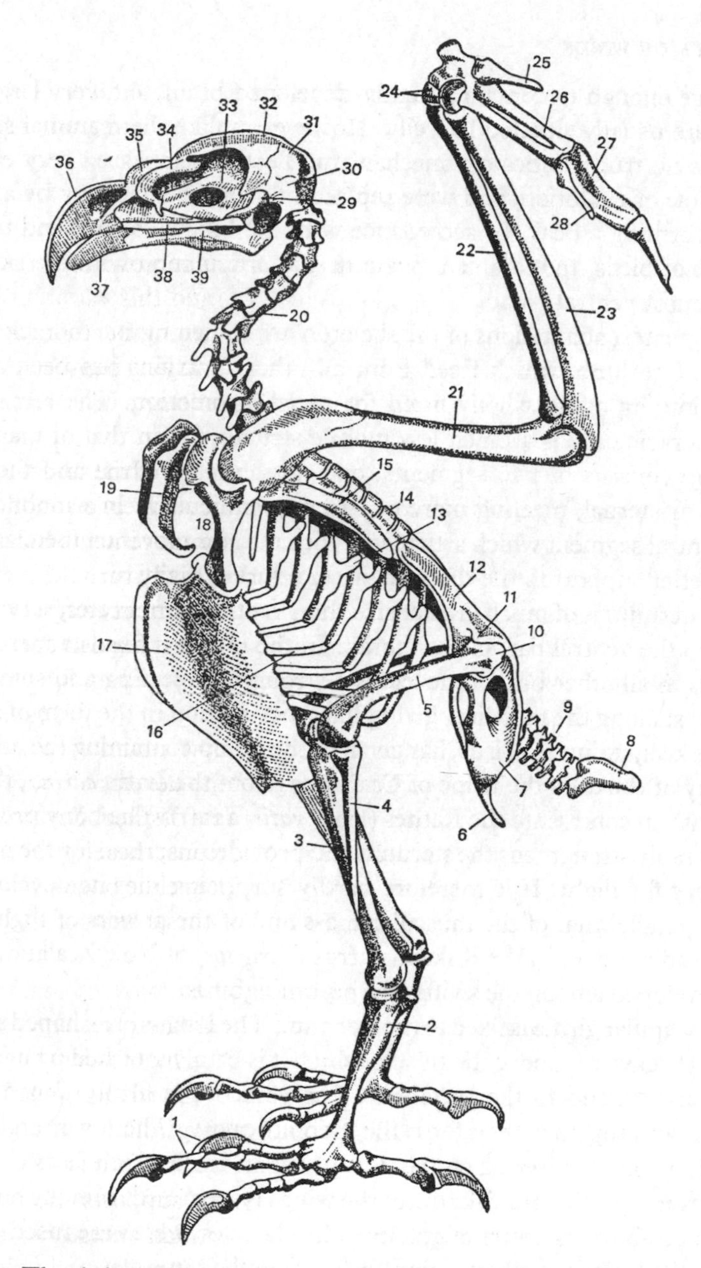

Figure 1. The skeleton of a raptor. 1. toes. 2. tarsus. 3. tibia. 4. fibula. 5. femur. 6. pubis. 7. ischia. 8. pygostyle. 9. caudal vertebrae. 10. trochanter. 11. antitrochanter. 12. ilium. 13. uncinate process. 14. thoracic vertebrae. 15. scapula. 16. sternal rib. 17. sternum. 18. coracoid. 19. clavicle. 20. cervical vertebrae. 21. humerus. 22. radius. 23. ulna. 24. carpal bones. 25 first digit. 26. metacarpal. 27. second digit. 28. third digit. 29. quadrate. 30. squamosal. 31. parietal. 32. frontal. 33. orbit. 34. lachrymal. 35. nasal. 36. upper mandible. 37. lower mandible. 38. jugal. 39. articular.

box large enough to contain a highly-developed brain, and very large eyes which are usually situated laterally. However, unlike the mammal skull, it contains no true masticatory mechanism. Teeth disappeared very early in the course of evolution, and were replaced during the Tertiary by a horny sheath forming a beak. In accordance with the bipedal stance and upright carriage of birds, the foramen magnum and articulation with the skull are in an almost ventral plane.

The greatest adaptations of the skeleton are shown by the thoracic girdle and the fore-limb, which freed from all other functions has been able to evolve into an organ wholly fitted for aerial locomotion. The remarkably rigid thoracic cage is a much less mobile structure than that of mammals. Each rib consists of two segments, one dorsal or vertebral and the other ventral or sternal, meeting more or less perpendicularly in a mobile joint. The ventral segment which articulates with the sternum is ossified, and this gives better support to the thoracic viscera during flight.

The sternum is of much greater size than in other vertebrates, serving as a shield to the ventral part of the thorax. In the ratites it forms a mere plate, whereas in all other birds it develops a strongly-projecting and sometimes huge crest along the mid-line, giving it a cross-section in the form of a letter T. This keel, unique to birds, has given to the group containing the immense majority of the class the name of Carinates (from the Latin *carina*, the keel of a boat), in contrast to the Ratites (from *ratis*, a raft). The bony projection considerably strengthens the sternum and provides insertion for the muscles necessary for flight. It is therefore hardly surprising that its development should parallel that of the muscular mass and of the powers of flight. The keel is rudimentary in the Kakapo parrot (*Strigops*) of New Zealand, and is best developed among the swifts and hummingbirds.

The scapular girdle is fixed to the sternum. The long sabre-shaped scapula extends backwards above the ribs, to which it is firmly attached by ligaments which also extend to the vertebrae. The coracoid is firmly joined to the scapula, and together they form the glenoid cavity. The lower end of the coracoid is rigidly inserted on the sternum at an angle which faces outwards and forwards, so that the fulcrum of the wings is separated from the body and positioned above its centre of gravity. The clavicles, which are fused distally to form the furcula in all but a few birds, act in the same way.

To this girdle are jointed the wings, formed mainly of the upper arm and forearm. Unlike those of pterodactyls and bats, the fingers form only a relatively small though important part of the wing. The humerus is rather short, with a greatly enlarged articular region fitting the glenoid cavity, and carries strong crests for firm muscular insertions. As in all vertebrates, the forearm is made up of radius and ulna. At its far end is the profoundly-modified articulation of the hand. Instead of flexing in the dorso-palmar sense, this works more or less laterally so that the angle of the hand to the forearm can be opened and closed, allowing those parts of the wing supported by the hand to fold like a fan towards the forearm when the wing is at rest. This modification is brought about by the reduction and profound alteration of the bones of the wrist. The hand too is greatly changed, and the number of fingers is characteristcially reduced. The carpal and metacarpal bones are partly fused together. The fingers are reduced to three, of which the thumb is minute while the two others, better developed, form a hand of flattened segments which, though rudimentary, is essential to flight.

The proportions of the various sections of the wing relative to one another vary widely according to the type of bird. In some strong and rapid flyers like the swifts and hummingbirds, the humerus is short and the forearm and hand longer. Others, and especially the albatrosses, show the opposite arrangement. All the segments tend to lengthen in gliding birds, and to shorten in those with rapid wing-beats.

The joints between the sections of the wing are highly specialized. The articulation of the humerus in the glenoid cavity is very free, so that the bone can move in any direction. On the other hand the elbow-joint works only in the plane more or less parallel to the surface of the wing, which opens like a pair of compasses, and the same is true of the carpal joint. Thus the wing can fold on itself only in a single plane, but can pivot as a whole in every direction. This arrangement makes the whole wing very mobile, yet to some extent clamps the various sections together to withstand the thrust of air during flight.

The skeleton is thus wholly and profoundly modified for flight. Despite its strength it is incredibly light, chiefly because the bones are hollow, penetrated by diverticula of the respiratory sacs connected to the large bronchi (see Chapter 6), during embryonic development or (as with the humerus) after hatching. The bones are thus full of air which is in communication with the respiratory sacs. The extent of this pneumatization varies with the kind of bird – though slight among small birds and certain aquatic ones (penguins and gulls), it is considerable among large birds capable of sustained flight such as the raptors and albatrosses. There is no direct relation between degree of pneumatization and flying ability, though they may be linked through more subtle correlations.

Pneumatization gives the whole skeleton greater strength for the same mass of bone, allowing a considerable economy in weight. In many birds the spaces are braced by bony trabeculae, whose highly elaborate arrangement recalls that perfected for the stringers of aeroplanes. In both cases the need is for maximal stiffness with minimal weight, and it is interesting that birds have solved the problem in much the same way as the aeronautical engineers.

The musculature

In order to pass through the air rapidly, economically and persistently, the body must be formed according to the laws of aerodynamics, and its fore-and-aft shape must be held almost constant during flight despite respiratory and digestive movements. The middle and hind parts of the spine form a rigid dorsal axis, so that a musculature of strong extensors is not needed to keep the body extended. The ribs are capable only of very restricted respiratory movements, which also implies the absence of a complex musculature. The axial musculature, especially the dorsal part, is therefore greatly reduced in birds.

The chief muscles of the head are those which act on the lower jaw. Since in most birds this does not act as a true grinder, these muscles are greatly reduced. The numerous vertebral apophyses, sometimes even bearing the rudiments of ribs, allow the insertion of muscles in multiple fasciae. This arrangement gives the head and neck a flexibility and a mobility in all planes of space which are unequalled among other vertebrates. At the same time it gives the strength and speed of action needed by many birds for procuring their food – for example by herons which spear their prey (and in flight fold their long necks, in contrast to storks and cranes which stretch theirs out in front).

The musculature which moves the limbs, and especially the fore-limb, is remarkably well developed. The arrangement of the muscles and the disposition of muscular masses are highly characteristic. They must be streamlined for travelling through the air and their weight must be concentrated near the centre of gravity, while large movements of heavy masses could waste energy in overcoming inertia. To satisfy these requirements, the muscular masses have been transferred towards the centre of the body, and send out towards the periphery and the extremities only slender muscles and sinews, of reduced weight and volume. Thus birds' muscles have a very characteristic conical shape.

There are about fifty well-developed muscles which move the wings and are inserted on the sternum and its appropriately-shaped keel. Three are especially important in enabling the spread wing to beat the air.

The first and most important is the *pectoralis*, which is broadly inserted on the sternum and keel at one end and on a strong crest of the humerus at

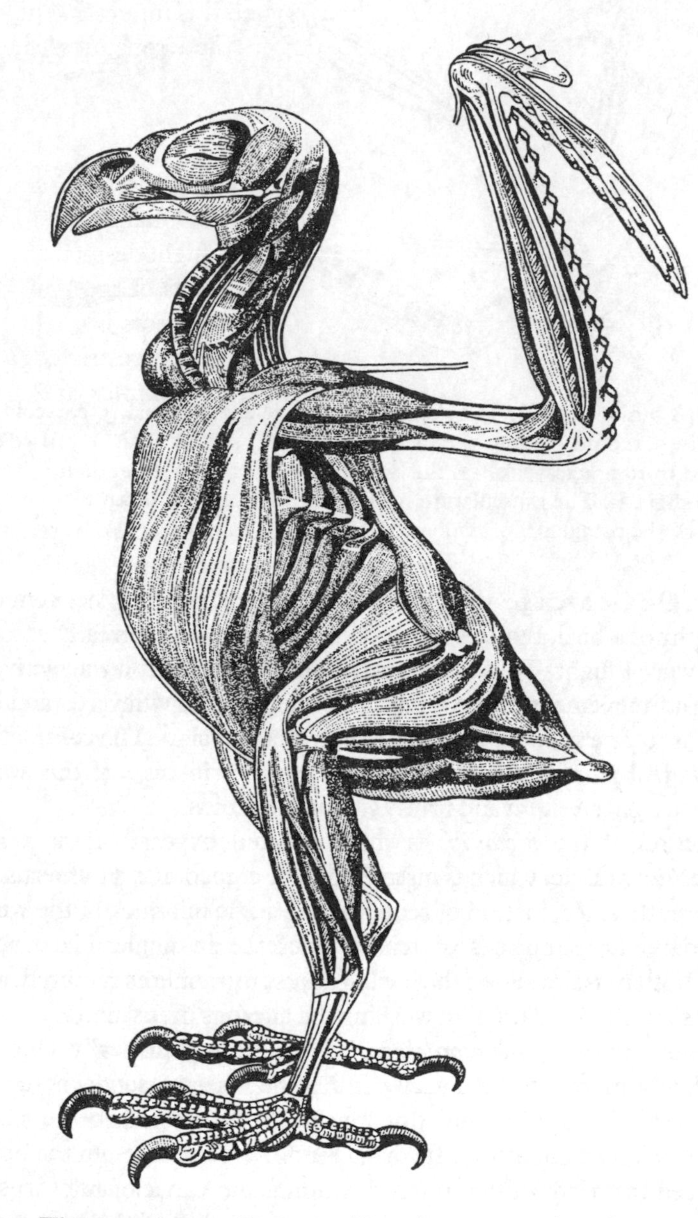

Figure 2. The musculature of a raptor. Note the concentration of muscles near the trunk, by their attenuation in the distal limb segments.

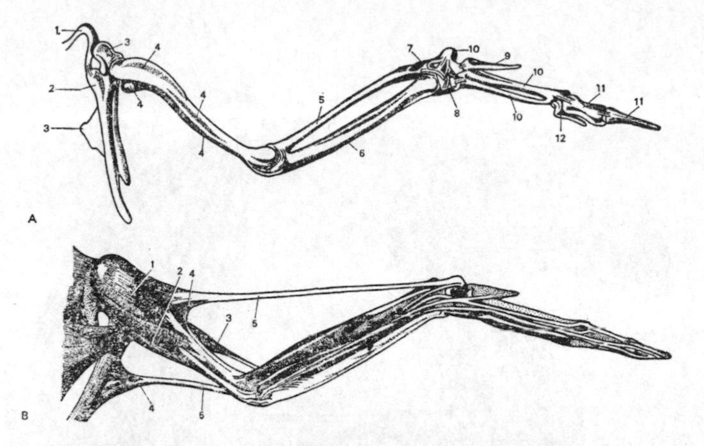

Figure 3. A bird's wing. A. The skeleton. 1. clavicle. 2. scapula. 3. coracoid. 4. humerus. 5. radius. 6. ulna. 7. radiale. 8. ulnare. 9. second digit. 10. fused first. second and third metacarpals. 11. the two phalanges of the second digit. 12. fourth digit. B. The musculature. 1. deltoid. 2. triceps. 3. biceps. 4. tensors of the patagium. 5. tendons of the patagial tensors.

the other. On the average, this pair of muscles make up 15.5 per cent of the total weight of a bird, but this common proportion varies greatly according to the power of flight. These strongest of all the muscles play an active part in the rapid movement of the wings, drawing them downwards and backwards so as to give support as well as forward propulsion. The contraction of such powerful paired muscles would flatten the rib-cage, if this was not braced by a rigid sternum and firmly spread coracoids.

The action of the *pectoralis* is supplemented by that of the *posterior coracobrachial* muscle, which is inserted on the coracoid and humerus. This acts differently since, instead of reducing the angle of attack of the wing by drawing down its leading edge, it tends to increase this angle. The concerted action of both muscles allows the bird to adjust the angle as required, and to control its own flight like a pilot working the ailerons of his aircraft.

The wing is raised by the *supracoracoid*, acting antagonistically. One might expect such a muscle to lie dorsally and to be inserted somewhere on the thoracic part of the spine, but this muscle is transferred to the sternum where it is inserted anteriorly. Its distal tendon passes through the opening left between the heads of the coracoid, scapula and clavicle, and turns back like a rope over a pulley to insert on the upper side of the humerus. Its contraction thus raises the wing, through what in mechanics would be called a transmission. This muscle is far smaller than the pectoralis. It is usually one-tenth the volume of the latter and sometimes even less, but in humming-

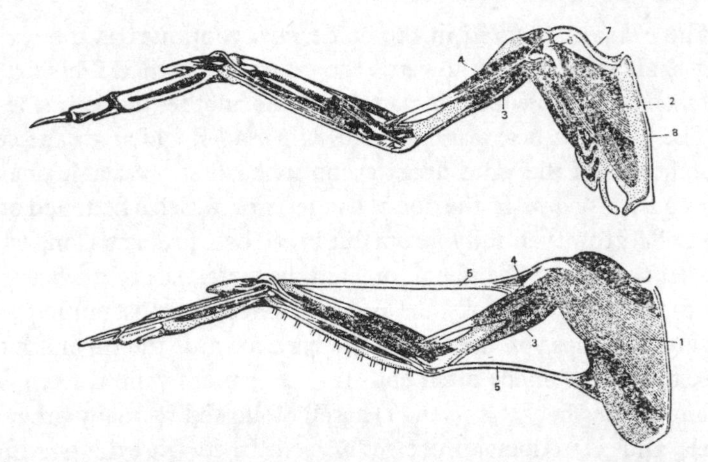

Figure 4. The wing musculature of a domestic goose *Anser*. 1. pectoralis major. 2. pectoralis minor. 3. triceps. 4. biceps. 5. tendons of the patagial tensors. 6. coracoid. 7. clavicle (part of the wishbone). 8. sternum.

birds on the other hand, it is as much as three-fifths the size of the pectoralis, since with their unique mode of flight these birds need to lift their wings as rapidly as they bring them down. The supracoracoid is usually well-developed in birds with a fast take-off.

These muscles are complemented by others, situated in the scapular and dorso-humeral areas, which form the controlling organs of the wing. Others in the limb itself move the various segments and produce the warping which is so important in controlling flight. The front limb of a bird is thus actuated by a highly-developed musculature, whose homology with the muscles of other vertebrates is difficult to establish.

The feathers

As a homoiothermic vertebrate, a bird needs an insulating layer to protect it from heat-losses; as a social animal, it needs visual signals; and as a flying animal, it needs aerodynamic supporting surfaces. All these needs are satisfied by highly-specialized developments from the skin – the feathers. Their structure and appearance alone are sufficient to characterize this class of vertebrates. They are more complex than hairs, which serve almost solely to protect mammals from cooling. A feather, formed from the more or less mineralized product of proteins, keratin, makes its first appearance as a dermal papilla. Once formed it is wholly separate from the bird's body, since unlike a hair it is a lifeless structure which is merely anchored in the skin.

Except for certain ornamental plumes, all feathers are built on the same plan. The *calamus*, planted in the epidermis, continues as the *rachis*, the primary shaft which tapers towards the tip of the feather. This carries the *barbs*, secondary axes which lie parallel to one another on both sides of the shaft. These in turn bear parallel series of *barbules*, which are the complex structural units of the vane. Inserted on its barb at a variable angle, each barbule consists of a shaft, the dorsal flange, from which a flattened scale, the basal lamella, grows ventrally. From this broad base projects a long whip-like segmented *pennulum*. The pennulum and the distal part of the basal lamella may carry fine spines or hooks, the *barbicels*. These are most numerous on the distal barbules, those on the side of the barb towards the tip of the feather. In these the basal lamella often ends in one or more ventral teeth, and the pennulum carries hooklets (*hamuli*) basally followed by many other ventral barbicels, and often dorsal processes as well. In contrast the proximal barbules, on the side towards the base of the feather, are much simpler, though they usually bear ventral teeth; their pennulum, which is often very long, is more or less smooth and almost entirely devoid of barbicels. The complex projections, which vary greatly from one species and one type of feather to another, hook the adjacent barbules together very efficiently, and so give the feather its combination of toughness and flexibility.

The largest feathers, the pennae, are implanted in the fore-limbs where they shape the wing (*remiges*), or on the last caudal vertebrae where they form the tail (*rectrices*), while the much smaller contour-feathers grow from and clothe the body. These two types of feather play fundamentally different parts in flight.

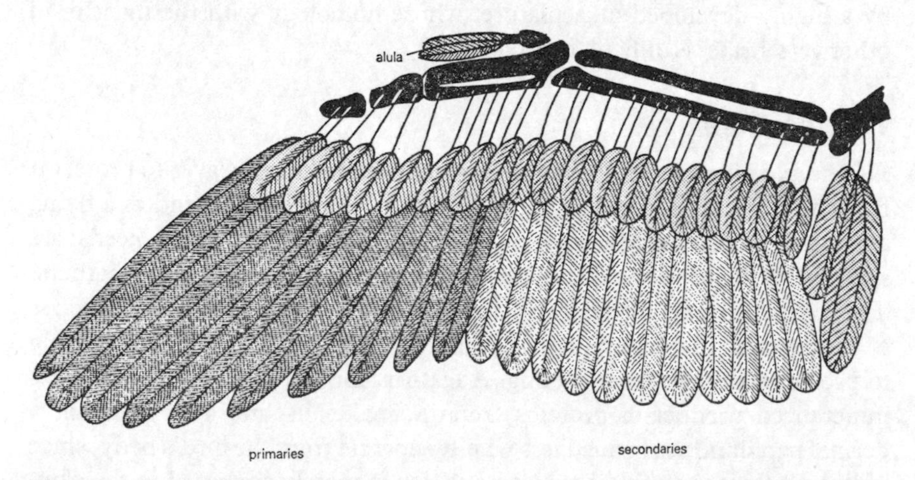

Figure 5. The implantation of the primary and secondary remiges in the skeleton of the wing – the primaries in the hand and the secondaries in the ulna.

The contour-feathers are not evenly distributed (except in the penguins and ratites), but localized in certain tracts, the *pterylae* separated by bare areas the *apteria*. A bird usually carries a great number of feathers. This varies between species and even between individuals: 1,100 to 2,800 among small passerines, 2,000 to 2,500 in woodpeckers and doves, 6,500 in gulls, 12,000 in ducks and 25,000 in swans. In contrast, the ruby-throated hummingbird *Archilochus colubris* has only a thousand feathers. Of the large variations within species, the commonest are seasonal fluctuations related to thermal insulation. Small birds have more feathers for each unit of their weight than large ones. The weight of the plumage is a significant fraction of a bird's total weight: according to Brodkorb, a Bald Eagle of 4,082g had 486g of feathers, or 12 per cent of the whole. By Turcek and Wetmore's measurements, the weight of the plumage varies from 3–4 per cent in penguins to 12 per cent in tits, and more, with the mean at 6 per cent.

The contour-feathers serve mainly to protect the bird's body from unfavourable conditions and to ensure its temperature regulation. They are not involved in lift, but do have an important aerodynamic part to play in modelling the shape of the bird, since its outline depends much more on the plumage than on the underlying anatomy. It is these feathers, overlapping from front to back like roofing tiles, that ensure a smooth flow of air and avoid turbulence which would hinder the bird's passage through the aerial fluid.

The feathers directly involved in flight are of course the remiges and rectrices. Though built on the same general plan as the contour-feathers, they differ in that each has a well-defined outline which is characteristic for the individual penna and the species. They are much larger, broader and more elongated than contour-feathers, and tougher though still pliant with fluted shafts. Their barbules carry a great number of hooklets and other barbicels, which cling with the greatest tenacity. If the vane of a feather is pulled apart it finally gives way, two barbs parting and leaving a gaping slit. However, it is only necessary to bring the edges of the slit together again and smooth them, for the vane to re-form, the hooklets of the barbules having caught up on one another once more.

This arrangement is very advantageous, resulting in unbroken flying surfaces which bear the thrust of air, and yet are flexible and recover easily from damage. If the bird strikes a branch or other obstacle the vane of a feather may be split, the attachment of the barbules being torn apart; yet a rapid preen with the beak will soon bring them together and restore the feather to its undamaged state. If the surface of the feather was continuous, an accidental tear would not be automatically repaired in this way, especially since a feather is not a living organ.

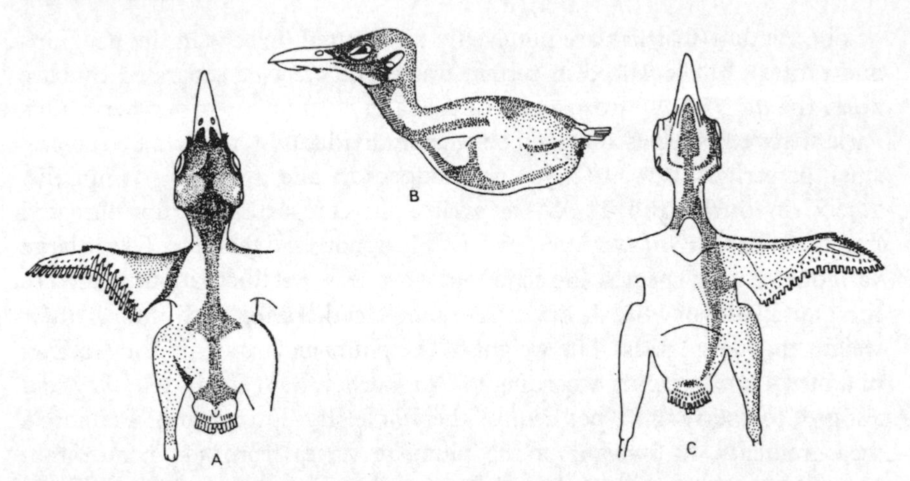

Figure 6. The pterylosis of *Aphelocoma coerulescens insularis*, a jay from the southern United States. A. Dorsal. B. Lateral. C. Ventral.

The shape and stiffness of the pennae vary, according to the needs imposed by the mode of flight of the species. The pennae are not scattered at random and their number is rigidly fixed. This is especially true of the remiges, which form the plane of the wing and are inserted on the bones of the hand and arm. In large birds which fly well, the ulna shows similar kinds of bony tubercules for the insertion of the calami. The remiges are named according to their levels of insertion. The three or four borne on the thumb form the *alula* or bastard wing. The *primary* remiges, inserted on the bones of the hand and the middle finger, are the most important of all since they contribute most to flight. Their number varies between nine and twelve, but is constant at ten for most birds. The *secondaries*, inserted on the forearm, vary more in number, from nine among small passerines to forty in albatrosses. Lastly the *tertiaries*, which are inserted on the upper arm (humerus), have the shape of pinions only among the long-distance flyers, and in other birds are like elongated contour-feathers.

Above and below the wing, the remiges are covered and protected by a strictly-determined sequence of overlapping contour-feathers, the *tectrices* or *wing-coverts*. These form a greater and a lesser series above and below, each covert accompanying a particular remex. The wing-bend is covered in small feathers, also directed backwards so that they smooth the airstream over the leading edge.

Except at their bases, the remiges overlap one another, the outer vane of one partly covering the inner vane of the next one outwards. Thus the remiges and their coverts together form a continuous flexible jointed plane, except towards the tips of the primaries. These are separated to different degrees, giving considerable aerodynamic advantages in some conditions of flight. A cross-section through the wing shows that the leading edge is thick and rigid because of the bones and tendons, whereas the trailing edge is thin and flexible since it is formed of nothing but the tips of the feathers. For greatest efficiency the wing is arched, convex above and concave below. This increases the positive pressure on the under surface and the negative pressure on the upper surface, and so improves the lift.

The first (outermost) primary is normally the shortest, and sometimes even vestigial, while the next few increase in length up to the tip and then decrease in a regular way. This can be expressed in a wing-formula, important as a specific character in some groups, which quantifies the general shape of the wing and the pointedness of its tip. Closely-related species may show marked differences in formula, while distantly-related ones show convergence, in accordance with their powers of flight and similar ecological adaptations.

Together these highly-specialized feathers thus form a remarkably well-adapted organ of support and propulsion. They must however be regularly renewed at the moult, which for the flight feathers takes place every year. During the moult, at the site of each feather-bud the worn feather is pushed out of the follicle by the growth of the new one. The shaft of the old feather is continuous with the new sheath and marked off from it merely by a constriction, which lets the worn feather fall when the new one is sufficiently keratinized.

The endocrine control of moulting is very complex, and involves most of the ductless glands. The thyroid plays a crucial part, since one can abolish the moult by removing the gland, or bring it on by injecting thyroid hormones. The effects of the sexual glands are not clear, since castration does not prevent the moult, besides which this can occur at any stage in the gonadial cycle: some birds continue the moult while breeding. All that one can say is that the gonads act as inhibitors, and play a part in controlling the development of nuptial plumage, without truly causing the prenuptial moult. The suprarenal cortex may act indirectly to inhibit the action of the sex hormones on the pituitary; and in turn the anterior lobe of the pituitary plays an important part through its thyrotropic hormones. This gland, the command-post of the body, certainly originates the moult by activating the pituitary-thyroid system. The natural moult of birds is thus controlled by a combination of endocrine rhythms, with the pituitary-thyroid system predominating (Assenmacher 1958).

In many birds the moult is annual, in which case it begins after the breeding season. In others there are two moults to each annual cycle, one before and one after breeding. The prenuptial moult is often only partial, not involving the flight feathers. In birds of temperate climates the moults are regular, synchronized with a well-defined breeding season. The same is true in tropical climates with contrasting seasons, where the moults, usually two a year, are apparently determined by the end and the beginning of the rainy season (as Miller has shown for the Rufous-collared Sparrow *Zonotrichia capensis* in Colombia), rather than by variations in photoperiod which are very slight at low latitudes.

The replacement of contour-feathers scarcely affects a bird's powers of flight, but dropping pennae presents serious problems. These feathers do not fall randomly but in a definite order. In most birds the primaries fall one by one, beginning with the innermost, while the moult of secondaries begins at several centres at once. This established order of succession ensures that most birds lack only a few pinions at a time, with the gaps symmetrically distributed, so that the aerodynamic ill-effects are minimized. The duration of moult varies, being short among small passerines and plovers, and long in albatrosses, petrels, pelicans, gannets and raptors. In eagles the moult can last almost the whole year.

On the other hand, certain birds moult all their flight feathers at once, so that they become entirely flightless and are very vulnerable. Some of these birds undertake moult migrations to favourable habitats where they are secure from predators. Wild ducks – Gadwall, Shoveler, Pintail and Wigeon – after nesting in northern Europe and Asia, assemble to moult in uncountable flocks at the mouth of the Volga during the middle of July. Their moult ended, they migrate again in the first fortnight of August, scattering in all directions. In the same way certain high lakes in Tibet serve as refuges for ducks on their way from Siberia to winter in tropical Asia. A similar moult migration is undertaken by the Shelduck of north-western Europe. During July these assemble in large flocks, especially on the marshes of the south-eastern part of the North Sea, the Heligoland Bight, and in smaller numbers at several points on the coasts of Denmark, Holland, France and Great Britain. There they moult their remiges, which all fall together, and leave once more for their homelands. Great flocks of Great and Short-tailed Shearwaters have been seen to gather off Greenland and in the Bering Strait respectively, where they lose all their remiges and are flightless throughout the moult. These birds have all precisely synchronized the loss of their remiges with their presence in areas where the loss of flight is least inconvenient.

In other respects too, migratory birds must change their plumage at times related to their travels, since they need all their powers of flight in order to perform long-distance migrations. The majority moult before migrating, leaving their breeding areas in fresh plumage. Though in some the contour-feathers moult during the migration, this scarcely affects their flying ability.

The plumage is thus a remarkable instrument of flight. It is a non-living tissue, not irrigated with blood and to some degree separated from the living body. This is especially important to the reduction of weight, since a comparable surface filled with blood would be very much heavier. Furthermore it would entail an unacceptable loss of heat, increased during flight by its continual movement through the air.

Gliding and soaring flight

We can examine how this mechanism works without going into details. The comparisons with aeroplanes and other flying machines, which are so tempting to make, are misleading because birds differ in one basic respect: their lifting planes and propulsive system are not functionally separated, but a single working surface carries out both functions at once, so that it operates under very special conditions. Both lift and propulsion through the air involve the expenditure of energy, which may come from the air itself as a result of its own movement, or from the actions of the bird in stirring the air. In consequence there are several types of flight.

Many birds can fly for a long time without flapping their wings. Like man-made gliders, they keep themselves aloft by using the horizontal component of the wind, variations in windspeeds and upcurrents. The principal characteristic of a glider is what engineers call the *lift-drag ratio* (or gliding ratio). High ratio is especially important in a glider since it reduces the slope of the descent and thus minimizes the loss of altitude, and the same is true of birds. Vultures, large raptors, pelicans and storks have very high ratios, and can glide in still air without losing much height. The loss of altitude has been measured on windless days, and found to be one metre for a horizontal travel of 8·5m in the Fulmar and of 22m in the American Black Vulture *Coragyps atratus* – performances which are inferior to those of the best sailplanes, which lose something like one metre in fifty or even better.

Soaring birds can fly without flapping and not only maintain their altitude but even increase it. For this they need either an ascending wind, or a horizontal wind of variable velocity. Terrestrial soaring birds rely mainly on thermal upcurrents for gaining altitude. In certain open habitats, especially semi-deserts and steppes, the soil is heated by the sun and warms the lowest layers of air, which rise because their density is thereby decreased. The resulting huge columns of rising air are widely used by soaring birds, notably vultures and large raptors, which maintain height by flying in concentric circles without flapping. They glide from one thermal to another, losing height which they regain when they find themselves in another upcurrent. These aerial performances are especially common in the tropics, where wheeling vultures, pelicans and various birds of prey are very characteristic of savannahs and steppes. Topographic relief also causes air to rise. Mountainous country, and sometimes a single crest, can deflect the wind which strikes it, and so cause upcurrents. Despite turbulence, birds in mountainous country, and especially raptors, use these upcurrents with great ease. Certain orographic lines are used by birds of prey on migration, so that the long mountain chain of the Appalachians, which traverses the eastern United States from north-east to south-west, is followed by great concentrations of raptors, especially buzzards, profiting from the upcurrents caused by even these low mountains.

Terrestrial soaring birds as a whole have broad wings, often almost rectangular, ending in diverging rectrices which leave wide slots open between them. This arrangement is especially well adapted to making use of slowly ascending air currents, since it increases the lift of the wings and allows slow forward travel, with stalling prevented by the wing-tip slots.

Many sea birds too soar excellently, with the albatrosses and shearwaters as undisputed champions. They cannot use thermal upcurrents, since these occur at sea only in exceptional circumstances. (This explains why terrestrial gliders avoid sea passages during migration: storks go round the Mediterranean by way of Asia Minor and the Straits of Gibraltar, rather than pass over the sea.) However, thanks to the fineness of their aerofoils, they can use local updraughts caused by the wind reacting against the waves. Thus they often fly along a wave and then cross to another, and so on in a regular rhythm. Another method is to gain speed and turn into the wind, when the bird climbs at the expense of speed. Changing course again, it loses the altitude it has just gained, in order to regain its speed. Thus by repeatedly changing its direction of flight it combines the advantages of the two situations: keeping aloft without flapping while tracing out circles or zigzags, it is actually drifting with the wind. Finally, sea birds know how to use the fluctuations in wind speed which often occur. To stay up without wasted effort, they need only fly downwind when the wind drops and upwind when it gusts, changing course and trimming the set of their wings. The paths traced by shearwaters, albatrosses and even gulls show that they follow the ever-changing reactions

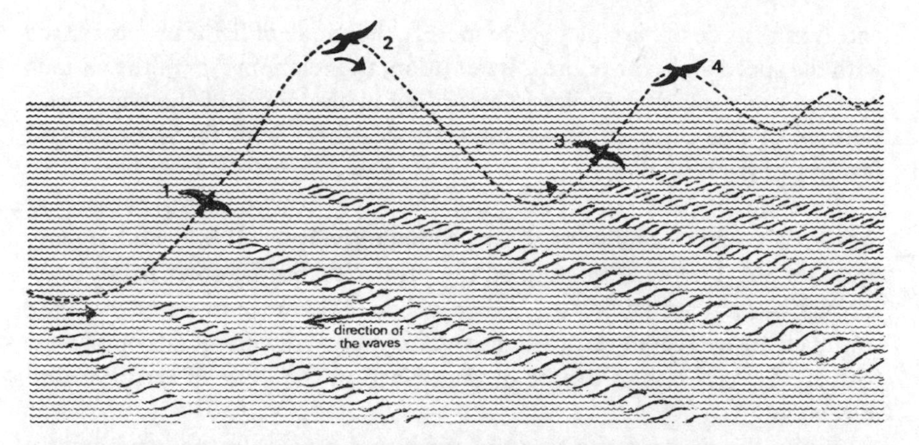

Figure 7. The path of a frigate bird in gliding flight: the bird makes use of the variations in wind speed by following curves which increase in amplitude with the strength of the wind. 1 and 3 headwind, 2 and 4 sidewind.

of the wind on the sea, anticipating its alterations in velocity. Marvellous sensitivity to aerodynamic conditions keeps them instantly informed of the shifts of the wind. Perhaps this requires special sense organs, which the Procellariiformes may have in the unusual structure of their nasal fossae (see p. 359).

Marine soaring birds have long narrow and pointed wings, in strong contrast to those of terrestrial ones. This conformation gives high lift with minimal drag at rather high wind speeds and also enables these birds to make use of rapidly gusting winds.

Despite the economy in energy it implies, soaring without flapping does demand a physical effort on the part of the bird, which has to keep its wings widely spread against the thrust of the air tending to lift them above its back. It must counteract this force by an effort of the pectoral muscles, while endlessly shifting its wings in order to trim its flying surfaces. The gliding bird is continually balancing with its tail, warping its wings, and turning this way and that. The success of the soaring birds shows that they have acquired that perfect familiarity with the conditions in their airy environment which this way of life demands.

Flapping Flight

Powered flight in birds is flapping flight, in which the wing performs an oscillating action which produces lift and forward propulsion simultaneously. The wing beats regularly, taking hold of the air and working like an oar. A multitude of variables act to modify the mode of flight in accordance with the speed and type of bird. In addition, the relationships of the various parts of the wing alter endlessly according to the phase of the wing-beat. Because of this complexity, we can only deal here with the most general features of flapping flight.

Normally the basal part of the wing, the upper arm and forearm carrying the tertials and secondaries, beats comparatively shallowly and mainly provides lift. The wing tip with all the primaries, on the other hand, beats through a wide arc and forms the propeller. The movements of the wing simultaneously create a vertical component (lift) and a horizontal component (thrust). Several phases can be distinguished in the cycle of reciprocating movement.

During the propulsive phase the wing, and especially the hand and its remiges, is carried forwards and downwards. It turns, partly because of the greater rigidity of the leading edge supported by the skeleton on the limb, so that its upper side faces forward (pronation), and thus presents the greatest surface to press upon the air. During the next phase the wing returns backwards and upwards, folding towards the body at the joints of wrist and elbow (at least in small birds, whereas in large ones the basal part of the wing remains more or less fixed). The primaries take up an almost vertical position, and each turns on its axis so as to let the air pass between them and so diminish the drag.

During its forward movement the wing thus acts like the blade of an airscrew in biting on the air and driving the bird forwards producing both lift and propulsion, their proportion depending on the angle of attack of the wing controlled by the play of muscles. Its backwards movement is usually unproductive, during which the kinetic energy of flight may actually fall. The instant that the bird begins to lose height, because its lifting surface is not giving enough lift, it compensates with a powerful wing-beat. In many passerines this produces a very characteristic undulating flight, in contrast to the level flight of larger birds whose better-developed secondaries and tertials give them sufficient lifting surface to support them through the cycle. Woodpeckers especially are noted for their sinusoidal flight, dives with wings almost closed alternating with strong beats which regain the lost height. However, other small birds fly directly, their inertia and streamlining allowing them to part the air unchecked.

If the primaries were rigid their angle of attack would be too great, but their flexibility and the spreading-out of their tips are very important in the dynamics of flight. A long pointed wing can twist as a whole, while in a short wide one the primaries, often emarginated, spread out so that each one

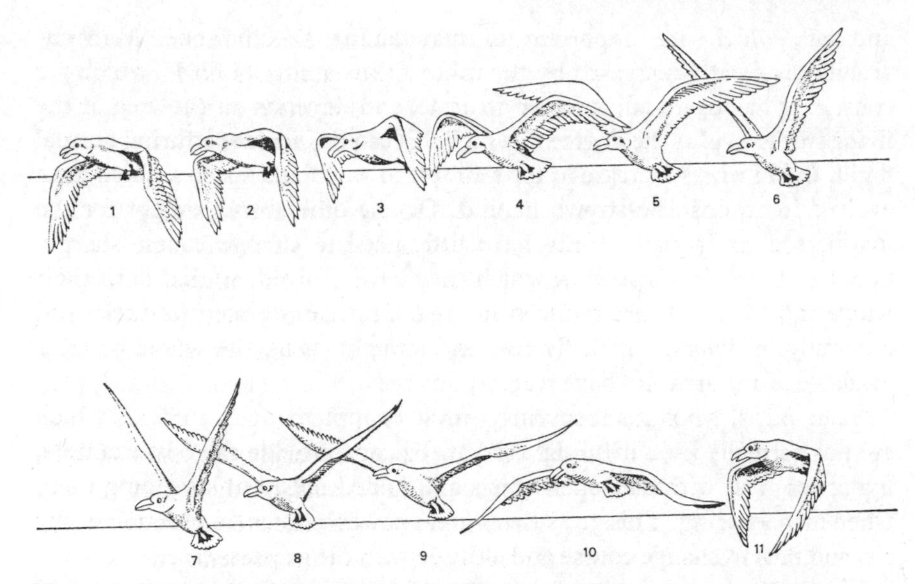

Figure 8. Successive wing positions of a Herring Gull *Larus argentatus* during flapping flight (from a film).

can twist to produce the same effect. During the backwards movement they turn almost perpendicular to the wing, like the slats of a venetian blind, thus letting the air pass between them and so offering the least possible drag. The *alula* also helps to prevent loss of speed. It controls the stream of air over the surface of the secondaries, ensuring smooth flow instead of the turbulence which would result if the leading edge lacked this deflector.

Some small raptors (such as the Kestrel) terns, kingfishers and small passerines can almost hover, flying for a moment on the spot while searching for prey or examining places and objects. To do this the bird holds its body vertical with its tail extended downwards and spreads the base of its wing out vertically, facing into the wind, while the tip beats backwards and forwards almost horizontally, reversing its angle so that it lifts on both strokes.

Flapping flight is made more complicated by the bird's need to steer, changing its height and course. During flight in a straight line and at a constant altitude, its weight acting downwards, and the resultant of the air reaction acting upwards, are in equilibrium along a vertical line. In order to manoeuvre, the bird has to throw these two forces out of line. The weight always acts through the bird's centre of gravity, but the reaction of the air can be moved forwards, backwards or to either side by trimming the wings and tail, which are important in maintaining this balance. Manoeuvrability is greatly increased by the inherent instability of birds, which are constantly having to adjust their attitude, and depends on the area of the flying surfaces, plus the degree to which this can be increased during normal flight. Broad wings increase it, for example in woodland species which have evolved in an obstacle-strewn habitat. On the other hand, except for the marine soarers, aquatic birds have little need to change course sharply because of the clear spaces in which they have evolved, so that both their wings and their tails are reduced in area. This can be seen in ducks and especially in divers, which fly fast and straight using the whole of their available lifting area and have scarcely any tails.

Some birds, when manoeuvring, employ supplementary surfaces which are not normally used in straight flight. Falcons provide the best example, flying very fast with half-open wings and folded tails, but spreading them when manoeuvring. This gives them great power to alter the reaction of the air, and thus to change course and altitude with the greatest ease.

Finally we may note that manoeuvrability is increased by the speed of which birds are capable, and by powerful muscles which can act strongly and rapidly on the flying surfaces. Flight involves complicated muscular actions which are co-ordinated by the cerebellum, always large, well-developed and richly-innervated in birds. As in other vertebrates, the cerebellum works in conjunction with the internal ear, a balancing organ vital to animals which move in all three planes of space.

Take-off and landing

Though birds need to make special efforts to launch themselves into the air, most of them can do this without a forward run, taking off vertically before they climb away at an angle. A bird begins its take-off with a jump, first crouching with bent legs and spread wings and then springing up while beating down its wings to get a grip on the air. In a series of very fast and powerful movements, the distal part of the wing acts like a paddle which is recovered for each stroke. The hand swivels at the wrist through more than a right-angle, so that its under-surface beats the air on the downstroke and its upper surface on the upstroke. Thus the whole wing-beat, including the recovery, is propulsive, which explains why birds which need to take off quickly have unusually strong supracoracoid muscles. During take-off the wing acts like the rotor blade of a helicopter, but its reciprocating actions means that its pitch has to be reversed at each stroke to keep the thrust acting upwards. Like any plane moving through the air at a large angle of attack, the wing causes eddies during its complex movements, and through the inertia of the air it encounters at each stroke the turbulence generated by the previous one, which augments its lift. The initial vertical ascent is soon transformed into a steep climb, the bird pitching progressively forwards while continuing the same movement of its wings, until it has established normal flight.

Some birds, like aeroplanes, are unable to leave the ground until they have attained horizontal flying speed. Large vultures, especially when weighed down by a heavy meal, shuffle clumsily forwards before launching themselves into the air. Albatrosses choose exposed nesting sites so that they may run into the wind beating their wings. Certain water birds run over the surface with giant strides, and the divers cannot take off from the land at all because their legs are too weak and placed too far back. Take-off is much easier from a high perch, and this is almost (though not quite) essential to birds like swifts whose long wings would otherwise strike the ground.

To land, a bird has to reduce its forward speed almost to zero, and since it then falls out of the air it can brake only at the instant before touching down. This it does by rearing back into a nearly vertical position with head high and tail spread, the wings beating forwards to produce very effective braking. It often returns to the vertical flight used in take-off, by keeping its body vertical and its wings beating back and forth in an almost horizontal plane, so that it maintains lift without forward movement and has only to let itself down on to the chosen perch. Another method of losing forward speed in landing is to glide in below a high perch and swoop up to it.

In landing the legs are stretched forwards and allowed to bend progressively on impact, acting as buffers to dissipate the kinetic energy of the bird's forward movement. This imposes a considerable load on the undercarriage which has to be strong enough to absorb and stand the shock. Ducks touch down and plane on their breasts like seaplanes, thus protecting their bellies and relatively weak breasts from the impact. So do divers, which are unable to reduce speed because of their small wings and short tails, and must set down at speed.

Types of flapping flight

Though flapping flight is fundamentally the same among all flying birds, it varies very much in detail from one kind to another. These variations are produced by the diverse proportions of the wing. As a result wings can be placed in two categories, though with many intermediate states.

The first type of wing is characterized by great width and small length; it

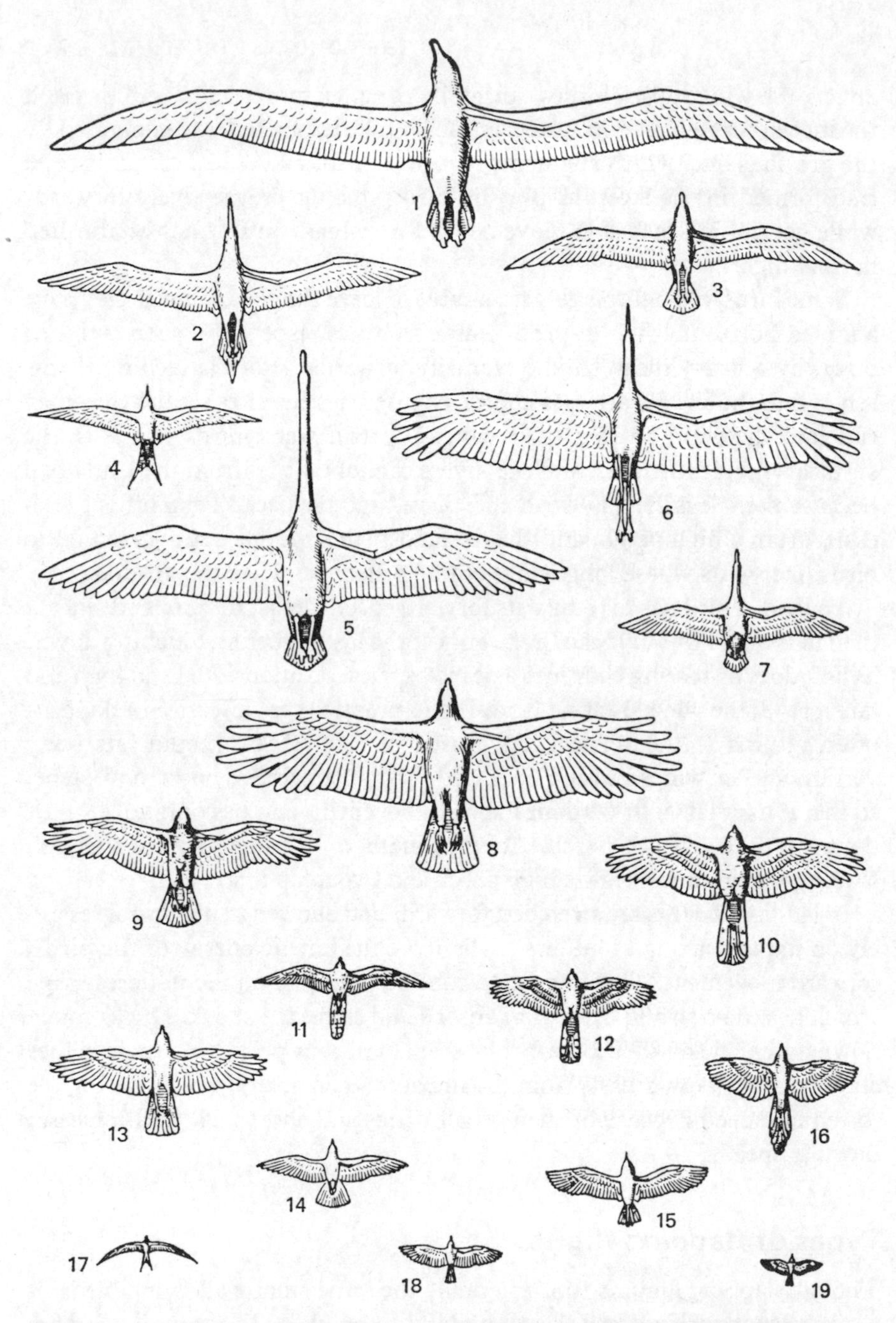

Figure 9. Outlines of birds in flight, seen from below. The proportions of the various segments of the wing skeleton – the humerus, radius and metacarpals – are shown diagrammatically on one side of each sketch.
1. albatross. 2. gannet. 3. gull. 4. tern 5. swan 6. stork. 7. duck. 8. eagle. 9. buzzard. 10. goshawk. 11. falcon. 12. sparrowhawk. 13. crow. 14. pigeon. 15. lapwing. 16. magpie. 17. swift. 18. thrush. 19. sparrow.

is rounded or elliptical, often strongly convex, and usually ends in widely-separated primaries which give a 'fingered' look in flight. This type is found in game birds, woodpeckers, and some pigeons and passerines: birds which mostly fly slowly with rather rapid wing-beats, but can manoeuvre with ease, even at low speeds without loss of lift, because of their large lifting area. That their take-off is equally easy is shown by the Pheasant, which despite its weight rockets up on the power of its strong pectoral muscles. However, these birds have little endurance, and therefore do not make long sustained flights. The nocturnal raptors also have wings of this form, large and with deeply digitated tips, so that they can cover the ground deliberately and adroitly at a slow beat, as they must to avoid obstacles and capture prey in the dark.

The second type is much longer and proportionately narrower, with the tip pointed rather than digitate, usually much less convex above. Most birds with wings of this type, such as the waders, swifts, swallows, most bee-eaters, falcons, and some passerines, fly fast and easily, but cannot easily remain airborne at low speeds. Their speed and the supplementary lifting surfaces which some of them can bring into action (for example by spreading out wings which are partly folded in normal flight) give them great manoeuvrability, while their wing-beats are often slower. However, some birds have wings of this type without being notable flyers. For example ducks, and divers, have small narrow wings for their weight, making up for small area by beating very rapidly, but are scarcely aerobatic. Although they fly fast thanks to rapid wing-beats, the auks are incapable of precise manoeuvring and despite their strong pectoral musculature they take off with difficulty.

There are fairly constant relations between wing dimensions and other morphological features such as weight (Greenewalt 1960b). The relation between the bird's weight and the area of its wings, or the loading for unit area, is particularly interesting. The following is a selection from extensive data (from Magnan 1922, *Ann. Sci. nat.* (19) 5: 125–334, and Poole 1938, *Auk* 55: 511–17).

Species	*Weight* (g)	*Wing-area* (cm^2)	*Wing-loading* (cm^2/g)
Great Northern Diver *Gavia immer*	2,425	1,358	0·6
Pied-billed Grebe *Podylimbus podiceps*	343·5	291	0·8
Wandering Albatross *Diomedea exulans*	8,502	6,206	0·7
Cory's Shearwater *Procellaria diomedea*	572	1,280	2·2
Storm Petrel *Hydrobates pelagicus*	17·4	100	5·7
Ascension Frigate-Bird *Fregata aquila*	1,620	3,240	2·0
Cormorant *Phalacrocorax carbo*	2,115	1,967	0·9
Heron *Ardea cinerea*	1,408	3,590	2·5

Species	*Weight* (g)	*Wing-area* (cm²)	*Wing-loading* (cm²/g)
White Stork *Ciconia ciconia*	3,438	4,951	1·4
Mallard *Anas platyrhynchos*	1,408	1,029	0·7
Whooper Swan *Cygnus cygnus*	5,925	3,377	0·6
Grey Lag Goose *Anser anser*	3,065	2,697	0·9
American Black Vulture *Coragyps atratus*	1,702	3,012	1·8
Griffon Vulture *Gyps fulvus*	7,269	10,540	1·4
Golden Eagle *Aquila chrysaetos*	3,712	5,382	1·4
Buzzard *Buteo buteo*	1,072	2,691	2·5
Peregrine *Falco peregrinus*	1,222·5	1,342	1·1
Kestrel *Falco tinnunculus*	245	708	2·9
Capercailie (♂) *Tetrao urogallus*	3,361	1,412	0·4
Partridge *Perdix perdix*	387	433	1·1
Crane *Grus grus*	4,175	5,553	1·3
Great Bustard *Otis tarda*	8,950	5,728	0·6
Snipe *Gallinago gallinago*	95·5	244	2·5
Dunlin *Eriola alpina*	44	126	2·8
Herring Gull *Larus argentatus*	850	2,006	2·4
Common Tern *Sterna hirundo*	118	563	4·8
Mourning Dove *Zenaidura macroura*	130	357	2·7
Barn Owl *Tyto alba*	279	1,163	4·2
Swift *Apus apus*	36·2	165	4·5
Chimney Swift *Chaetura pelagica*	17·3	104	6·0
Greater Spotted Woodpecker *Dendrocopos major*	73	238	3·3
Ruby-throated Hummingbird *Archilochus colubris*	3	12·4	4·1
Swallow *Hirundo rustica*	18·35	135	7·3
Carrion Crow *Corvus corone*	470	1,058	2·2
Magpie *Pica pica*	214	640	3·0
Great Tit *Parus major*	21·45	102	4·7
House Wren *Troglodytes aedon*	11	48·4	4·4
Blackbird *Turdus merula*	91·5	260	2·8
Whitethroat *Sylvia communis*	18·65	87·1	4·7
Starling *Sturnus vulgaris*	84	190·3	2·3
Song Sparrow *Melospiza melodia*	22	86·5	3·9
Chaffinch *Fringilla coelebs*	21·15	102	4·8

However, the relations which have been determined are not strictly comparable between birds of greatly differing weight, since the weight differs according to the cube of the linear measurements while the area of the wing varies according to their square. In taking account of the corrected relation, one discovers that within certain groups the wing area divided by the 2/3 power of the weight yields a practically constant value, expressed in the formula $A = cM^{2/3}$, where A is the area, M the weight, and c the corrected loading per unit wing area. Between families however the value of c, which reflects the mode of flight, varies greatly, averaging 7·5 among ducks and 20 among raptors, and results in marked differences in speed and in manoeuvrability. Small birds generally have a better reserve of lift than larger ones because of their distinctly more favourable wing-loadings. Large birds may therefore encounter difficulty in special conditions, as when raptors lift prey to dismember it or carry it to their young.

From these facts it is clear that the larger and heavier a bird is, the larger *proportionately* its wings must be and the heavier they in turn must be, in order to withstand the thrusts imposed by their own action. It has been calculated that the thickness of birds' wings increases as the power of 1·34 of their length, so that their weight increases very rapidly with increasing span. Thus in a Griffon Vulture of 7269g a wing weighed 1599g or 22 per cent of the whole, while in a Carrion Crow of 470g one weighed only 74·7g or 16 per cent, and in a Spotted Flycatcher of 14·35g as little as 1·8g or 12·5 per cent. The unavoidable need to strengthen the wing means that its inertia would greatly exceed a functional limiting value if the bird increased beyond a certain size. There is no doubt that the condors have reached the upper limit of size for flying birds: the few birds which are heavier have totally lost the power of flight. This provides an especially clear example of a general biological law, determining the optimal and the limiting sizes for a given type of animal, though these sizes differ greatly between types of very different organization.

Another very interesting topic is the frequency of wing-beats, which obviously varies very much between different types of birds. The following table gives some values (from Blake 1947, *Auk* 64: 619–20, and Meinertzhagen 1955, *Ibis* 97: 11–114).

Species	*Beats per second*
Great Crested Grebe *Podiceps cristatus*	6·3
Cormorant *Phalacrocorax carbo*	3·9
Capercaillie *Tetrao urogallus*	4·6
Pheasant *Phasianus colchicus*	9·0
Coot *Fulica atra*	5·8
Heron *Ardea cinerea*	2·5
Mallard *Anas platyrhynchos*	5·0
Wigeon *Anas penelope*	5·1
Mute Swan *Cygnus clor*	2·7
Black Kite *Milvus migrans*	2·8
Peregrine *Falco peregrinus*	4·3
Turnstone *Arenaria interpres*	4·0
Ringed Plover *Charadrius hiaticula*	5·3
Lapwing *Vanellus vanellus*	2·3
Black Guillemot *Cepphus grylle*	8·0
Puffin *Fratercula artica*	5·7
Herring Gull *Larus argentatus*	2·8
Wood Pigeon *Columba palumbus*	4·0
Mourning Dove *Zenaidura macroura*	2·45
Belted Kingfisher *Megaceryle alcyon*	2·4

Species	*Beats per second*
Blackbird *Turdus merula*	5·6
Starling *Sturnus vulgaris*	5·1
Magpie *Pica pica*	3·0
Rook *Corvus frugilegus*	2·3

Some birds, especially ducks, seem to compensate for their reduced wing-area by more rapid wing-beats. Otherwise the frequency is more or less proportional to the wing-length, remaining almost constant for a given bird, under any normal conditions and at any speed. Changes in speed are not produced by altering the rhythm of wing-beats, but by adjusting the angle of attack. This suggests that the wings act as mechanical oscillators whose frequency is determined by several constant factors, especially the weight and inertia of the various parts of the wing and its musculature.

Certain birds which have evolved from strongly flying groups have more or less lost the power of flight through secondary regression. Their wings, normally constructed but weak and with reduced musculature, serve only for very short flights or even none at all. In certain birds such as the penguins the wings are completely transformed, and in others such as the cassowaries and kiwis they are vestigial. Such more or less complete regression follows from very special ecological niches, or from isolation on islands.

The hovering flight of hummingbirds

The flight of hummingbirds is most highly specialized, fast for their size, with frequent and almost instantaneous changes in direction and speed. Above all they can hover perfectly still, as when they are drinking from flowers, and even uniquely fly backwards. The skeleton of the fore-limb is highly modified, with remarkably short upper arm and forearm and inactive elbow joint, the hand forming the greater part of the wing, and the humerus forming a fixed V with the radius and ulna. The wing bones are moved by powerful muscles which make up almost 30 per cent of the birds' weight. The supracoracoid muscle which raises the wing is no smaller than those muscles which lower it. Therefore, a hummingbird's wing must act as strongly during the back-stroke as the fore-stroke.

Superficially, wings jointed only at the shoulder seem to work like oars, but in fact their movements are much more complicated. In hovering they are moved horizontally forwards and backwards. As they move forwards and downwards their leading edges are to the front and they take up a small positive angle of attack. As they recoil backwards they rotate through about 180°, so that their leading edges are to the rear and their upper surfaces

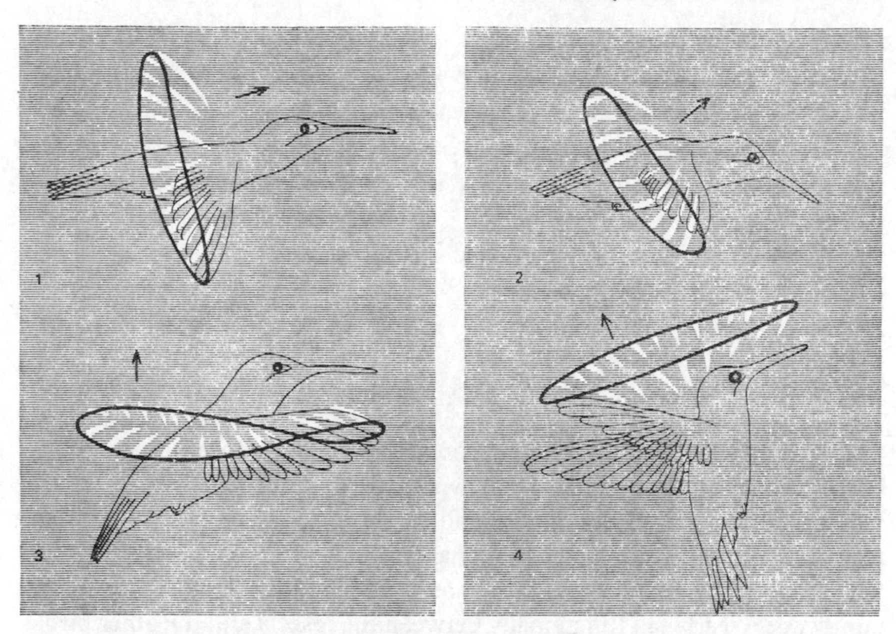

Figure 10. A hummingbird (Trochilidae) in different types of flight: 1 and 2 normal, 3 hovering, 4 backwards. In each mode the wingtip describes an almost elliptical path.

downwards. They then revert to the first position to begin the next forwards movement, the wing-tip tracing out a figure-of-eight during the cycle. This mode of flight allows hovering, and vertical ascent like that of a helicopter through delicate adjustment of the angle of attack.

For a hummingbird to fly forwards, the plane in which its wings beat tilts progressively from the horizontal, so as to produce both upward lift and forward thrust, until at the bird's maximum speed the wings are beating vertically. In backwards flight the body is held upright and the wings beat in a plane inclined backwards, their tips tracing out circles, so that lift is combined with backwards thrust. Although hummingbirds can fly backwards much less rapidly than forwards, this is a useful accomplishment. They can also change direction merely by altering the angle of attack of the wings, helped by adjustment of the tail. The changes of direction which these birds make in their very quick manoeuvres involve an astounding facility in rolls, and other manoeuvres in the rolling plane.

The wing-movements of hummingbirds are thus quite different from those of other birds, the wing acting like rotors except that its movement is reciprocating not rotary. It acts propulsively during both strokes, though the strongest propulsive thrust is during its forward stroke. The wing-beats are always distinctly faster than those of other birds, increasing in frequency from the larger to the smallest species. The following values are from Greenewalt (1962):

Species	*Wing-length from wrist to tip (mm)*	*Beats per second*
Eupetomena macrura	78	22
Melanotrochilus fuscus	80	25
Amazilia cyanura	53	41·5
Archilochus colubris ♀	44·5	52
Archilochus colubris ♂	38·5	70
Calliphlox amethystina	33	78

The rapid friction of the air acting on the remiges produces the humming characteristic of these birds, from which their flight can well be called vibrant. In this and in the rigidity and mode of action of their wings they resemble insects. When various characteristic relations, such as that between weight and wing-length, are examined this highly specialized family does clearly fall functionally between the insects and the other birds. Certain hawk-moths have a rate similar to that of hummingbirds, with which they are often confused by inexperienced observers.

Formation-flying

Long-distance flights, such as the daily journey between roost and feeding-grounds (undertaken for instance by some ducks and some waders) and especially migratory flights, involve considerable expenditure of energy. When the greatest economy in energy is necessary, birds fly in groups. Indeed, many large species fly in true formations, each bird in a definite position and at a definite distance from its neighbours. Some fly in line astern, each bird following directly behind the one ahead of it. This formation is seen especially when cormorants are returning to their fishing-grounds. Other birds (such as waterfowl, pelicans, cranes, and even some waders like the Golden Plover) fly in reversed V formation, the point foremost. It is noteworthy that this formation is never seen among gliding birds, but only among large birds with regular flapping flight.

Despite the contrary opinion of some observers, it seems that formation-flying can be explained on aerodynamic grounds. A beating wing generates an ascending eddy at its tip, of which a following bird can take advantage if it positions itself at the correct distance and so gains support from the up-current. Observation of a flight of geese or swans in formation, or a photograph showing their wing-beats frozen at an instant in time, demonstrates a remarkable synchronization between the birds of a single formation which cannot be accidental. In echelon flight, as in the arms of a V formation, each bird loses some of the benefit from the favourable eddies set up by both wings of the bird in front. However, the members of the formation are better placed to judge the distances which separate them, and so to avoid a collision, which would be dangerous to birds the size of ducks. Though the beating wings of birds work very differently from the fixed wings of aeroplanes, the result is comparable. This helps to explain why the leading bird falls back after a while to rest at the rear of the line, or a bird switches from one limb of the V to the other to change the wing supported by eddies.

Flight performance

Birds as a whole enjoy highly developed powers of flight. Thanks to their profound anatomical modifications these vertebrates have gained complete mastery of the air, where they manoeuvre with perfect ease, and with a power which in many small species is astonishing.

The ability to fly is innate in birds, which take wing as soon as their remiges have grown. The young birds sometimes seem to devote themselves to exercises, but these are merely locomotor movements which have not yet become co-ordinated. Some leave the nest still unable to fly, and must creep about and flutter clumsily in the undergrowth, yet once they have developed sufficiently their flight is as perfectly controlled as that of adults. Moreover some birds, such as swifts and swallows, launch themselves into the air as soon as they leave the nest. Such birds, whose true environment is aerial space, usually spend a long time in the nest so that they may develop completely before they begin to fly.

Birds are able to cover considerable distances, especially during migration. The most spectacular example is the crossing of the Gulf of Mexico by the minute Ruby-throated Hummingbird *Archilochus colubris*, which weighs only 3g and seems quite unfitted to cross 800km of water without landing. The Bristle-thighed Curlew *Numenius tahitensis*, which annually returns from Alaska to Polynesia, flies without a break between the Aleutian Islands and Hawaii, covering 3300km. At an average rate of 26 metres and two wing-beats per second, this bird must take 35 hours in the crossing and beat its wings 252,000 times (Stresemann). The Arctic Tern *Sterna macrura* covers at least 17,000km during its post-nuptial migration.

Birds also cover considerable distances daily within their own territory, in searching for food during the nesting season. The Blue Tit, which visits its nest 30 to 40 times an hour to feed its young, covers 100km during the day. This daily journey amounts to several hundred kilometres in swallows, while swifts which are on the wing from 16 to 18 hours a day cover a thousand kilometres (Steinbacher). This locomotor activity involves considerable expenditure of energy.

The flight speeds of birds are very variable: normal speed is used in ordinary travel, while a burst of high speed is used to escape an enemy or pursue prey. Migration speeds are somewhat intermediate, with birds flying at average speeds above those used in normal transit flight. The table gives some information on speeds recorded.

Species	*Speeds (km/hr)*	*Flight conditions*
Quail *Coturnix coturnix*	48–91	migration
Killdeer *Charadrius vocifer*	45–88	
Sandpipers, *Calidriinae*	175	
Long-tailed Duck or Oldsquaw *Clangula hyemalis*	86–115	
Mallard *Anas platyrhynchos*	74–96	
Geese *Anser* species	40–90	
Whooper Swan *Cygnus cygnus*	62–70	
Turtle Dove *Streptopelia turtur*	61–72	migration
Sparrow Hawk *Accipiter nisus*	41	migration
North American buzzards	42–67	migration
Peregrine *Falco peregrinus*	63–290	
Ruby-throated Hummingbird *Archilochus colubris*	43	
Swift *Apus apus*	61–97	
Swallow *Hirundo rustica*	35–97	
American Robin *Turdus migratorius*	27–51	
Blue-headed Wagtail *Motacilla flava*	46–48	migration
Starling *Sturnus vulgaris*	37–74	normal
	69–78	migration
Rook *Corvus frugileus*	52–72	migration
Chipping Sparrow *Spizella passerina*	24–32	
Chaffinch *Fringilla coelebs*	34–46	normal

The mean speeds of pigeons are about 60 to 70 km/hr, those of swifts and waders about 90–100, and those of ducks about 100. Most small passerines scarcely exceed 50–60 km/hr, but some such as crows and starlings fly faster. The Peregrine travels at 100–140 km/hr. Speed-records among birds are held by swifts which can attain 200 km/hr, and by the Peregrine which, in a swoop with wings almost plastered to its sides, reaches 290 km/hr, the highest speed recorded for a bird and no doubt for any animal.

Birds travel at very various heights. On migration they usually fly at moderate heights. Except under special circumstances, small birds seldom exceed about a hundred metres, though some crossings are made at altitudes up to 1,500 or more rarely 3,000m, even by small passerines such as Linnets. Under exceptional circumstances birds can fly at much greater heights. Some fly over the highest mountains, such as the condor met at up to 6,000m in the Andes, or the Chough and a few raptors at similar heights in Asian mountains. During migration geese, ducks and waders regularly cross the high ranges of the Himalayas, and the altitude record is held by geese seen at Dehra Dun in India, flying at 9,000m. Flight at such heights presents severe physiological problems, because of the low temperature and rarefaction in the high levels of the atmosphere. Oxygenation is difficult because of the low air pressure, just when considerable effort is needed to make good the loss of lift in the rarefied air. The bird must also be able to withstand the rapid climb from low to high altitudes. The contrast between mountaineers who laboriously assail the peaks of the Himalayas, and birds which fly over them with ease, is particularly striking.

Chapter 3

Terrestrial and Aquatic Locomotion

THOUGH they have gained the mastery of the air, birds have not thereby lost the ability to move on land, while some can still swim. It is true that some strictly arboreal birds like the swifts have become aerial to such a degree that they can move about on the ground only with difficulty. On the other hand others can walk as well as they can fly and move with great ease over the ground, and some do not even fly in seeking food or escaping their enemies. Others again can travel over land, swim adeptly, and fly with ease. On land and water the organs of locomotion are the hind limbs or legs, highly modified from the primitive vertebrate type. We shall briefly examine their adaptations before considering how birds walk, climb and swim.

Pelvic girdle and legs

The pelvic girdle is profoundly modified so as to increase the rigidity of the posterior axial skeleton and the strength of the hind limbs. The bones which form it are very widely fused to coalescent sacral vertebrae. The ilium, ischium and pubis are solidly combined into a single bony structure, strong despite its light weight. The pubic bones remain free at their caudal ends, forming two slender rods turned forwards, allowing the passage of eggs which are large in comparison with the abdomen. Only in some ratites, notably the Ostrich, do they fuse together in a pubic symphysis.

The leg, though built on the general vertebrate plan, has diverged widely from it. The femur is short and wholly hidden in the muscular mass of the trunk, articulating with a longer tibia. Beside the tibia lies a vestigial fibula which is fused to it for most of its length, except at the upper end where it forms a separate part of the knee joint. Here there is a kneecap which is usually small, but in the divers is highly developed being fused to the tibia below the joint and forming a projection on which important swimming muscles are inserted. A bird's tibia is really a complex bone in which the proximal tarsal bones are fused, so that it is properly called the *tibiotarsus*.

Figure 11. A Golden Eagle *Aquila chrysaetos:* A. in flight, B. perched. In each diagram the appropriate system of supporting bones is shown in black: in flight, the fore-limb, shoulder girdle, ribs and sternum; when standing, the spinal column, pelvic girdle and hind-limb. The position of the centre of gravity is indicated.

Similarly the next segment of the leg, often called simply the tarsus, is really the *tarso-metatarsus*, since it is formed by fusion of the distal tarsal bones with the metatarsals. In penguins this fusion is incomplete, as in *Archaeopteryx*, but in all other birds three metatarsals are joined into a single bone, showing their separate origin only in the triple articulation at their lower end. Together with the toes, this segment makes up what is commonly known as the 'leg' of a bird, since the tibia is hidden by plumage and the femur in flesh. Except in a few species in which the legs are feathered (owls and sandgrouse) they are sheathed in horny scales. These are usually most closely arranged on the front of the tarsus, where they overlap like tiles or are even fused into a single plate (the booted or ocreate condition of some passerines).

Birds' feet never have more than four toes, the fifth having been lost during evolution. In most birds three toes point forwards while the first toe, or hallux, which is sometimes less developed, points backwards (a condition known as anisodactyle). Apart from the last phalanx, which carries a curved claw, the hallux consists of one phalanx, the second (inner) toe of two, the third (middle) toes of three and the fourth (outer) toe of four phalanges. Their most highly developed condition is seen in some climbing birds.

As adults, birds are wholly digitigrade, resting only on their phalanges. However, their young are plantigrade and rest on their tarsi, sometimes provided with callosities which later disappear. The Marabou Stork and some other birds still rest in this attitude when adult.

The hind limb is moved by strong muscles which together form a triangular sheet, its base stretching from edge to edge of the pelvic girdle and its tip inserted on the knee. The anterior muscles of the thigh (notably the *ilio-tibialis anterior* or *sartorius*) pull the femur forwards and more strongly extend the lower leg, straightening it on the thigh. The posterior muscles pull the femur backwards, increasing the angle which it makes with the pelvic girdle. Thus simultaneous contraction of these two sets of muscles puts the two first segments of the leg into extension, a strong movement which is important for the jump before take-off. The opposing muscles tighten to retract these segments back on each other. The angle of the tarsus on the tibia opens and closes under the influence of muscles in the lower leg, of which the most important is the *gastrocnemius*. This pulls on the tarsus to open the angle at the heel, by way of a joint which works like the kneecap with a cartilaginous tendon passing over the end of the tibia. Very complicated tendons run from muscles in the thigh and lower leg, along the tarsus almost to the tips of the toes, which are flexed by some of these tendons and extended by others. There is no true muscle tissue in these lower segments of the leg, which are therefore light and resistant to cold.

Except for the femur the segments of the leg are long, and the angles they make with one another can be greatly changed. They are arranged in a zigzag, and can fold up on one another or stretch out straight. It is notable that, from the femur to the tarsus, the extensor muscles are better developed than the flexors. Their powerful contraction opens the angles between the segments and stretches out the whole limb. However, at the level of the foot the flexors are better developed, and since these close the foot they also help to lengthen the limb. This is important in walking and especially in take-off, when the contraction of these powerful muscles all along the limb suddenly increases its length, throwing the bird upwards as if by the release of a spring. During the landing these muscles can correspondingly absorb the residual kinetic energy and resist the violent impact. In addition, the powerful flexors of the foot give a bird a strong grip of the branches on which it perches. It is less important for the other segments to be folded back, closing the angles between them, since the weight of the bird itself assists in this movement.

Walking, running and hopping

Though most birds can travel on the ground, some as a rule do not venture there. This is especially true of the swifts, hummingbirds (except a few Andean species) and woodswallows (Artamidae), which are ill adapted to walking because of their weak legs with short tarsi and feeble toes, and also because the length and mode of action of their wings makes take-off from the ground difficult. Swifts hang on to vertical faces, which they do all the better because their first toe can be turned forwards, so that the foot with all four toes bearing recurved claws forms a sort of climbing-iron with multiple hooks. This pamprodactyle arrangement is shown also by the mousebirds (Coliidae). Passerines which perch also descend only rarely to the ground, where their relatively short legs (and especially tarsi) seriously handicap them. Some water birds such as divers and grebes, whose legs are placed far back on their bodies, travel on the land with great difficulty and are therefore rarely found there.

On the other hand many other birds move easily on solid ground, those which run well having long legs and especially tarsi. This is particularly true of passerines, among which one can pick out the running and ground-living species by the proportions of their tarsi and toes. In the best adapted, the pipits and larks, the claws are long and almost straight.

Ground-travelling birds move in two different ways, each accompanied by anatomical modifications to the hind limb which can be detected even within the passerines (Rüggeburg 1960). Those which are adapted especially to life on the ground walk and run, moving their legs alternately. Their bodies are usually long and carried horizontally, their toes long but with short claws, except for the first toe which carries a long and almost straight claw. This mode of progression has the advantage that it involves only small oscillations of the centre of gravity, and so is very economical of energy. It is therefore the gait of the truly terrestrial birds, which run from danger instead of flying and never willingly perch in trees. It is seen also in some small passerines such as larks, pipits and the Snow Bunting, in larger ones such as the Starling and the Crows and of course in other terrestrial groups such as the game-birds. As birds walk they make a characteristic balancing movement of the head, which seems to be thrown back and then forwards at each step. This movement, though it must give some mechanical advantage, is mainly important in keeping the head fixed in space for as long as possible, and then moving it

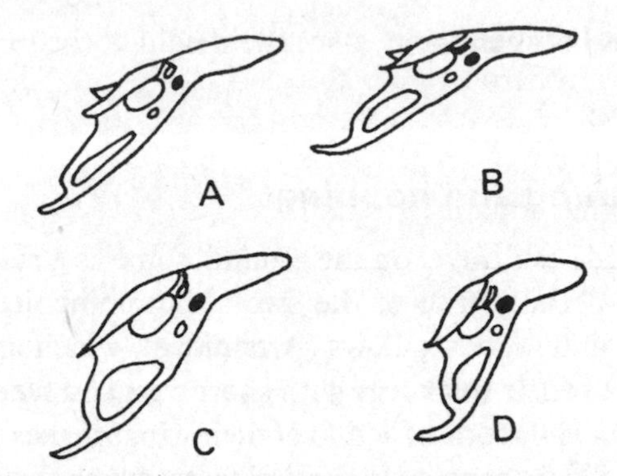

Figure 12. The pelvic girdles of various birds, seen in profile. A. lark *Alauda;* B. Snow Bunting *Plectrophenax;* C. nuthatch *Sitta;* D. warbler *Sylvia.* In the first two, which are walking birds, the girdle is less raised and the *foramen oblongum* narrower and more elongated than in the perching nuthatch and warbler.

quickly forwards to its next fixed position. This gives better visual perception of the environment, by momentarily fixing the projected image on the retina, than if the head was moving throughout the step. These synchronized head-movements are particularly noticeable in birds with highly developed lateral vision, and those with narrow sectors of high visual acuity.

Other birds, all of them small passerines, move over the ground by a very characteristic hopping, the series of jumps being interrupted by pauses. This gait is only possible to small birds, since for larger ones it would involve disproportionate expenditure of energy. They can move fast on the ground for a rather short time, while most of their life is spent in trees to and from which they take flight at the least alarm.

The flightless ratites show an advanced specialization of the legs towards a solipede condition like that of horses. Rheas, emus and cassowaries have retained three rather short and stubby toes. In ostriches the inner toe has disappeared, while the outer one is greatly reduced and is used only to improve the balance. The leg thus ends in a highly developed middle toe with a strong claw, which really carries the whole weight of the bird and also serves as a formidable weapon. The length and power of their legs give these birds a fast sustained trot over the expanses of hard bare soil which constitute their habitat in semi-deserts and steppes. They beat with their flightless wings only to help keep their balance. Bustards, which share the same open habitats, also have long legs ending in three short toes, well adapted to fast running. So does the Secretary Bird, a curious raptor with the gait of a wader, very long tarsi and short toes (the hind one reduced though it carries a stout claw), which chases snakes on foot through the African savannahs.

In contrast, the toes of birds which walk in wet and swampy country tend to be enlarged, as in herons, rails and gallinules. They reach their greatest development in the jacanas, whose slender toes are enormously long and end in long straight claws, an adaptation to walking over the floating leaves of water-lilies and other aquatic plants on which these birds spend most of their time.

Though terrestrial locomotion is not of course nearly as rapid as flight, some birds can run fairly fast: the Emu at 50 km/h, the Pheasant and Road-runner *Geococcyx californianus* at about 34 km/h. At such speeds they can well escape from their enemies and capture their prey.

Perching

Because of their largely arboreal habits, many birds perch on branches, though some seem poorly adapted to this attitude. This is especially true of large waders, herons, egrets, ibises, and above all ducks – though despite their inappropriate-looking webbed feet some of these, such as the whistling ducks, are definitely perching birds. The legs of many truly arboreal birds, especially among the passerines, are better adapted for taking a firm grip of branches. Such a leg usually consists of a tarsus so short as to hamper movement on the ground, but ending in real pincers which can grip a branch tightly. The hind toe is generally well developed and opposed to the middle toe so that together they form the two jaws of this vice, while the lateral toes improve the balance by taking a grip on either side.

This trend is carried further in the birds which were formerly classified as Scansores or climbers, an artificial group based on purely adaptive characters and comprising the parrots, woodpeckers and other Piciformes, trogons and cuckoos. In these birds one of the lateral toes is turned backwards, so that each jaw of the pincers consists of two toes. It is the fourth toe which is turned (*zygodactyle* arrangement) in all except the trogons, whose second toe has undergone this evolutionary change (the unique *heterodactyle* arrangement). The leg tendons are specialized so as to ensure simultaneous control of the toes under muscular contraction. In the woodpeckers, which are the most highly developed climbers, the second and fourth toes are flexed by anastomoses of the tendon which flexes the hind toe, so that all three close simultaneously like the two jaws of a pair of pincers. The peculiar arrangement in turacos (Musophagidae) and the Cuckoo-Roller *Leptosomus discolor* (Madagascan Coracii) is noteworthy, the fourth toe being reversible so that it can be turned forwards or backwards at will.

In all birds perching is made much easier by the arrangement of tendons in the hind limb, which is such that flexion of the heel joint between the tibia and tarsus automatically causes flexion of the toes, which close and tighten on the branch. The bending of the leg, produced automatically by the weight of the body when the muscles relax, thus closes the foot, so that contraction involves hardly any work and muscular energy is conserved. This adaptation is especially important in allowing birds to cling tightly to branches and so remain in position while they sleep. It is reinforced by a very peculiar specialization of the tendons at the level of the toes. Their undersurface is strewn with innumerable tiny points, which under the weight of the bird and the tension of the toes catch into the transverse ridges with which the sheath of the tendon is provided. Thus while the bird is perched the tendons are jammed, further economizing in muscular effort (Welty).

Climbing

Many arboreal birds can climb along steeply sloping branches and trunks, but only a few have overcome the problem of climbing vertical trunks.

The treecreepers (*Certhia*) have solved the problem without any very marked morphological adaptation. The sharp recurved claws with which their feet are provided ensure them a firm grip on the bark, and they climb by successive jumps, supporting themselves by pressing their long pointed rectrices against the trunk. The nuthatches (*Sitta*) are much more skilful, for they are the only birds which can climb with great ease not only up but down trunks, always headfirst. Unlike other climbing birds which always cling vertically, nuthatches move obliquely, hanging on with the upper foot and pulling against it with the lower so that the claws cannot slip. Their short tails are not used for support.

The Dendrocolaptidae, a very diverse passerine family in the South American forests, explore tree-trunks in the search for insects hidden under the bark, over which they move by clinging with their long toes and claws. The shafts of their tails are stiffened yet flexible so that they can be used as props against the trunks. This stops the birds from rocking backwards, but they do not need a strong support since their way of seeking food involves only moderate force. Woodpeckers, on the other hand, must be firmly braced for their task of drilling into wood, and using their specialized feet with two toes in front and two behind they have developed a much more effective posture. Their strong and very sharp claws are curved round in an arc of a circle. The contraction of the flexor muscles make the claws rotate about the joints of the last phalanges, so that having penetrated the bark they are locked against one another, fitting together like some hooking device. The bird closes all the angles of its hind limbs so as to reduce the overturning couple, and clings tight against the tree. It uses its tail, pressed against the trunk, not to prevent itself sliding down but to oppose the force which tends to pull it from the tree and topple it backwards. A woodpecker's tail has strengthened shafts which are tough yet flexible, well adapted to their function as firm props against the trunk.

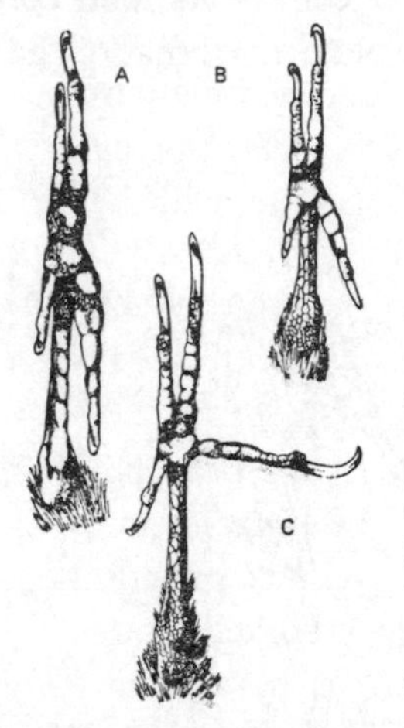

Figure 13. Plantar surfaces of the feet. A. aracari *Pteroglossus*, a percher. B. woodpecker *Celeus elegans*, a percher and climber. C. another woodpecker *Dryocopus pileatus*, a climber. The toes are shown in their normal action positions.

The zygodactyle arrangement of woodpeckers' feet is not a direct adaptation to climbing tree-trunks, but is much more relevant to the habit of perching on branches (Bock & Miller 1959). The toes which are turned backwards are scarcely used in climbing, as is shown also by the evolutionary tendency in some groups of woodpeckers for the hind toe to regress, becoming rudimentary or even vanishing (as in *Dinopium* and *Picoides*). In contrast a lateral support is very useful in balancing, and is provided by the outer toe which in many woodpeckers (such as *Dendrocopos* and *Picoides*) can turn into a lateral position perpendicular to the line of the trunk.

Because the demands made by moving up vertical trunks and by boring into wood are not altogether compatible, woodpeckers show a kind of

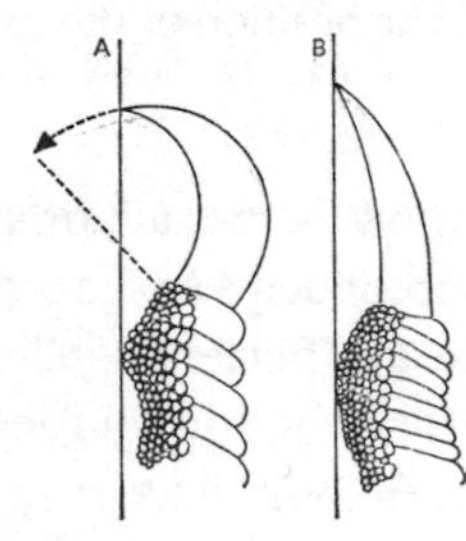

Figure 14. A. the curved claw of a climbing bird. B. the relatively straight claw of one which does not climb. When the foot is applied to a tree-trunk, only claws of the climbing type can pierce the bark.

balance between the two requirements, some being better climbers and others more powerful drillers. This results in different ecological specializations. Thus in North America the Hairy Woodpecker *Dendrocopos villosus* climbs much more easily and with less waste of energy than the Three-toed Woodpecker *Picoides tridactylus*, but on the other hand is less well equipped to bore into wood. These adaptive differences are reflected in its greater locomotor activity and in its diet: only 45 per cent of its food consists of insects extracted from wood, in contrast to 85 per cent of that of the Three-toed Woodpecker.

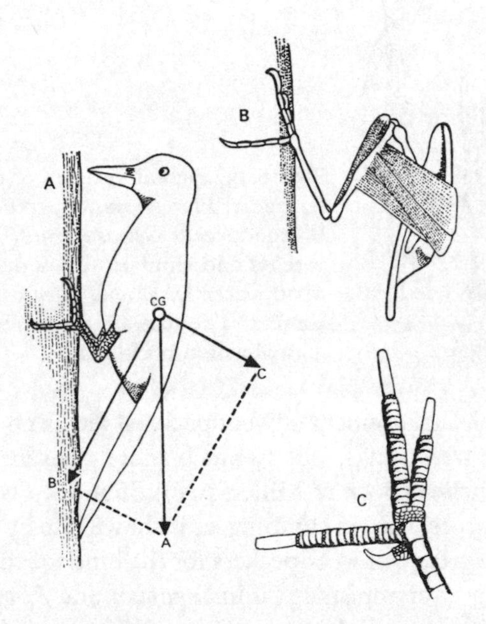

Figure 15. The mechanism of climbing in woodpeckers.
A. Diagram of forces which act on the climbing bird. A. the weight is analysed into two components, B. centripetal and C. centrifugal. These forces originate from the bird's centre of gravity (CG).
B. The pelvic girdle and hind-limb, with the flexor muscles of the leg (flexors of the tibia and tarsus only, those of the toes not shown).
C. The dorsal surface of a *Dendrocopus* foot, showing the position of the toes while climbing.

Birds which can climb along tree-trunks show some adaptations in common, though these are developed to very different degrees. Except in the nuthatches the caudal vertebrae and pygostyle are better developed, in accordance with the more powerful musculature and the special function of the tail. The tarsus is generally short and the toes long, except in certain woodpeckers in which a secondary regression tends towards a reduction even in the number of toes. The flexor tendons are modified in accordance with the important function of the feet.

Swimming

Many water birds can swim and even dive under the water. They float effortlessly because of their low specific gravity. This is given by their plumage, which acts like a natural lifejacket, since the water cannot penetrate between the unwettable feathers. The feathers of aquatic species have a specialized structure (Rutschke 1960). Their barbules are very long, especially at the level of the pennulum, and through torsion they lie in many different planes. They carry innumerable barbicels, which may make up a real network, and thanks to this structure they are especially impermeable. In addition the bird preens, spreading onto the surfaces of its feathers the secretions of the uropygial, or oil, gland. This, the only external gland on a bird's body, is located above the last caudal vertebrae, where it is buried in the mass of muscles which move the rectrices. It consists of two joined ovoid sacs, each (with a few exceptions) having an excretory duct through which it releases an oily substance which sometimes carries a distinct smell (and gives the Musk Duck its name), and is especially well developed among water birds. However, the secretion apparently does not work by waterproofing the feathers, but acts only to keep them in good condition with all their elements in place. The secretion is very important, since birds deprived of their oil glands soon notice that their feathers are wetted every time they enter the water, and learn to shun it. If washed with a detergent, ducks soon lose buoyancy and actually sink. The impermeability of the plumage is also very important in thermoregulation, since it places between the bird's body and the highly conducting water an insulating layer which prevents heat loss.

True diving birds are liable to be handicapped by their lightness. Generally speaking, divers, grebes, cormorants, auks and diving petrels have distinctly denser bodies than other birds, because of their heavier skeletons with reduced development of the air-sacs (see Chapter 6). Before they dive they can sometimes be seen to sleek their feathers against their bodies, by means of a well developed cutaneous musculature, so as to expel air. It has been suggested that swallowing stones, needed for grinding their food, also helps to increase their density. The bodies of diving birds are more cylindrical and longer than those of birds which mainly fly. They are often wrapped in a layer of fat which acts as a heat insulator against the cooling of the water. Their feet are often placed far back on their bodies, where they are more effective for propulsion and easy manoeuvring. This configuration is especially marked in divers and grebes, which seem to thrust against the water in what is almost a crawl kick.

In swimming the feet, which are usually webbed, act like oars. In most swimming birds the web stretches between the three front toes, as in the gulls, ducks and geese; but it sometimes includes the hind toe, as in cormorants, gannets, frigate birds and tropic birds – members of an order to which this peculiarity once gave the name of Totipalmes. The movements of the leg in swimming are much the same as in walking, but with the foot more folded when it is being carried forwards. The toes are thus closed up so as to offer the least water-resistance, just as the tarsus is laterally flattened, in order not to impede the bird's forward movement. When the foot is vigorously thrust backwards the toes spread widely, stretching out the web. Except at top speed the feet move alternately, and they control changes of direction by thrusting unsymmetrically.

In order to dive, birds either tip into a 'duck dive' on the surface, or hurl themselves from a height like gannets, which plunge into the sea like arrows with bills outstretched and wings folded. Most birds swim underwater almost entirely by thrusts of their feet as on the surface, but some make equal use of their wings. In scoters the alula is spread out to act as a fin in directing the stream of water. In diving petrels and auks the whole wing is used for propulsion, and this has certainly affected its size and shape to the detriment of its efficiency in flight.

In penguins, the best-adapted swimmers among all birds, the wings are so highly modified as to be useless for flight, having become real fins or flippers like those of whales. These flattened, heavy, short wings are mounted on a very powerful scapular girdle. The very long sternum with a heavy keel allows the insertion of hugs coracoids and crescentic scapulars. The bones of the wing are flattened and very strong, without joints at the elbow and wrist, so as to strengthen and stiffen the wing against the reaction of the water. The flippers are covered in feathers which overlap like scales, without any differentiation of pennae. The bullet-shaped body of a penguin, protected by a thick layer of fat, is remarkably well streamlined, with the head set into the shoulders, the maximum girth one-third from the front end, and the legs pointing backwards along the midline and involved only in steering. These birds are comparatively heavy and can thus stay submerged with little effort.

Thanks to their flippers, penguins truly fly underwater by movements like those of aerial flapping flight. Like paddles the flippers exert a lifting component, but change pitch at each stroke like controllable propellers, so as to maintain a constant traction. Thanks to this propulsive system, penguins swim easily and fast underwater, momentarily reaching 45 km/h. Over long distances they 'porpoise', breaking surface in a series of leaps just like dolphins. They can plunge into the water from a height of several metres, their abdominal cavities being protected from the impact by their lengthened breastbones. They can also leap out of the water to land on their feet, attaining 1·20m in height and covering 3m horizontally.

In contrast, penguins move much less easily on solid ground. They walk with a studied gait, using their flippers as balancers in a very characteristic waddle which gives them the appearance of clumsy little men. When they wish to move faster they lie prone on the snow or ice and toboggan, pushing themselves along with their feet, their flippers working like oars to help them along. Despite their awkward walk penguins can cover distances of over 100km. Their tracks have been found several hundred km from the antarctic coast and at more than 1,400m above sea-level.

Diving birds can reach considerable depths, although most do not usually go below 5 to 10 metres. Divers have been caught in fishing nets at 55m, diving ducks and cormorants at 40m. Their dives do not usually last more than one or two minutes, but in experiments ducks have survived being submerged for fifteen minutes, which shows that they are anatomically adapted to resist pressure and physiologically adapted to stand shortage of oxygen.

Chapter 4

The Colours of Birds

WHEREAS mammals live in a world of smells, the study of birds' sensory apparatus (Chapter 7) shows that they live in a visual world. Many of the messages by which they communicate between themselves confirm that vision plays the principal part. They benefit from the combination of excellent sight with colours which are lacking from mammals other than primates. One cannot help thinking that a combination of characters of coloration, especially in the plumage, has developed in step with their sensory capabilities. Apart from some groups of insects, coloration has certainly attained its greatest diversity among the birds appealing to our sense of beauty, but serving a utilitarian purpose as the means of expression in most social and sexual contacts between the birds themselves.

These colours are produced in various ways. Some originate rather simply from *pigments*, substances of high power to absorb light. These chemical compounds absorb a part of the incident light, filtering it and reflecting only a part of the spectrum, so that white light is transformed, and the pigment-bearing surface takes on a colour which depends on the nature of the compound and its absorption spectrum. Other colours result from more complex optical effects produced by microscopic structures, through reflections in heterogeneous media or through optical interference. The fine structure of the feather always modifies the coloured appearance.

Melanins

The melanins are certainly the most widely distributed pigments in the animal kingdom and are met with throughout the birds, being responsible for the blacks, greys, browns and yellowish beiges which they so commonly show. Melanins also play a part in the production of so-called 'physical' colours, in which their screening and absorptive properties are fundamental, since it is at this level that the optical interference characteristic of metallic colours is produced.

During the growth of a feather, before it sprouts, the chromatophores containing the pigment gather in the Malpighian layer. Their threadlike processes introduce the granules into the developing barbs and barbules, where they are laid down in layers between those of keratin, as much in the barbs as in the barbules. Although the density of these grains of pigment largely determine the colour of the feather, their arrangement and the structure of the feather also play a part.

Melanin pigmentation is under the control of the endocrine system and especially of the thyroid. Under the influence of sexual hormones, the eumelanin in the male plumage is often replaced by phaeomelanin in the plumage of the female. The quantity of melanin, and therefore the depth of coloration which it produces, is also related to the humidity of the bird's environment. It has long been known that desert birds are pale-coloured, which has been regarded as cryptic coloration, matching the mineral substrate. This influence of the environment is clear when birds from desert habitats are compared with the geographical representatives or close relatives from humid areas. Systematic studies carried out on the Great Grey Shrike *Lanius excubitor* in North Africa have shown that the gradient in deepening coloration of the plumage precisely follows the variations in rainfall (Jany), as in the tits *Parus montanus* and *P. palustris* across the palaearctic region (Kniprath). It has also been shown experimentally that humidity favours the intensification of melanic pigmentation. Diet affects the colour and depth of pigmentation.

Carotenoid pigments

The pigments which are responsible for most yellow and red colours in birds belong to the large family of carotenoids. Their colouring properties and absorption spectra depend on their chemical structure.

Vertebrates, and especially birds, cannot synthesize carotenoids, which they must obtain in their diet. Thus for example flamingos reared in zoological gardens lose their pink colour unless they are fed with certain crustacea rich in carotenoids, which the crustacea themselves have partly derived from the microscopic algae which synthesized them. Birds show important differences from mammals in their carotenoid metabolism. They seem able to assimilate only the xanthophylls or oxygenated carotenoids, from which they produce their plumage pigments, while mammals can assimilate only deoxygenated carotenoids. Experiments have shown that canaries entirely deprived of carotenoids end by becoming white, and that return to their yellow coloration only follows the feeding of xanthophylls, while the true carotenoids have no effect (Völker 1934).

Carotenoids make their appearance dissolved in fatty inclusions within the cells of the feather bud, where they accumulate and soon form a highly coloured concentration, just as the cells begin to keratinize. The keratin thus take the colour of the pigment which is dissolved within itself, unlike the melanins which form granules between the layers of keratin.

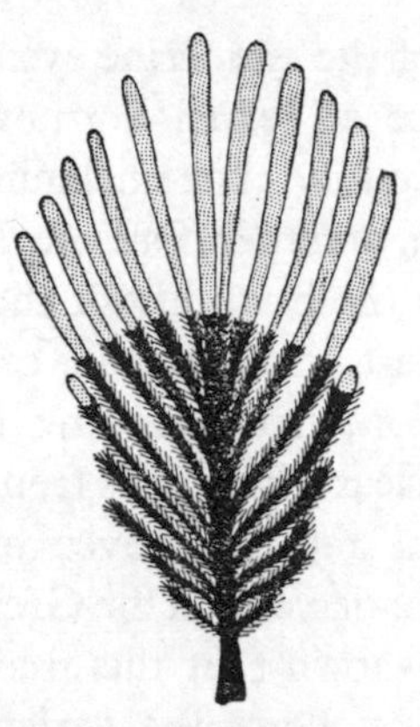

Figure 16. A neck feather from a Cut-throat *Amadina fasciata*, enlarged about 20 times. The proximal part of the feather contains melanins (black), the distal part carotenoids (stippled). These pigments are similarly separated in the Goldfinch and other species.

It is noteworthy that carotenoids are laid down especially in the barbs of the feather. When their concentration increases beyond a certain value, the barbules disappear completely, so that under the microscope the pigmented barbs look like minute, highly coloured sacs. Carotenoids are deposited only towards the tip of the feather, while towards the base the basic structure is preserved and the colour is usually blackish from melanins. Since they hinder the differentiation of barbules, it is easy to understand why carotenoids are not found in the flight feathers, where mechanical necessity demands a strength and flexibility incompatible with the loss of these fixing devices.

Carotenoids are absent from certain families of birds. They are responsible for all true red and yellow colours, except for those metallic colours which occupy this part of the spectrum. They are equally present in green and violet feathers, where they are superimposed.

Turacin and other pigments

The Turacos, a tropical African family of the Cuculiformes, are mostly characterized by a spot of deep carmine on each wing, produced by a very special pigment. This can be extracted from the feathers of the living bird, without the trouble of grinding them up. Experiments have confirmed that washing in slightly alkaline water dissolves this pigment. *Turacin* is a porphyrin, a member of a group of compounds of which haemoglobin is the best known. Where the complex structure of the haemoglobin molecule contains an atom of iron, that of turacin contains one of copper and is unique in the animal kingdom.

Much of the plumage is green in many turacos, at least those of the genus *Turaco*. In contrast to the situation in most birds, this colour is not produced by the specialised structure of the feather but by a true green pigment, *turacoverdin*, whose composition is still unknown. The turacos are thus clearly distinguished by their biochemistry.

Some of the pigments responsible for the coloration of plumage have not long been known, and ornithologists were greatly surprised to discover that many plumages, when exposed to ultra-violet rays, show a very characteristic and usually reddish fluorescence. This phenomenon, shown by pigeons, bustards, some ducks, turacos, owls, nightjars and various passerines, is due to the presence of porphyrins, especially coproporphyrins. The distribution of these substances is not understood, since it does not seem to be connected with systematic position, diet nor way of life. One might suppose some link with nocturnal habits, because of the abundance of these compounds among nightjars and owls, but they are just as concentrated in the feathers of the typically diurnal bustards. Possibly porphyrins are a kind of excretion, voided when the feathers are moulted.

Finally we may note the existence of other unknown pigments, neither carotenoids nor porphyrins, in the feathers of certain birds. These are fluorescent in some parrots, but not in others such as the Budgerigar.

Colours due to scattering

Birds do not possess a true blue pigment. Kingfishers, rollers, some bee-eaters and cotingas, and many passerines, show very extensive blue areas. These result from the reflection and absorption of white incident light within heterogeneous media.

Except for those produced by interference, blue colours in birds are produced in this way, within highly specialized feathers. Their barbs show a particulate structure known as Tyndall blue, or 'Blaustruktur' to German ornithologists, while their barbules are not involved in the coloration. Frank (1939) and Schmidt & Ruska (1962) have shown that the barbs of these feathers are essentially formed of an outer layer of refractive keratin, beneath which are one or more ranks of comparatively large thick-walled 'box cells', which appear under the microscope to be pierced by innumerable micro-vacuoles. The ultramicroscope shows the walls to be full of tiny holes 0·1 to 0·25 μin diameter, separated from one another by keratin walls seldom more than 0·01 to 0·15 μ thick. Since the walls of the box cells measure 2 to

4μ, from 10 to 30 microvacuoles fit within this thickness, rising to about 100 when the walls are thicker. The absorption of the red parts of the spectrum and reflection of the blue takes place in this foamy sponge. Beneath this layer of chambers are others filled with melanin, which contribute to the coloration only by absorbing the stray rays, especially red ones which have passed through the layers described above.

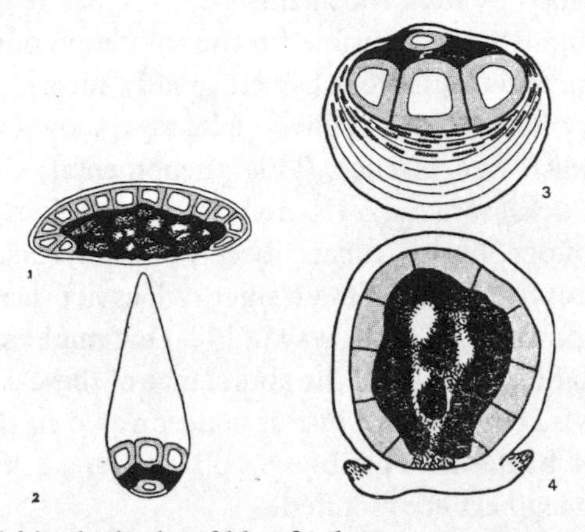

Figure 17. 'Box cells' in the barbs of blue feathers.

The arrangement of these various elements varies between species and areas of plumage. This allows differences in hue, saturation and brilliance to be explained in terms of chromatic purity and the admixture of incompletely 'filtered' components of the spectrum. Some feathers, as in tanagers of the genus *Calospiza*, are of an impure diffuse blue, and spectral analysis shows that the light reflected from their plumage contains a dominant blue, but one which is far from pure being mixed with other wavelengths. Microscopic examination shows that their barbs have few and thin-walled box cells, so that there are not many microvacuoles and the reflected light is relatively poorly 'filtered'. In contrast, where the blue is very intense there is a greater volume of box cells and hence of the heterogeneous medium, so that the composition of the reflected light is more profoundly modified and the colour of greater spectral purity.

Except for special cases such as that of the turacos, green feathers owe their colour to the same phenomenon, since green can be produced by a mixture of blue and yellow. They have the same structure as blue ones, except that the generally thicker layers of keratin enclosing the barbs are impregnated with yellow carotenoid pigment. The intensity and hue of the colour produced depend partly on the feather-structure which produces the blue effect, and partly on the carotenoid content of the feather, while the structure of the keratin layers influences its brilliance. In certain tanagers which show golden reflections, the regular arrangement of these layers allows a play of light which gives a glossy appearance to the plumage.

Violet colours, except for those resulting from interference and those of fruit-doves in which a carmine-violet carotenoid pigment is responsible, are produced in a similar way. A red carotenoid such as zooerythrin is diffused through the superficial layers and gives a red component, which combines with the blue effect to produce violet. In some birds, such as certain Ploceidae (the Gouldian Finch *Poephila gouldiae*) and kingfishers, (the back of the Ruddy Kingfisher *Halcyon coromanda*), the same effect is produced by reddish-brown melanins.

White coloration is produced by diffuse reflection of light, in feathers devoid of pigments.

Interference colours

Some birds are distinguished by a bloom, or even a really metallic glitter which makes them glow like jewels. While other plumage colours do not vary with the angle of the incident light nor with the angle at which they are

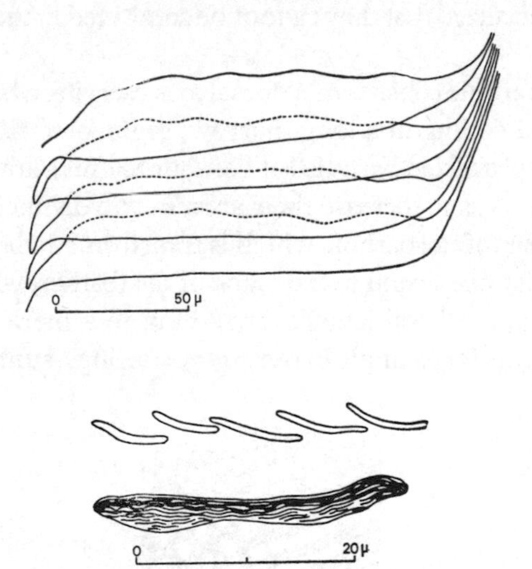

Figure 18. Part of a dorsal feather from the South American hummingbird *Aglaeactis cupripennis*. Above, a series of four successive external barbules from the iridescent area. Note the width of the lower lamellae, which overlap very widely.
Below, the cross-section of an iridescent barbule, greatly magnified.

viewed, these do change according to the conditions of observation, their hue shifting towards blue as the angle of view decreases or the angle of illumination increases. What are called iridescent colours are incontestably the most brilliant and luminous of all those shown by birds.

These colour effects are very widely distributed, sometimes appearing as simple metallic gloss which glows at certain angles. This is true of crows for example, whose black plumage has a characteristic shine. Other birds show a much more intense brilliance, for which naturalists reserve the term metallic. Such colours are sometimes shown only on a very limited area of plumage, as in ducks whose wings are ornamented with a sharply defined patch known as the speculum, with a species-specific coloration in blue, green or gold. Among birds of paradise, humming birds, sunbirds and others, metallic feathers cover larger areas forming remarkable ornaments, especially throat and crown patches. In others again, such as the most brilliant humming birds, some American icterids and the glossy starlings of Africa, the whole plumage shows a metallic glitter which is often very intense.

Such brilliant coloration, never attained by pigmentation, is due to optical interference which takes place in highly specialized barbules. The barbs play no significant part. The minute structures which give rise to the interference are so highly specialized that they cannot be combined in the same anatomical unit.

An ordinary barbule consists of a dorsal axis carrying a basal lamella on its internal side and ending in a long, narrow, segmented flagellum, the pennulum. The unspecialized barbules of contour feathers are greatly altered in metallic feathers so as to increase their area, in two distinct ways. In some it is the terminal part of the barbule which is transformed, the pennulum being widened into a flattened band in the plane of the feather, while the basal part is much reduced, the basal lamella dwindling to a mere small strip. This arrangement occurs for example in the glossy starlings, sunbirds and trogons.

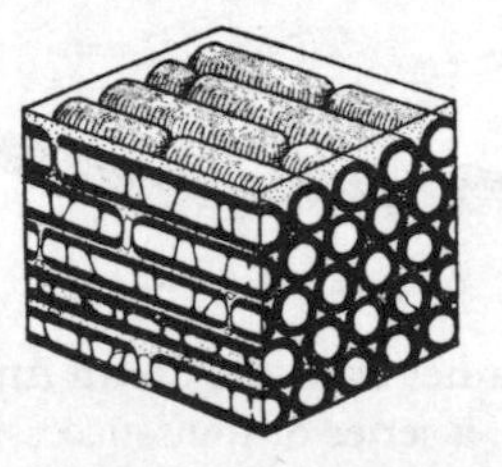

Figure 19. Sections, transverse on the right and longitudinal on the left, of the superficial layer of a barbule from a Trogon. Melanin black, keratin stippled, air-filled spaces white.

Elsewhere it is the basal part of the barbule which is expanded, the pennulum remaining filiform and unspecialized. The basal part is enlarged and widened so as to produce an extensive surface, the basal lamella being so considerably developed that the neighbouring barbules of a single barb overlap to form a continuous plane. This arrangement is found in several glossy cuckoos, e.g. the Emerald Cuckoo, and the hummingbirds. The most highly evolved of the latter show a specialized structure in the development of a superior basal lamella from the dorsal axis. This supplements the original lamella, covering it entirely and substituting for it as the organ of reflection. Thus here too an extensive reflecting surface is formed, more or less in the same plane as the 'useful' part of the feather, and allows the greatest possible scope to the optical phenomena.

Interference colours have been studied especially in hummingbirds (Dorst 1951, Schmidt 1952, Greenewalt, Brandt & Friel 1960), because of the brilliance and variety of their ornaments. Experiments have conclusively shown that these make use of interference phenomena, and the shining feathers contain no trace of pigments apart from quite ordinary melanins. Study of the spectral reflectivity of these feathers yields curves which in general show very sharp maxima, corresponding to the colours shown by the feathers, on either side of which the reflection falls off sharply. Such colours are often very pure, sometimes attaining a purity of 0·90 and approaching that of spectral colours, as those produced by other optical phenomena never do.

Metallic plumages vary enormously in colour, since the wavelength and purity of light reflected from the interference systems depends upon the fine structure of the reflecting organelles, the separation between their walls, their refractive index, and no doubt also on their arrangement and that of the keratin layers. Unlike other types of coloration, that due to interference extends across the whole spectrum, so that these colours are as varied as they are brilliant.

Colour-effects in plumage

We have seen that the colours of birds are produced by a variety of physical causes. The plumage of a single bird often contains feathers whose colours are due to very different mechanisms. Sometimes different types of coloration are even shown by a single feather, as in the throat of the sunbird *Cinnyris senegalensis* which is metallic blue overlaid with brilliant red. Each feather of this ornament has a blackish melanic basal area, a middle part with flattened barbules specialized into organelles producing optical interference, and the tips of the barbs devoid of barbules and coloured red with lipochromes.

The coloured appearance of feathers is almost infinitely variable, thanks to the different types of coloration and to variations in the structures in which they arise. The macroscopic structure of the feathers plays an equally important part, since the various parts which bear the colouring elements themselves vary considerably and this has a marked effect on the appearance, especially the brilliance of colouring. Some birds of paradise may seem to be dressed in velvet, because the short feathers spring perpendicularly out of the skin, as in the throat patch of *Paradisea apoda.* Other birds in contrast have feathers whose components are pressed close together and arranged so as to expose the maximum area, giving a lacquered effect as in some feathers of pittas. The velvety, shiny, silky or hairy appearance due to the arrangements of the feather's components greatly affect the colours and their brilliance.

The plumage of many species, especially among the pigeons, parrots, raptors, cranes, owls, woodpeckers, and some passerines such as the woodswallows (*Artamus*), is dull-looking as though it had been powdered. In fact this appearance is due to a real powder resulting from the disintegration of certain feathers, especially in patches of specialized powder down. The powder particles, with a diameter of about 1μ, reflect only part of the light, diffusing it and accentuating the blue end of its spectrum. This results in the matt finish characteristic of such birds.

It is appropriate to include here those colorations known as 'cosmetic' or 'adherent', produced not by the feather itself but by substances fixed upon it which give it a colour different from its own. These added colours arise especially from the oil gland secretions, which birds spread over their plumage and which may be coloured. Thus in hornbills the properly white parts of the plumage are yellowish, coloured by carotenoids from the uropygial secretion, while some gulls, terns and pelicans are pink for the same reason. These colours fade and disappear from skin after death. Other false colours are produced by substances foreign to birds, of which iron oxide is the most widely distributed having been detected on about 120 species, especially the Bearded Vulture and ducks. Particles from the soil, or from suspension in water, attach themselves to the barbules and to their microscopic organelles. Soot and industrial pollutants are likewise apt to colour the plumage of birds (Berthold 1966).

Colour changes

Feathers are periodically replaced by moulting, once or twice a year. As a rule the plumage can change its colour and texture only at these times, in accordance with the bird's sexual condition, resulting in the juvenile and adult, winter and breeding plumages. Some authors (including Staples) have believed that substances can pass through the keratin itself, which is especially permeable to fats. These substances would not carry fresh pigments into the feather, but would be capable of repairing those already there, especially the carotenoids, and of altering the structure of the feather so as to allow the development of its microelements. The least change in the critical structures could produce important changes in coloration, especially in the interference colours. This hypothetical process, aptosochromatosis, deserves study, though it can be predicted that the effects it may produce are only minor in comparison with those which take place at the moult.

Light can certainly change the colours of feathers, since certain pigments, especially carotenoids, are very sensitive to light. The Green Magpie *Cissa chinensis* and some bee-eaters change from green to blue as a result of change in their yellow pigment, the blue colour due to scattering being resistant to the action of light. Wear, beginning at the exposed edges of the feathers, can also alter the appearance of birds considerably. In some the fresh feathers are edged with a different colour which, since this part of the feather is the most conspicuous, greatly affects the general appearance of the plumage. As the edges are worn away, the coloration changes with the increasing exposure of the more central parts of the feathers. This is true of the male House Sparrow, whose black throat patch is smaller in winter because the feathers have wide pale edges. This kind of change is fairly common among birds, which can thus appear under two different guises while undergoing only one moult a year.

Plumage coloration in the bird's life

Apart from the flight feathers, the most important function of plumage is as a protective facing against damage, and especially against thermal effects including solar radiation. Besides this the patterns produced by coloured areas, and the colours themselves, act as visual messages, sign-stimuli of the greatest importance in the whole behaviour of birds: in recognition between and within species, recognition of prey and enemies, social behaviour, recognition of the sexes and nuptial displays, care of the young, and so on.

The nature, texture and coloration of the plumage has also long been used by systematists in establishing the classification of birds. However, it is proper to be extremely cautious as far as this concerns the delimitation of large groups, despite the opinion of some authors. Feathers, being superficial are subject especially to the influence of the environment and the way of life of the bird, whose ecology has more effect than its systematic position. However, this does not mean that differences in coloration are of no systematic value. Some groups are characterized by particular kinds of pigmentation. Certain structures, such as the double basal lamellae on the barbules of hummingbirds, appear only within well defined families, while colour characters are very important at the level of genera and species. Coloration is important in evolution since it plays a part in most bird behaviour, and especially in reproductive behaviour, and so provides opportunities for selection. However, the ecological significance of plumage characters cannot be denied, since many of them clearly reflect environmental influences. The distribution of colour types across climatic and biological zones is strikingly regular. Tropical forest avifaunas are much richer than others in birds with large areas of brilliant colour, whereas in temperate countries such vivid colours are usually confined to small areas of the plumage. This is especially true of metallic coloration, which in temperate avifaunas are virtually confined to the specula of ducks: birds such as the glossy starlings and hummingbirds, whose colours are entirely due to interference, are typical of hotter climates.

The hypothesis that this regularity is related to metabolic rates may be set up for testing. According to this hypothesis, tropical forest birds live at a higher metabolic level, life being easier for them thanks to a favourable physical environment and an abundance of food. This allows them not merely to meet their 'normal' energy requirements, but even to spend a surplus in synthesizing more intense pigments and in elaborating external structures for their display. Dementiev's hypothesis about desert vertebrates is crucial to the argument. Most of these animals are lightly pigmented, a fact which has long been interpreted in terms of cryptic coloration, as an adaptation making them less conspicuous to their enemies in that pitiless light. However, Dementiev suggests that the prime cause of their loss of pigment is a reduction in vital activity, a slowing of metabolic processes, and especially a limitation in the oxidation of melanin precursor. In contrast, animals of the wet tropics attain a high metabolic rate thanks to their much more favourable environment, and have thus been able to develop 'luxurious' biochemical processes. Intense oxidations yield considerable amounts of melanic pigments, while carotenoids too are much more abundantly represented. The latter are the first pigments to disappear under less favourable circumstances, and are scarcely shown by desert faunas.

These hypotheses deserve to be tested, since they seem to agree with what we know of birds and their evolution. Furthermore, the greater density of species in tropical forest habitats calls for the development of distinguishing marks between species and between the sexes. More generally, the very favourable conditions of tropical forests and the abundance of food ensure that competition is much less keen there than elsewhere. Under such conditions even those specializations which at first sight seem most useless can develop and initiate adaptive radiations.

Chapter 5

Food and Feeding Habits

THE digestive system of birds is highly developed and differs markedly from that of mammals, its form being controlled by two requirements. The first of these is for maximum lightness, needed for aerial life. The head is very much reduced in size compared with the total bulk of the bird, and does not provide space for a powerful chewing apparatus like that of mammals. Teeth, possessed by the earliest birds, soon disappeared, and the bill and mouth serve only to hold on to food and sometimes to tear and break it up. Its reduction to finer particles is left to other organs within the trunk of the bird, near the centre of gravity.

The second requirement of the digestive system is for mechanical and chemical efficiency. Birds need to take in large quantities of food and ensure its rapid digestion, so as to transform it into metabolites with the least delay. In order to be light enough to fly they do not accumulate significant reserves, and it is only the flightless birds which lay down permanent deposits of fat. A bird therefore needs to eat well and often in order to provide the indispensable fuel for its high-efficiency engine. Birds have succeeded in occupying extremely varied food niches, at almost all trophic levels, through their highly diverse diets.

Bill and buccal cavity

The jaws of birds are covered by horny sheaths (*rhamphotheca*) which form the bill, and though devoid of teeth are especially strong and thick towards their tips and edges. At the base of the bill, parrots, diurnal raptors and pigeons have an area of soft skin, the *cere*, which is often brightly coloured and in the middle of which the nostrils open.

Like the whole skull, the skeleton which supports the bill is light. The articulation of the jaw is like that of reptiles, with the quadrate bone involved in multiple articulations and allowing a great mobility to the whole buccal region. In many birds the upper mandible shows a movement on the cranium (*kinesis*), permitted by a naso-frontal hinge or a cartilaginous connection at this level, and by a bony connection between the premaxillaries and the quadrates. Contraction of the adductor muscles of the quadrates rotates these forwards, and the movement is transmitted to the front part of the upper jaw which is free to move upwards. In many birds, including flamingos, woodcocks, parrots, woodpeckers and hornbills, this movement is ample, whereas it is reduced or abolished in others (Beecher 1962, Bock 1964). The lower mandible is always very mobile, so that all birds can open their bills widely. The ability to raise the upper jaw as the lower jaw is lowered must be related to the capture of prey and its movement through the buccal cavity, since it allows the mouth to open in the centre line of the mandibles as they separate. It is also certainly involved in the strong pressure which some birds exert on their food in breaking it up, and in absorbing the shocks which the bills of woodpeckers, for example, must bear.

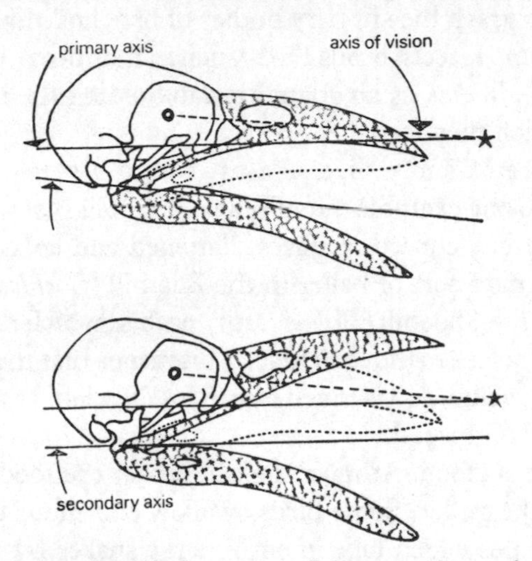

Figure 20. Diagrams showing displacement of the secondary axis of the bill in different birds, with (above) and without (below) kinesis of the upper mandible. The kinetic bill opens more widely, with less displacement of the axis.

The horny bill is made up of several plates, which in most birds fuse together so intimately that the structures are invisible. Only in the Procellariiformes do they remain separate, thereby retaining a primitive condition of which traces can be seen in ostriches, skuas and tinamous. In puffins the bill is covered during the breeding season by extra horny plates, which later fall off in a true moult. In other birds the bill does not moult, but grows continually to compensate for the wear at its edges, which is rapid in graminivorous and frugivorous birds.

Birds' bills appear in a variety of shapes and sizes and a study of feeding adaptations illuminates this enormous diversity. Among the most curious bills are those of the toucans and hornbills, which are huge and bulky yet incredibly light because they are quite hollow, containing vast air-filled cavities which are merely strutted by bony trabeculae. Other bills are enormously elongated, like that of the hummingbird *Docimastes ensifer* which is longer than the bird's body and justifies the name 'sword-billed'. Crossbills (*Loxia*) have mandibles which cross at their tips, while the New Zealand plover *Anarhynchus frontalis* has a unique bill laterally curved to the right.

The edges of the bill (*tomia*) are often sharp, especially in raptors which tear flesh with their bills, and which often have one or two 'teeth' on the tomia. These assist the hooked tip in holding and cutting the prey. Some fish-eating birds, especially the mergansers (*Mergus*), have denticulated tomia in order to grasp the slippery bodies of fish, and this specialization is also shown by some insectivorous birds such as motmots (*Momotus*) and the hummingbird *Androdon*, as an adaptation to holding the insects with hard carapaces on which they feed.

Bills may be remarkably diverse even within a single family, and the Ardeidae provide one example out of many. The bill is straight and pointed in herons and egrets, curved in ibises, flattened and enlarged at the tip in spoonbills. It forms a sort of bailer in the Boatbill (*Cochlearius*) and a huge swollen organ in the Shoebill (*Balaeniceps*), each of which may be placed in a monotypic family close to the Ardeidae. This means that the shape of the bill has no systematic value, and is much more closely related to diet and the way in which the food is grasped.

No bird chews its food. At most the bill breaks the food up into morsels which can pass the gullet. Some birds swallow enormous prey whole, such as the serpent eagles which take in entire large snakes by peristaltic movements of the oesophagus. Only in some graminivorous birds is the bill used to crush the hard coats of grain, or of fruit stones whose kernels are sought for food.

The bill is always provided with sensory organs, especially tactile corpuscles by which the bird can explore its environment and seek out its prey. The tip of the bill is especially well innervated in woodpeckers and ducks, while in waders it is soft and dimpled with sensory pits.

The tongue is hidden within the buccal cavity where it is connected only towards the distal end, originating from a complicated pharyngeal structure which includes the hyoid and mandibular arches. Its rich innervation comprises motor fibres distributed throughout its length, and sensory fibres towards its base where they play an important part in the sense of taste. Tactile corpuscles, especially developed in woodpeckers, occur throughout its length. The tongue is highly mobile thanks to a skeletal and muscular apparatus which varies from species to species, being especially well developed in the woodpeckers and in nectar-feeding birds.

Birds' tongues are very diverse in shape, their variations being as great as those of the bill. In most birds the tongue is of the ordinary lance-like shape, keratinized at the tip and for almost its whole length, and bordered towards the rear by backwards-pointing cornified papillae. In the cormorants and other pelecaniformes, which swallow fish whole, the tongue is reduced to a mere vestige, while in penguins and certain other fish-eating birds it is highly keratinized and covered with backwards-pointing spikes. In contrast it is thick and fleshy in parrots, where it plays an obvious part in sound production and imitation of the human voice, and in ducks and geese, where its base carries small horny lamellae which join those of the bill to form a filter system. In nectar-feeding birds it forms either an elongated tube through which nectar is drawn from the corollas (in the sunbirds, honey-eaters and hummingbirds), or in contrast a sort of small brush with which the nectar is licked up (in the brush-tongued lories). In woodpeckers the tongue is specialized for collecting wood-boring larvae, after the bird has broken these out from their hiding places with its chisel-like bill.

At rest the tongue is enclosed within the buccal cavity which is hard and keratinized, only becoming soft towards the back of the mouth. Birds lack the soft palate which prolongs the vault of the bony palate backwards in mammals, and so they swallow in a different manner. The food is thrown to the back of the pharynx by a movement of the head, and simultaneous reflexes close the internal nares and move the larynx forwards to block the respiratory passages. The throat is usually sufficiently wide to allow bulky prey to pass through. To drink, a bird simply fills its lower jaw with water, and tilting its head backwards allows the liquid to flow under gravity into the oesophagus.

The buccal cavity is strewn with sensory corpuscles and with mucous glands. These glands are most numerous in graminivorous birds which take a dry diet, but correspondingly much reduced in aquatic birds and even absent from the Pelecaniformes. Similarly, salivary glands are developed to a very variable degree. They are especially large in the Jay and in woodpeckers. The floor of a woodpecker's mouth carries a pair of highly-developed glands (up to 7 cm long in the Green Woodpecker), which serve to lubricate the tongue and make it sticky for capturing insects, just as in mammalian anteaters. The salivary glands are even larger in swallows, and in swifts they increase still more during the breeding season, the mucous saliva cementing together the mud and vegetable matter of which their nests are made. This trend reaches its extreme in the swiftlets (*Collocalia*) of the Far East, whose nests stuck to rocky walls and inside caves are built largely, and even in some species entirely, of compressed and hardened saliva. These nests provide the basis for the famous 'birds' nest soup' so much valued in Chinese cooking.

The gut

Although the digestive tract of birds is formed on the general vertebrate plan, and conspicuously preserves certain reptilian peculiaritics, it is notable for its unique features.

OESOPHAGUS AND CROP

The oesophagus is wide compared to other vertebrates, especially in those birds which swallow large prey – the divers, auks, gulls, cormorants, herons, raptors, owls, and kingfishers. In birds whose prey is hard and sharp, its internal surface is strongly keratinized.

Besides acting as a duct, the oesophagus serves for the temporary storage of food. In many birds, such as the cormorants, part of its length is always widened for this purpose, and this specialization is carried further by the

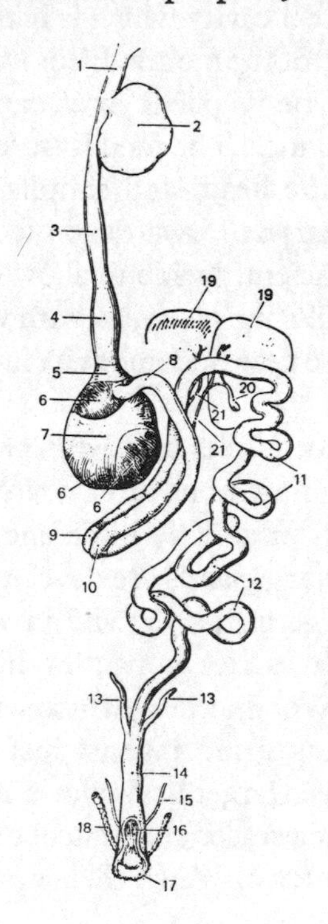

Figure 21. 1. oesophagus. 2. crop. 3. crop. 4. proventriculus. 5. cardiac orifice. 6. muscles of the gizzard. 7. tendinous plate. 8. pyloric orifice. 9. small intestine. 10. pancreas. 11. jejunum. 12. ileum. 13. caecum. 14. large intestine. 15. ureter. 16. bursa of Fabricius. 17. cloaca. 18. vas deferens (in the male bird). 19. liver. 20. gall-bladder. 21. hepatic ducts. 22. cardiac diverticulum.

raptors in which the oesophagus is strongly dilated. This foreshadows the formation in some birds of a clearly separated pouch, the *crop*, which in pigeons, parrots and game-birds is a huge extensible sac, supported by the clavicles when it is full. The crop is essentially a temporary reservoir for food. It is an organ for the rapid storage of large amounts of food, in birds which are at the mercy of their enemies while they are feeding. This is true, for example, of pigeons searching on the ground for grain, when they are threatened by mammalian predators.

The crop is truly important in digestion only in the Hoazin *Opisthocomus hoazin* of the amazonian forests, in which the oesophagus has differentiated such a huge pouch that part of the breastbone is deformed to make room for it. This muscular keratinized crop acts like a true stomach, grinding up the tough leaves on which this strange bird feeds. Two further pouches are attached to it, swelling from the posterior part of the oesophagus before this opens into the true stomach, which is contrastingly small. Thus the functions of the stomach are relegated to the crop, in connection with the specialized diet of this bird.

The crops of pigeons are highly specialized for the production of 'milk', a cheesy substance rich in proteins (10–15 per cent) and fats (20–30 per cent), though poor in sugars. This is produced by the proliferation of cells from the lining epithelium, which is controlled by the pituitary hormone prolactin – the same hormone as releases the secretion of milk in mammals. The secretion of 'milk' begins in both parents on the eighth day of incubation and continues until the sixteenth day after hatching. The milk is regurgitated and fed to the squabs in an adaptation unique among birds, the pigeons being thus the only birds to rear their young at least in part on a secretion from their own bodies.

STOMACH

The stomach region of birds is always made up of two very distinct parts, the anterior one glandular and the posterior one muscular. Comparison with reptiles shows that only the muscular part corresponds to the reptilian stomach, and that the glandular part is a new development. Mechanical and chemical functions are carried out in two clearly separated pouches, the chemical process beginning before the mechanical one, in contrast to the order in mammals.

The glandular stomach or *proventriculus* is a cigar-shaped sac whose walls are rich in glands. It is developed to a variable degree depending on the diet of the species, being largest in the Procellariformes, gulls, cormorants and herons, all birds which eat large prey. The glands sometimes occupy the whole wall of the proventriculus, as in many passerines, and are sometimes concentrated in certain areas which may even form a kind of diverticulum. The glands secrete juices rich in proteolytic enzymes, especially pepsin, which break down the proteins in the food. The glandular stomach is under very precise nervous control, which adjusts the amount of secretion to the quantity and quality of food which it contains. This organ provides a staging point before the muscular stomach, where the food is impregnated with gastric juices and begins to be digested. Sometimes it also serves for food storage, thus supplementing the crop.

The *gizzard*, or muscular part of the stomach devoid of digestive glands, also shows many very strict adaptations to particular diets. Occasionally it forms nothing more than an extensible pocket with reduced musculature, in which the food already impregnated with secretions of the proventriculus undergoes further digestion. This arrangement is found only in purely carnivorous birds, the raptors, owls and fish-eaters. In contrast the gizzard of vegetarian, and especially graminivorous birds is a very muscular pouch forming an extremely powerful grinding organ, with muscles radiating from two symmetrical centres to form opposing discs. The cavity between them is lined with a thick membrane of keratinoid substance (*koilin*), which protects the underlying muscle from bruising and tearing by hard bodies. Herbivorous and graminivorous birds assist the action of the gizzard by swallowing

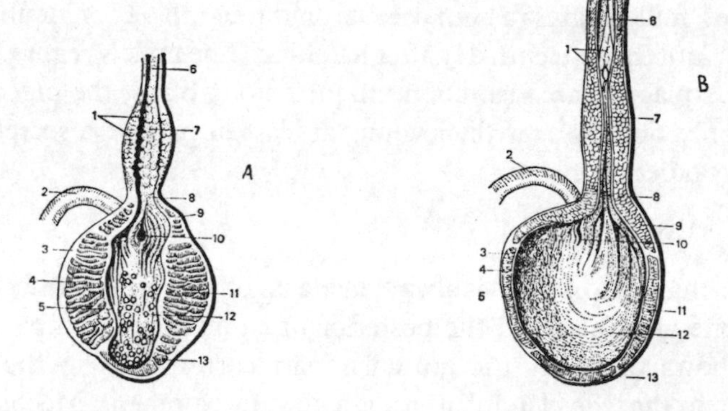

Figure 22. Gizzards:
A. of a graminivorous bird. 1. wall of the proventriculus. 2. duodenum. 3–5. muscles of the gizzard wall. 6. oesophagus. 7 and 8. epithelium of the proventriculus. 9, 11 and 13. muscles of the gizzard. 10. cardiac orifice. 12. cornified layer.
B. of a carnivorous bird. 1 and 7. proventriculus. 2. duodenum. 3–5, 9, 11 and 13. muscular wall of the gizzard. 6. oesophagus. 8. cardiac orifice. 10. pylorus. 12. epithelium.

pebbles which lodge in the gizzard for a long time and often accumulate to form a considerable bulk, converting the gizzard into a much more effective ballmill. As early as 1752, Réaumur showed that a turkey could grind in its gizzard a metal tube which would resist a force of up to 36kg. Also it is well known that ostriches, rather unusually, swallow foreign bodies which have nothing in common with their food. One swallowed a copper candlestick which was found to be twisted when it was retrieved from its stomach, demonstrating the power of that organ.

The gizzard also acts as a barrier against indigestible fragments in the food, such as spines, bones, feathers, hair, insect carapaces and shells. Sometimes its pyloric part is even provided with long fibres which form an efficient filter, preventing hard bodies from passing through the posterior opening of the stomach. Indigestible particles are regurgitated and expelled by way of the mouth in the form of *pellets*. Raptors and owls do this in a very regular rhythm, and analysis of the pellets allows reliable study of their diets.

Functionally the gizzard replaces the buccal cavity and teeth of mammals, though its truly digestive function is more important than theirs, since it is here especially that proteins are broken down by enzymes secreted in the proventriculus. This is particularly important in carnivorous birds, in which the mechanical action plays a smaller part in digestion than the chemical one.

INTESTINE

Birds' intestines are less clearly differentiated into distinct segments than those of mammals. However, although this part of the alimentary canal undergoes large peristaltic movements the disposition of the loops remains much more constant because the intestine is more firmly held in place. This is an aspect of that stability which the demands of flight impose on all organs.

As among mammals, the intestine is relatively long in herbivorous and graminivorous birds – as well as in piscivorous ones in which it is slender with a narrow lumen – while it is shorter in frugivorous and especially in carnivorous ones. Larger birds have relatively longer intestines. Nevertheless, this variability is much less than among mammals, since in birds it is the anterior parts of the alimentary canal which are mainly concerned in fundamental adaptations.

As in other vertebrates, the intestine is the seat of digestion and absorption. The liver, which is especially bulky in piscivorous and insectivorous birds, is proportionally larger than in mammals, and like theirs plays an essential part in all the metabolic processes. It consists of two lobes which pour out their secretions through separate ducts, the one leading from the caudal lobe sometimes bearing a bladder. The bile secretions play the same part as in mammals, although the bile is acid and not alkaline. The bulky pancreas, enclosed within the duodenal loop, is the more important because the action of its starch-reducing enzymes is not preceded by that of salivary secretions as in mammals. This essential organ also secretes other enzymes which break down fats and proteins.

A pair of caeca of very variable development opens into the small intestine, at the level at which in mammals it runs into the large intestine. These caeca are absent from Procellariiformes, raptors, parrots, and kingfishers, and much reduced in hummingbirds and swifts. In contrast they are highly developed in herbivores, especially ostriches, grouse and tinamous, in which their area is often increased by extensive lobulation. They play a complicated part, being involved in the absorption of water and of proteins while serving most importantly as the site of a true microbial digestion. Their development and size are strictly correlated with the proportion of cellulose in the diet.

The rectum, which is relatively short and is important in the absorption of water rather than of nutrients, leads to the cloaca into which the excretory and sexual ducts also open. The cloaca is divided into several chambers (the *coprodeum*, *urodeum* and *proctodeum*) by transverse flanges. In young birds a diverticulum, the *bursa of Fabricius*, opens towards the middle of its dorsal wall. This plays no part in digestion, but is of a lymphatic structure analogous to that of the thymus. Like the latter it retrogresses as the gonads develop and (except in certain ratites) disappears with maturity. Its function is unknown but cannot be important, since removal of the bursa has no effect.

Digestion

The digestive systems of birds, highly evolved anatomically, are also remarkably efficient physiologically. Each bolus of food passes notably quickly through the alimentary canal, and faster still in frugivorous and especially in carnivorous birds. A shrike digests a mouse in three hours, while a waxwing digests berries in sixteen to forty minutes and a thrush in thirty minutes. Passerines in general take an average of one and a half hours for food to pass through the alimentary canal, whatever their diet. This allows them to eat a large amount of food in relation to their size, taking it in frequently and metabolizing it rapidly. This rapid transit is ensured by the great mobility of various parts of the canal. Intense peristalsis, of up to 75 contractions an hour in the fowl, takes place in the oesophagus and crop, while the gizzard makes a rotary contraction two or three times a minute. The intestines too are swept by waves of peristalsis running towards the rectum. These mechanical movements are accompanied by a rapid chemical breakdown of the food, under the influence of very active enzymes. A high digestive yield is attested by the rapid growth of young birds. For example a young stork can increase in weight by 170g through eating 500g of fish and frogs.

Recent research has shown that the process of microbial digestion also goes on and is especially active in the Tetraonidae: grouse are known to feed largely on the needles of conifers, a diet rich in substances which the birds' own enzymes cannot break down.

Diets

Birds have adapted to all diets, and feed on every thing on earth. Many birds take a mixed diet, merely showing a more or less marked preference for a particular kind of food.

Corvidae feed as much on plants as on insects, worms, small vertebrates, carcases and all sorts of rubbish. Gulls feed on fish and other marine animals as well as on very diverse rubbish accumulated on beaches, and on insects and their larvae and worms. Other birds with much more specialized diets being dependent on strictly specified kinds of food, are known as *stenophages*. Some raptors hunt only a well-defined class of prey; cormorants and gannets take only fish; many pigeons feed solely on grain; swallows and swifts chase only flying insects; while cuckoos are entirely insectivorous, and especially take hairy caterpillars rejected by other birds.

Between these two categories are the birds of mixed diet. Many birds show no more than moderate dietary preferences. Some change their diet during the year, and from being stenophages at one season become polyphages at another. Often the young birds take a diet different from the adults, in relation to the demands of growth, as in graminivorous passerines whose young are fed especially on insects.

The dangerous Portugese Men o' War, pelagic siphonophore coelenterates, are preyed upon by certain albatrosses which are naturally immune to the highly potent venom of these animals. The food of birds thus provides evidence of their remarkable adaptive faculties and allows the differentiation of very diverse ecological niches, as a few examples will make clear.

VEGETARIAN DIETS

Many birds are strict vegetarians, most often exploiting a strictly defined category of plants, while others take vegetable matter as part of a largely animal diet.

Some, such as larks and other passerines, bustards and cranes, feed on herbs, especially their young shoots. Rooks and crows take sprouting grains, which they skilfully dig up with their long powerful bills. Many geese are herbivorous in the strict sense, actually grazing grass. Magellan Geese (*Chloephaga*) of the southern part of South America feed mainly on the grass of the pampas. It has been estimated that seven geese eat as much as a sheep, which justifies the hatred of the breeders who accuse them of competing with their flocks.

The brent geese (*Branta*) of the northern hemisphere show a marked preference for the fields of sea-grass (*Zostera*) and algae (*Enteromorpha* and *Ulva*) uncovered by the tide. Swans eat so much aquatic vegetation that they are useful in keeping ponds clear. Other birds, especially some coots, feed on submerged vegetation (e.g. *Myriophyllum*). All these birds have strong bills, often with cutting edges for biting through the stalks.

Few birds eat leaves, with the exception of the Hoazin (*Opisthocomus*). In contrast buds, and especially flower buds, are sought after by tits and many other passerines, such as Bullfinches (*Pyrrhula*) which are therefore accused of harming fruit trees. The bills of these birds most often have strongly cutting edges.

Black Grouse and their allies feed largely on the shoots of juniper and heather, especially in the winter, but also on the needles of conifers which are scorned by other birds. Willow Grouse (*Lagopus*) feed in winter mainly on the shoots of dwarf willow (*Salix*), which form 94 per cent of their diet in the arctic tundras.

Fruits provide a rich food supply which is especially lavish in the tropics, so that it is not surprising that many birds there have become frugivorous – some groups of pigeons such as the fruit pigeons (Duculinae) and fruit doves (Ptilinopinae); the turacos, parrots, trogons, toucans, barbets, and colies (mousebirds); and many passerines such as the cotingas, tanagers and dicaeids. The Oilbird *Steatornis caripensis* feeds solely on pulpy fruit, whose remains form a thick carpet to the caves where this peculiar bird nests. None of these groups have expanded out of the tropics, no doubt because they are so dependent on a diet of fruit that they must live in areas capable of providing it throughout the year. However, the temperate regions are occupied by species which are partly frugivorous, such as the Turdidae (thrushes and smaller related forms), Sylviidae (warblers) and Parulidae (American wood-warblers), and to a less extent the crows and orioles. Many other birds readily make up a mixed diet with fruit, even among the raptors: the Palm-nut Vulture (*Gypohierax angolensis*) takes the fruit of the Oil Palm.

A diet of fruits with soft pulpy flesh requires no special adaptations of the bill, and in fact frugivorous birds show every type from the fine bills of warblers to the huge ones of toucans and hornbills. At the most one can call attention to the adaptation of the Pacific fruit pigeons, which can markedly stretch the base of the bill like the jaws of a snake, so as to swallow huge fruit. They can thus take fruit up to 5cm in diameter, such as those of the nutmeg to which these birds are partial, and their intestines are wide enough to pass the stones of these fruits.

Other birds have become graminivorous, exploiting the seeds of a great many plant species and showing specializations in accordance with their physical properties. This preference is understandable because of the high energy content of seeds, and the fact that they can be found even in cold countries, and at the most unfavourable times of year. Graminivorous diets impose much more marked morphological and anatomical adaptations than frugivorous diets, because of the hardness and dryness of such food. These adaptations are particularly clear among the passerines, in which a graminivorous diet may be a secondary adaptation, since apparently these birds are primitively frugivorous and insectivorous (Ziswiler 1965, 1967). Some passerines, such as larks, tits, tanagers, caciques and starlings, are graminivorous only occasionally or for part of the year. In contrast others, such as the finches, buntings, weavers and estrildid weavers, make seeds their main or even their sole food. This diet has produced a series of special adaptations, which are most conspicuous in the bill and alimentary canal.

Whereas birds which eat seeds only occasionally swallow them whole in their husks, leaving them to be broken open entirely by the gizzard, true seed-eaters husk them with their bills. This involves a considerable mechanical effort, demanding a stronger bill than those of other birds.

Birds use two methods of husking seeds, and can be divided into two groups on this basis. The first group take a seed between the cutting edges of the bill and slice it, the lower jaw making very pronounced backwards and forwards movements. This method is used by all the Fringillidae and by *Erythrura* among the Estrildidae, especially on the seeds of dicotyledons. The other group positions the seed within the bill, wedged against the ridged palate, and breaks it by powerful pressure of the lower jaw, acting upwards. This method is used by the buntings, the American seed-eaters (Emberizinae), the weavers, and most of the estrildid weavers. It is applied especially to the seeds of monocotyledons and notably those of grasses. Birds adapted to this method of opening seeds have generally more robust bills, without cutting edges but with more highly differentiated palates, bearing ridges, plates and tubercules. Their jaw articulations and musculature also differ from those of birds which cut seeds with the edges of their bills. In all seed-eating birds the seed is husked with the tongue after the bill has done its work.

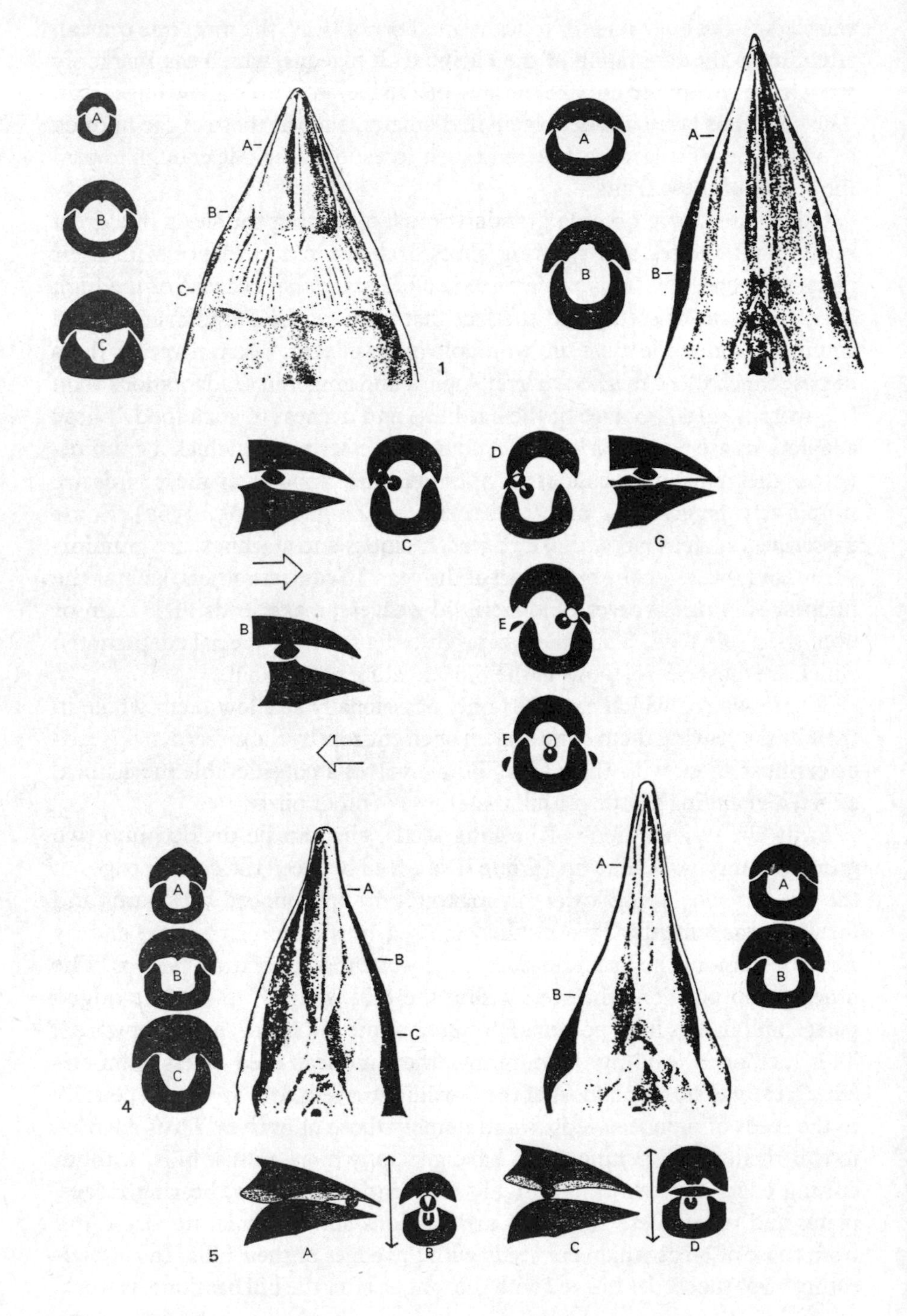
A
B
C
1
A
B
A
B
C
D
E
F
G
A
B
C
4
A
B
A
B
C
D
5

The Hawfinch and other true grosbeaks (*Coccothraustes* and related genera), which feed partly on kernels taken from very hard nuts, are remarkably adapted to this diet, as is shown by the conformation of their skulls and bills which are constructed so as to exert considerable forces (Mountfort, Sims). Comparison of the Hawfinch's skull with that of the Chaffinch, a bird belonging to the same family and of similar size, shows that the former is 2·5 times as large and four times as heavy and solid as the latter. Within each jaw are two furrowed knobs or anvils, between which the nut to be cracked is placed. The adductor muscles of the lower jaw are highly developed and capable of exerting considerable force. It has been calculated that the force needed to break the stones of cherries is 27 to 43kg, and of olives as much as 48 to 72kg. That the Hawfinch can crush these stones in order to extract their kernels proves the rigidity of its skull and the power of its musculature.

Similar arrangements are shown by other birds, notably the Nutcracker *Nucifraga caryocatactes*, whose palate is likewise provided with a projection against which fruit stones are wedged to make it easier to break them open. Large parrots, such as cockatoos and macaws, feed largely on stones with very hard shells. The African Grey Parrot *Psittacus erithacus* feeds largely on the fruits of oil palms, breaking their woody stones. The crossbills (*Loxia*) are largely adapted to taking conifer seeds from between the scales of their cones, for which the tips of their bills are uniquely crossed. The divergence of the tips allows the bill to be inserted between the scales and used to prize them apart. However, this adaptation is not indispensable since the Pine Grosbeak *Pinicola enucleator*, related to the crossbills, has a stout but normally shaped bill yet feeds on the seeds of conifers, as do certain

---

Figure 23. 1. Palate of the Hawfinch *Coccothraustes coccothraustes*, and cross-sections of the bill at levels A, B and C.
2. Palate of the Chaffinch *Fringilla coelebs*, and cross-sections of the bill at levels A and B.
3. Diagram showing the technique used by the Goldfinch and related species to break open and husk seeds. The seed is broken, A and B (in profile), C (in cross-section). The seed is husked, D–F (in cross-section), G (in profile).
The white arrows show the direction of movement of the lower mandible. The white area in the profiles of the upper jaw represents its lateral wall.
4. Palates of two ploceids, *Hypargos niveoguttatus* on the left and *Emblema picta* on the right, and cross-sections of the bills at the levels shown.
5. The technique used by an astrild (Ploceidae) to open up a round seed (A, B) and an elongated one (C, D).
American hummingbirds (Trochilidae).

woodpeckers and tits. Furthermore, the diet of crossbills is not as rigid as was thought, since they can feed on other fruit – especially during migrations which take them far from any coniferous forests.

Some birds which feed partly on seeds lack a bill adapted for cracking tough seed coats, and use instead the trick of wedging the fruit into a crevice in the bark and then hammering it with the bill. This is true of woodpeckers, which choose sites (sometimes called anvils) to which they regularly return with fruits or pine cones, in order to open them and remove the kernels or seeds. A nuthatch similarly takes a hard shelled stone, and having jammed it into a cleft drills it with its narrow conical bill, and tits often do the same.

Strictly graminivorous diets also involve modifications of the alimentary canal. While the proventriculus and intestine differ little from those of insectivorous birds, this is not true of the gullet and the gizzard. This part of the alimentary canal is provided with mucous glands, whose arrangement varies between groups. So does that of the very powerful musculature of the gizzard, which is most highly developed in graminivorous birds.

These birds are thus well equipped to take, tear open and digest seeds. Most are specialized for collecting the seeds of particular plants, whose properties and the nature and strength of whose seed coats determine the choice. Every species feeds by preference on seeds of a size adapted to its bill, and which it can husk most economically in the shortest time (Hespenheide 1966).

Plants thus provide in their fruits and seeds an abundant source of food. The birds in their turn perform various services for the plants on which they feed, contributing largely to their dissemination in a great variety of ways. For example, birds are liable to distribute certain hooked or sticky seeds which get caught up in their feathers.

A more active means of dissemination is carried out only by those birds which eat hard seeds or fruit, which they pick at one place and carry elsewhere to be eaten more conveniently. During this conveyance several fruits may fall unnoticed to the ground, to germinate and thus perpetuate the species. This is true of the Grey Parrot, which takes the fruit of the oil palm some distance before eating them. It drops some of them along the way, and thus spreads the oil palm through the African forests.

A still more effective way of scattering seeds is that of frugivorous birds which eat the fruit containing them. The pulp is digested, but the seeds merely pass through the alimentary canal and are voided with the faeces, their germinating power preserved by the husks which protect the seedling from the digestive juices. Indeed, these woody containers are often softened and attacked by the enzymes, which thus facilitate sprouting. The classic example is the mistletoe, whose berries are sought during the winter by numbers of birds, especially by the Mistle Thrush and by warblers. Its seeds are very soon voided (which explains why dissemination of the mistletoe takes place within a small area), and if one falls onto a branch to which it sticks it germinates to form a new parasitic plant. Warblers do not even swallow the berry whole, but carry it to a branch to tear it open, leaving the seed which becomes stuck and is thus truly sowed by the bird. This association between mistletoes and birds holds throughout the world, the Loranthaceae being disseminated in tropical Asia and Australia by the dicaeids, in America by the tanagers, and in Africa by the turacos and barbets. Though this association can scarcely be obligatory, it none the less favours the plant and thus in turn the bird, which seems to plant its own vegetables.

Another important plant resource, though one limited to the tropics, is provided by nectar, the sugary liquid secreted by the flowers of certain plant families. This contains dissolved sugars and is therefore sought by anthophilic birds. Species which feed on nectar show alimentary specializations, except for some tanagers and finches which drink it only occasionally. Nectar-feeding birds are found among the hummingbirds (Trochilidae) and honeycreepers (Coerebidae) in the new world, and among the honey-eaters (Meliphagidae), sunbirds (Nectarinidae), Hawaiian honeycreepers (Drepanididae) and the parrots known as lories (Trichoglossidae) in the old world. Except among the lories, the bill is generally elongated and hides a tongue which is profoundly modified into a tubular organ, extensible by means of a highly developed hyoid apparatus. The long and very narrow tongue of hummingbirds is supported basally by a part of this apparatus, and by a pair of paraglossals which are partly ossified and partly cartilaginous. Along the middle of its base runs a gutter-like groove, which becomes more marked anteriorly, separating the tongue into lateral halves. These are hollowed out dorso-laterally to form two tubes, with walls curled over on themselves but not entirely closed. The end of the tongue is split in two and thus forms two tubes, which are open along their sides and frayed at their tips. The tongue is very strongly protractile thanks to a well developed hyoid apparatus and to muscles (*m. geniohyoideus, m. stylo-hyoideus*) which wrap round the cranium and insert at the nostrils from behind. When this tongue is poked from the bill into the corolla of a nectar-bearing flower, it does not as one might expect work like a 'straw' through which drink is sucked, but fills itself by capillarity (Weymouth *et al.* 1964). However, it seems likely that reduced pressure produced by movements of the throat does play some part in the flow of nectar. The tongue of sunbirds forms a long narrow tube by rolling up on itself longitudinally. That of honeyeaters, similarly rolled, ends in four lobes which vary in shape from species to species, but are always extensively frayed to form a bundle of 'hairs', with which the nectar can be licked up and then sucked through the tubular part.

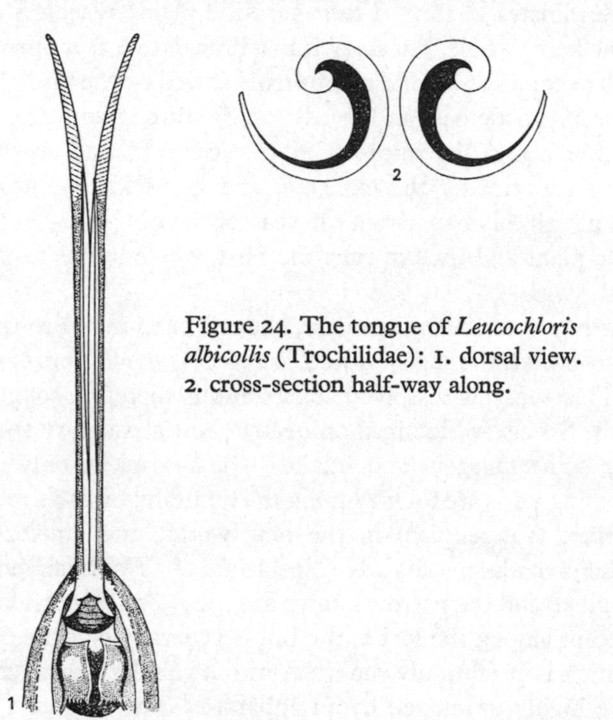

Figure 24. The tongue of *Leucochloris albicollis* (Trochilidae): 1. dorsal view. 2. cross-section half-way along.

In contrast the tongue of lories is differently formed. It is not curled, and retains some of the thickness of other parrots' tongues, but ends in a dense bundle of 'hairs' forming a true little brush for licking nectar. It is reasonable to suppose that this structure is an adaptation to the shape of Australasian nectar-bearing flowers. In contrast to many African and South American plants, the sources of nectar for sunbirds and hummingbirds, which have flowers with tubular corollas, those of most of their Australasian counterparts are in the form of widely open cups. This is true of the eucalyptus and other Myrtaceae which are so abundant in the Pacific flora. A 'brush' like that of the lories seems better adapted to such flowers than a 'pipette' like that of hummingbirds, and it is noteworthy that the honeyeaters, which occupy the same area, are also provided with a kind of brush supplementing the tubular structure of the tongue.

Nectar-bearing flowers are assiduously visited by nectarivorous birds,

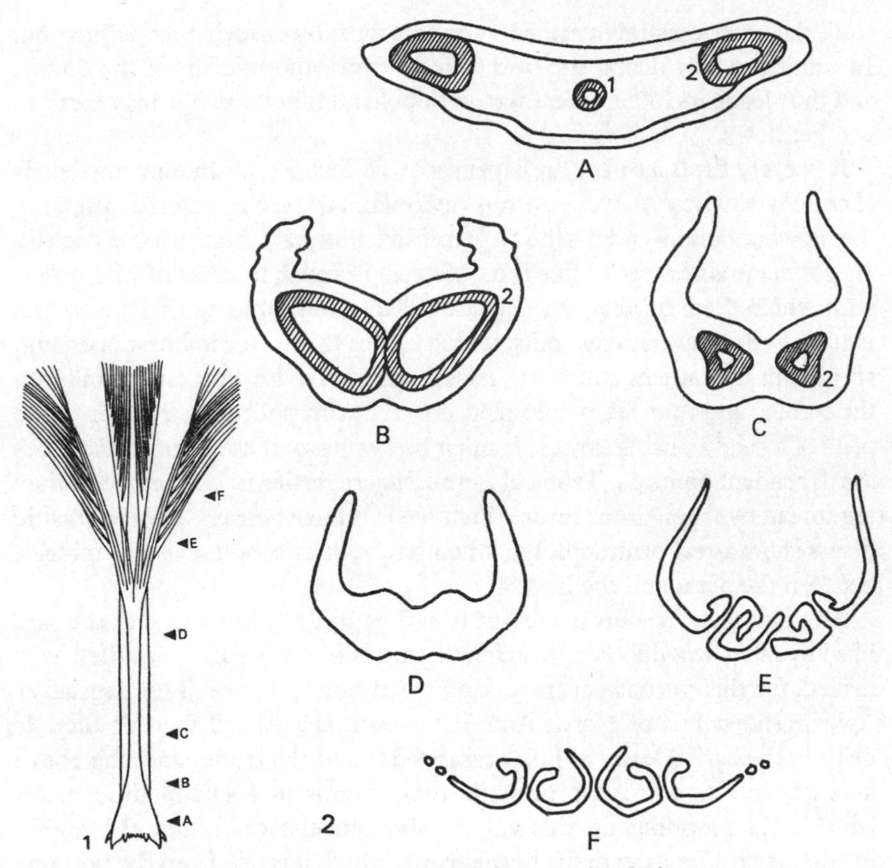

Figure 25. The tongue of a honeyeater *Meliphaga fasciogularis:* 1. dorsal view. 2. cross-sections (A–F) successively nearer the tip. 1 os entoglossum, 2 paraglossals.

which flock from a great distance to a tree in flower. These animals with a rudimentary sense of smell are not attracted so much by the scents of flowers, like insects, as by their colours. Red, and to a lesser degree yellow flowers attract birds more often than those of other colours. In order to take nectar, most birds perch near the flower they are to 'plunder'; but the hummingbirds being able to hover behave like hawk-moths, and thus pass in a flash from flower to flower. Nectarivorous birds are not rigidly dependent on this diet, since they also feed on the great numbers of insects which like themselves come to feed on nectar-bearing flowers, catching them in flight. Their diet is therefore most often mixed, the animal content satisfying their need for proteins which are not provided by nectar.

If the food-plant provides the nectarivorous bird with a preferred foodstuff, the latter certainly performs services for it by ensuring its pollination. In coming to take nectar the bird strikes against the stamens of the flower, and thus loads its bill and feathers with pollen grains which it may carry to another flower.

However, birds are the indispensable pollinators of Indian mistletoes (*Loranthus*), whose flowers do not open without strong external pressure (being what botanists describe as 'explosive flowers'). Sunbirds, especially *Nectarinia asiatica*, seek after these flowers in order to draw off the nectar with which their tubular corollas are filled. They squeeze the tips of the mature buds between their bills, which causes the flower to burst open and spread out its stamens and pistil. At this instant the bird thrusts its bill into the corolla, and thus has its plumage covered with pollen for transfer to the pistil of a neighbouring flower. It must be emphasized how much mistletoes are dependent on birds. Tropical *Loranthus* are fertilized by them, and they are spread by frugivorous birds which feed on their berries. These parasitic plants show a real ornithophily, which has been interpreted as a symbiosis between the plant and the bird.

All other plant resources are put to use by birds. This is true of the sap, liked by many woodpeckers which lick it up as it flows from holes they have drilled, for this purpose or in seeking wood-boring larvae. The Sapsucker *Sphyrapicus varius* of North America is even specialized for this diet. It drills in the bark a series of holes arranged round the trunk, one ring above another, and its scarcely protractile tongue ends in a sort of little brush suitable for mopping up the sap. It also eats the cambium, the highly nutritious proliferating tissue of the trunk, which it strips from the bottoms of the holes. Even fungi are eaten by birds, as seems to be shown by analyses of the stomach contents of the Red-breasted Pygmy Parrot *Micropsitta bruijnii*, from New Guinea, which has been seen to eat fungi growing on rotten wood.

CARNIVOROUS DIETS

Many other birds are secondary consumers, feeding on animal prey and so occupying higher ranks on the food chains. Analysis of the carnivorous diets of birds show how these animals have diversified so as to exploit all edible resources. Exploitation by birds is largely concentrated on the more important links in the food chains, insects, molluscs, fish and rodents, even though some species may have turned towards less abundant prey.

Insects make up a large biomass, and as producers and transformers of energy represent essential links in terrestrial communities, in which they occupy important positions. It is not surprising that many birds are specialized for their capture, in every kind of habitat. Except in woodpeckers, the capture of insects does not demand special adaptations. However, it is noteworthy that many insectivorous birds, especially flycatchers and tyrant-flycatchers, are provided with bristles on either side of the bill, which are often strongly developed. These highly specialized feathers, reduced to their shafts and having the appearance of hairs, probably have some sensory part to play in the close detection and capture of insects, since it is less probable that they act merely mechanically.

Birds of terrestrial habit, such as many Turdidae, wagtails, pipits, some Corvidae, rails and waders, explore the soil. Most of these birds take a mixed diet in which insects figure beside worms, grubs, fruit and seeds. They often scratch the ground and the layer of leaves and humus with their feet. This habitual searching technique of the game birds is seen also among passerines of much smaller size, such as the American towhees (*Pipilo*), which betray their presence by the characteristic noise they make in scratching about among the leaves. Storks and egrets are equally important consumers of insects, especially grasshoppers and crickets, to whose swarms they are attracted in large parties.

Many birds pursue their insect prey in flight, such as the swifts and swallows during the day and the nightjars at twilight and during the night. They have short bills, but with very wide gapes, which make it easier for them to seize their prey as they fall upon it at high speed.

Other insectivores hunt from a lookout point, perched in an exposed place from which they can keep watch over a wide area and fall upon prey which passes nearby. This is the technique of the flycatchers, many tyrant-flycatchers, todies, bee-eaters, jacamars and many kingfishers; and also of the trogons, although these are partly frugivorous. This hunting technique calls for deft and rapid flight, which makes up for lack of specialization of the bill. Some of these birds are specialists in the capture of particular prey, such as the jacamars which especially hunt butterflies, and the bee-eaters which take Hymenoptera to whose venom they are immune.

Insects which live among the foliage and on the branches and trunks of trees are sought after by other birds – such as tits, wrens, some tyrant-flycatchers, warblers, finches and tanagers – which hunt by careful searching, adopting various positions to do so. Some, such as the creepers and nut-hatches, hunt especially on trunks which they cover with great ease. Their thin elongated bills enable them to take prey deeply hidden in narrow clefts, while one of the Darwin's Finches *Camarhynchus pallidus* uses for this purpose a cactus thorn instead of its own short bill (see Chapter 28).

Woodpeckers put their ability to move about easily over trunks to good use in searching for insects and xylophagous larvae, which they hunt by excavating the tunnels in the wood. A woodpecker's bill is strong and rigid, straight and shaped like a chisel or pick. Taking a firm stance, the bird hammers repeatedly at the wood, using its head, which is moved by powerful muscles, as a mallet. In some species such as Sapsuckers (*Sphyrapicus*) only the neck takes part in this movement while the rear part of the body scarcely moves. In others such as the Three-toed Woodpecker *Picoides tridactylus* the whole body acts, by rearing back from the trunk and balancing on the feet, to give greater force to the blows. In this way the bill can attack even sound hard wood, excavating a tunnel chip by chip to expose the desired prey.

Once the insect tunnel is opened, the woodpecker inserts its tongue, which is highly modified and as extensible as that of a hummingbird. The hyoid horns are enormous, and together with the genio-hyoid muscles they curl round the skull and reach the nostrils. The contraction of these muscles thrusts the horns forwards and sticks out the tongue, which at rest is folded within the bill. In this way the length of the tongue in the Green Woodpecker can increase by 8cm. Supple yet resistant, it ends in a more distinctly cornified part which carries strong teeth and spikes, pointing backwards like the barbs of an arrow. Woodpeckers also have highly developed salivary glands, which lubricate the tongue and make it sticky. The bird inserts its tongue into the hiding places of insects or xylophagous grubs, and pulls them out by transfixing them with this harpoon-like structure. In North America, 45 per cent of the food of the Hairy Woodpecker *Dendrocopos villosus* is obtained by drilling, 30 per cent from the surfaces of trunks, and 25 per cent from other sites. These proportions are respectively 85, 10 and 5 per cent in the Three-toed Woodpecker *Picoides tridactylus* (Burt). Thus these birds are primarily forest and tree lovers. Others however have become terrestrial and hunt for insects in the earth, such as our Green Woodpecker *Picus viridis*, the American flickers (*Colaptes*) and the South African Ground Woodpecker (*Geocolaptes olivaceus*). Putting to use the extensibility, mobility and sticky coatings of their tongues, they catch ants on the ground, hunting them rather after the manner of mammalian anteaters.

Finally, it is appropriate to point out the very special diet of the honeyguides, a family of the Piciformes with remarkable habits which is especially widespread in tropical Africa. They feed particularly on Hymenoptera – wasps and bees, and on the larvae of these insects which they take from their hives. Very strangely they also eat wax, a fact noted by a Portugese missionary in 1569: he complained that honeyguides came to feed on the candles which he set out on the altar of his little church. Wax is eaten regularly, and is no doubt one reason for the birds' behaviour in guiding mammals to beehives.

Analysis of the wax during its passage through the alimentary canal shows that part of it is assimilated by the bird, which is rather surprising since wax is supposed to be indigestible by vertebrates. It is not yet known whether honeyguides secrete powerful and specific enzymes, or whether the breakdown of this substance is carried out by specialized intestinal bacteria (Friedman).

Small terrestrial molluscs are sought by many passerines which, like thrushes especially, swallow them whole. They grasp larger molluscs in their bills and knock them on stones so as to break open their shells.

Large marine molluscs such as mussels and cockles (*Cardium*) are widely eaten by eiders, scoters and various other marine ducks. These birds swallow them whole, since their bills are incapable of opening them, and their strongly muscled gizzards grind the shells to expose the digestible flesh. Other birds first open the shell to take out the soft parts. Gulls carry shellfish into the air and let them fall from a height on to a rock or a stone jetty, repeating this trick until the shell is broken. Other birds have bills shaped for opening shells. Oystercatchers (*Haematopus*) have chisel-shaped bills, elongated and very strongly compressed latcrally. They look for slightly open bivalves into which they put their bills, plunging them quickly in as deeply as possible, and then using them as wedges to force the valves apart.

The Everglade Kite *Rostrhamus sociabilis*, distributed over much of the hotter part of America, specializes in taking gastropod molluscs, especially the family Ampullariidae. It captures them in pools when they surface to renew their air supplies, and carries them in its feet to a tree. It then waits for movement by a mollusc, and as soon as one comes slightly out of its shell it seizes it with its bill, which ends in a long recurved needle-sharp point. The mollusc at once goes limp, destruction of its complex ganglionic system having paralysed its motor muscles. The bird shakes its head violently and so extracts the animal, whose shell falls to the ground to join the easily recognizable small heap, at the foot of the tree to which the kite returns with each capture. Another tropical American raptor, the Slender-billed Kite *Helicolestes hamatus*, which also has an upper bill with a sharp hook, shares the same stenophagous diet based on freshwater molluscs.

Some large wading birds are as specialized for the capture of molluscs. The principal food of openbills (*Anastomus*), which occupy the old world tropics, is large molluscs, especially Ampullariidae. These birds have long laterally compressed bills, with a large gap between the upper and lower mandibles, the edges of which carry horny denticulations. This arrangement must be considered as an adaptation for grasping a mollusc firmly and breaking up its shell before swallowing its body. The Limpkin *Aramus guarauna*, found from the southern United States to Argentina, also feeds on freshwater molluscs. Its bill also has a gap between the mandibles, though a narrower one than that of the openbills, and this too is an adaptation for holding a mollusc while the Limpkin breaks it open by knocking it against a stone or stump (Huxley 1960).

Small aquatic animals which make up the macroplankton are sought by a great many birds, which have developed special organs for catching them. The most curious of these birds are the skimmers (*Rhynchops*), close relatives of the terns, whose bills are long and strongly flattened transversely. The lower mandible is much longer than the upper, and ends in a real knife-blade. The bird habitually flies low over the surface of the water with rapid but shallow wing beats, keeping its bill open and cutting the water with its lower mandible. The pressure and flow of water over the bill, caused by the bird's progress, sweep minute prey into its mouth. The shock of the richly innervated tip of the bill striking a larger prey alerts the bird, which can then seize it. This method of fishing is used especially at twilight, when the plankton returns to the surface.

Many birds which eat plankton or other small animals have developed filters which allow them to take in water or mud mixed with the prey, and to separate them. However, methods of capture differ somewhat. Spoonbills have flattened bills with the tips enlarged into spoons. These are used to sweep the surface with a wide semicircular movement and to 'skim off' the surface layer, closing on any prey encountered. Ducks have flattened bills whose edges carry horny lamellae forming transverse ridges. Their thick fleshy tongues are also edged with horny outgrowths, and together these corresponding parts form an efficient filtration system, which is highly sensitive because it is provided with numerous sensory corpuscles. When dabbling, a duck takes in water and mud together with the suspended organisms, filters them, and swallows the prey while allowing the water and inorganic matter to escape. This sorting out is helped by splashing about with the bill. Flamingos have bills which work on the same principle, but are still more specialized, being very sharply bent so that they look broken in half. Hence a flamingo feeds in a very characteristic way, submerging its bill in shallow water by hanging its head upside-down. The lower jaw stays still, and it is the distinctly smaller upper jaw which moves. The bird tramples the mud with a rhythmic motion while turning round, and absorbs the minute animals thus disturbed, by filtering the water and mud in which they are suspended.

Some petrels, such as the prions, *Pachyptila*, of the southern oceans, also have a horny filter in their upper mandibles. In the most highly evolved of these species (*P. forsteri*) this filter consists of 150 horny lamellae 3·5mm long, arranged in a row and separated from the edge of the mandible by a space into which fits the edge of the lower mandible. This apparatus is completed by a fleshy tongue and an extensible throat pouch. These birds range over the surface of the sea filtering out the minute animals, especially crustacea, floating near the surface. They feed mainly at night when the plankton rises.

In aquatic habitats, whether marine or freshwater, fish form an essential link in the food chains. The capture of such mobile prey, whose fusiform bodies are the more difficult to grasp because they are slippery, demands highly developed organs of capture. Many fishing methods are adopted by birds. The least specialized in this respect, and yet the most effective, are certainly the penguins, whose bills carry not a single hook for retaining their prey. They are such excellent swimmers that they compensate for this lack by remarkable manoeuvrability, and can swallow several fish underwater before surfacing. So it is with the guillemots, auks and puffins, which can hold up to ten fish in their bills, each gripped by the middle of the body, and so carry them back to the nest to feed to their young.

Cormorants have long cutting bills ending in a strong hook, with which they hang on firmly to their prey until they get it to the surface and swallow it head first. So it is to a lesser degree with pelicans, whose huge and rather pliable bills are highly perfected tools for the capture of fish. Other fish-eating birds are armed with harpoons rather than with hooks, their thin strong bills ending in sharp points with which they transfix their prey. This is true of terns, which fly over the water and fall upon fish swimming near the surface, harpooning them with one of their needle-sharp mandibles. Darters fish under the water, their long pointed bills acting like harpoons propelled by sudden strong thrusts of their long snakelike necks. Herons fish almost from ambush, standing motionless or wading with slow steps through shallow water. In striking with their bills, they can correct for the visual error caused by refraction at the water surface. Some egrets spread their wings widely around their heads to avoid the reflection of the sun on the water, and thus fish within a circle of shadow which helps them to detect their prey. Kingfishers lie in wait for theirs from perches hanging over the water and falling upon them like lightning, using their narrow, strong and pointed bills either as harpoons or as forceps. They can dive and surface again by using their wings underwater.

Another adaptation for grasping fish is the development of denticulations, like those on pliers used for handling slippery objects. Thus mergansers (*Mergus*) have long narrow bills with regular denticulation along the edges, and these organs are the more effective since they are strongly hooked at the tip and since these birds' tongues are provided with horny bristles. Other birds have similar grips on their feet. Fishing raptors such as ospreys, sea-eagles and fish-owls (*Ketupa*) have sharply pointed recurved talons, and the underside of the toes and the pads of the feet bear pointed spicules which allow them to take a firm grip on fish. These birds fly over the water and fall upon fish swimming near the surface, which they lift into the air by means of these grips and carry away to break up in some quiet spot.

Many fish-eating species take frogs, some herons even showing a marked preference for such prey, while many birds hunt for reptiles, even larks having been seen to catch small lizards. While reptiles form a variable proportion of the diet of some birds, they are the principal prey of a few predators. Serpent eagles (*Circaetus*, *Spilornis*) and the South American Laughing Falcon *Herpetotheres cachinnans* habitually eat snakes which they swallow whole. The Secretary Bird *Sagittarius serpentarius* is highly modified, with the build of a wading bird. It stalks about the African savannahs hunting reptiles, especially large snakes.

Birds themselves form the prey of their fellows. Nestlings and eggs are sought by many plunderers, and make easy prey even for poorly-armed predators. Crows, jays, shrikes and many raptors plunder the nests of passerines which nest in trees or on the ground. Gulls, especially large ones like the Herring Gull *Larus argentatus* and the Great Black-backed Gull *L. marinus*, are dangerous predators for the broods of other sea birds such as puffins, auks and guillemots. The growth of populations of the Herring Gull, benefiting from protection in some parts of Europe, has thus led to serious dangers for other marine species whose position is more precarious. Great Skuas *Catharacta skua* feed largely on birds, small gulls, terns and young penguins, besides fish, rubbish and carcases of all sorts. These Laridae are true predators, attacking the young as much as the adults and their eggs, and are even cannibals on occasion. During their stay inshore in the Antarctic they feed especially on penguins, and become fish eaters during their spells at sea. The eggs and young of tropical birds are equally appreciated by many birds. In Africa the Harrier-Hawk *Gymnogenys typicus* plunders the nests of ploceids. According to recent observations in east Africa, Egyptian Vultures pick up stones in their bills and throw them at ostrich eggs, which they could not break unaided.

Some birds of prey are specialized for the capture of adult birds, which demands entirely exceptional powers of flight. Sparrow Hawks *Accipiter nisus* chase small passerines such as sparrows, finches and tits, which form the most conspicuous part of their diet. So do Eleonora's Falcons *Falco eleonorae* which take up positions on tall cliffs dominating the sea, and fall upon birds reaching the shore during migration. Oddly enough they only chase them over the water, and ignore those which have made landfall (Thiollay). Goshawks *Accipiter gentilis* hunt much larger birds, especially jays, pigeons, partridges, crows and pheasants which make up 60 per cent of their prey. Peregrines *Falco peregrinus* and related species especially pursue ducks, waders and small gulls, which they can capture thanks to their swift and skilful flight, diving on them like lightning. Most raptors seize their victims with their exceptionally strong talons, using the bill only to break up the carcase afterwards. Falcons fall on birds and strike them with their strong breasts, so that the victim falls towards the ground but is caught up in the falcon's talons before it reaches it.

Mammals too are prey to many raptors, some of which are specialized for their pursuit. Rodents are especially important in the diets of these birds, since they occur in very large populations. Small falcons such as the Kestrel *Falco tinnunculus* hunt rodents, as well as smaller prey such as insects. The Buzzard *Buteo buteo* is another great rodent-eater, taking field-mice and voles especially. The Rough-legged Buzzard *B. lagopus* of the northern tundras feeds on arctic voles and lemmings, a diet which it shares with the Snowy Owl *Nyctea scandiaca*. Rodents and shrews which are active at night also make up a very large part of the diet of other owls, which swallow them whole and later reject the indigestible waste in the form of pellets, whereas diurnal raptors usually eat only the digestible parts.

Eagles hunt larger prey such as marmots, rabbits, hares and small ungulates. African species take large rodents, rock dassies, small antelopes and the young of larger ones, and even viverids such as colonial meercats, while American eagles hunt prairie dogs (*Cynomys*). The larger species are specialized for the capture of stronger animals. The Harpy Eagle *Harpia harpiya* of the Amazonian forests hunts sloths and monkeys which it plucks out of the crowns of trees, its talons being certainly the most fearsome weapons possessed by any bird. The Monkey-eating Eagle *Pithecophaga jefferyi* of the Philippines, which is scarcely inferior to it in size and strength, is also a specialist in catching monkeys. In Africa these primates are preyed upon primarily by the Crowned Hawk-Eagle *Stephanoaetos coronatus*.

Finally, carrion is the food of poorly armed birds incapable of hunting living prey. Carrion eaters are found among the Corvidae, Laridae (especially skuas), raptors (kites and sometimes even eagles) and Procellariiformes (Giant Petrels), not to mention all the omnivores, while Marabou and Adjutant Storks feed largely on carcases. The most characteristic carrion feeders are the vultures of the Old World (classified with the Accipitridae) and of the New World (Cathartidae). These large birds – the Andean Condor, with a wing-span of 3m, is the largest living bird after the Wandering Albatross, while a North American fossil condor, *Teratornis incredibilis* must have reached a span of 5m – are poorly equipped to hunt for themselves. Their wings are well adapted for soaring but useless for pursuit, which demands swiftness. Furthermore their feet are weak and armed with hooked claws, ill-fitted to capture living prey. In contrast their bills are strong, often huge, with cutting edges and strongly hooked tips, well-fitted to open up the hide and tear the flesh off a carcase. Vultures have remarkable digestions even able to dispose of the bones of a large mammal in a few days, and its hide more quickly still.

Vultures find carrion while gliding at great heights. They then wait until the mammalian predators have opened the carcase, fed, and left. Lions in particular keep watch to prevent carrion-feeders from coming to their prey, and keep them at a distance. When all danger is past the vultures gather in mixed flocks, within which there is a hierarchy of species, largely in accordance with their size. In East Africa the dominant species is the Lappet-faced Vulture *Torgos tracheliotus*, before which the others scatter since it freely attacks them and drives them away. Next in the hierarchy is usually the White-backed Vulture *Gyps africanus*, then the White-headed Vulture *Trigonoceps occipitalis* and the Hooded Vulture *Necrosyrtes monachus* (Attwell). The dominated species wait until the others have surfeited themselves, and by this delay they profit from the rending of the hide, making the morsels of the carcase more easily available to their weaker bills. Vultures also compete with hyaenas, from which they flee, and with jackals and Marabous which in contrast they usually keep at a distance. They cut up the carcase with their bills, using their feet to hold the gobbets. Despite their strength, they cannot tear open thick hides but wait until the predators have rent them, or else make their entry through the natural orifices of the body. Their long mobile necks enable them to clean out the inside of a carcase in this way. Finally, though vultures only very occasionally hunt small antelopes and young ungulates, they eat insects and all sorts of rubbish, including even faeces. The Lammergeyer *Gypaetus barbatus* is also a carrion eater, though it is reputed sometimes to attack mountain animals when they are in precarious positions. It feeds especially on bones, which it takes from the carcases of animals killed by other predators (in Europe mainly wolves). It breaks them open by letting them fall from a great height on to slabs of rock, which from this regularly repeated practice become littered with bony fragments, and takes out the marrow.

Seasonal variations in diet

The adaptability of birds is also expressed in the variations in food preference which many show during the course of the year. This is especially true of those birds which exploit very diverse sources of food, whose diets thus adapt to existing conditions. Obviously the proportions of different foods taken reflects the availability and edible biomass of each category at a given moment. However, at any particular season birds seek selectively for special foods, whose part in the diet is therefore not entirely proportional to their availability in nature. These intrinsic seasonal variations, which are not directly related to variations in the availability of the foodstuffs themselves, reflect dietary preferences which fluctuate according to the season, and to physiological needs determined by metabolic variations. Thus mainly vegetarian birds take insect food more readily at times when insects are more abundant, and still more so when reproductive needs impose on them a diet rich in proteins (as is true of many of the Fringillidae). The Blue Grouse *Dendragapus obscurus* lives for most of the year on an almost exclusively vegetarian diet, made up of 98 per cent vegetable matter (of which 64 per cent is pine needles); but in August the proportion of animal matter exceeds 20 per

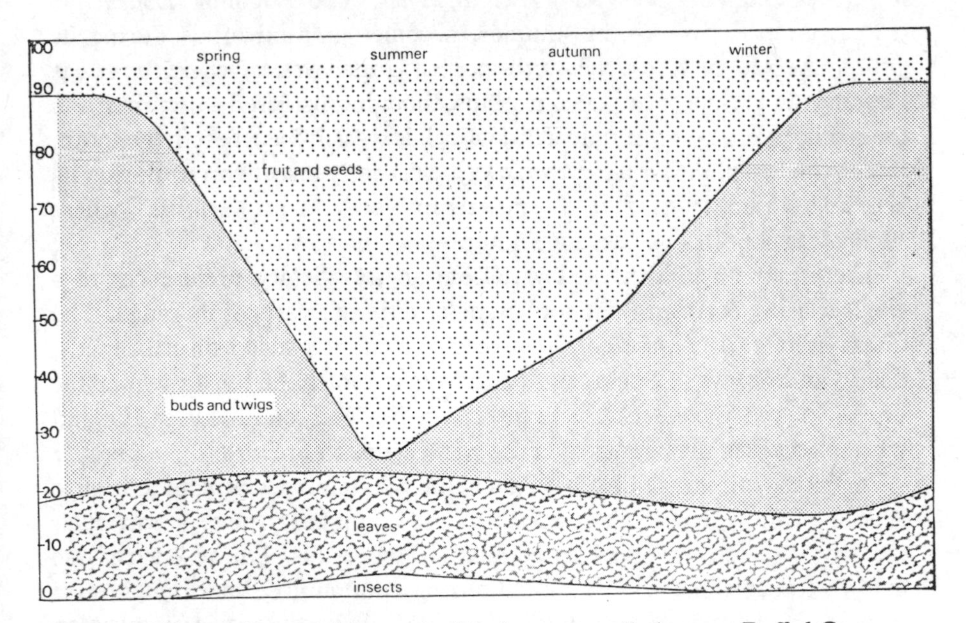

Figure 26. Proportions of different foods taken seasonally by 1093 Ruffed Grouse *Bonasa umbellus* of North America (ordinates indicate percentage by volume).

cent. In Finland the Great Spotted Woodpecker *Dendrocopos major* feeds on insects, and especially ants during the summer, and on conifer seeds during the winter. European thrushes (*Turdus*), which migrate relatively little and spend the winter on the Atlantic seaboard, have a mixed diet throughout the year; but during the summer their food consists mainly of animal prey such as insects, spiders, worms and molluscs, whereas in winter it is based on berries and fruits.

It is significant that the change to a greater proportion of animal proteins coincides with the beginning of the active reproductive phase, in temperate regions as well as in the tropics. For example, in the African savannahs guinea fowl (*Numida*) feed especially on grasses, but during the rainy season actively hunt for insects, apparently in relation to their breeding season.

Seasonal changes in the diets of migratory birds are also very obvious, when their preferences are compared at difficult phases of their annual cycles. Thus Rooks *Corvus frugileus* are mainly insectivorous during the summer, feeding on insects, grubs, worms and so forth, and more clearly vegetarian in their winter quarters. Animal food makes up 75 per cent of their summer diet, while in winter the proportions are reversed. The eastern European populations of rooks winter in the French plains, where they considerably augment the resident populations and are especially blamed for devastating the fields of growing wheat. The Bobolink *Dolichonyx oryzivorus*, insectivorous in summer, becomes graminivorous during its autumn migrations across the south-eastern United States, where the rice is ripening at the time of its passage. Thus, though considered a useful bird in the northern United States and in Canada, it is regarded as a pest in the south. So too the Starling *Sturnus vulgaris*, pre-eminently insectivorous during its breeding season, ruins olive plantations in North Africa (Tunisia), where some of its populations pass the winter.

There may also be marked variations in the dietary preferences of a single species, between the populations of different parts of its range. The Great Grey Shrike *Lanius excubitor* feeds on birds in Sweden, on arthropods, lizards and shrews in Spain, and on insects in North Africa. In winter the Great Spotted Woodpecker eats insects in England, and seeds in Finland where because of the harshness of the winters insects are rare.

In places where the edible biomass is limited, interspecific competition is keen, and the species which make up the community force a particular species to concentrate on a limited range of foodstuffs. The absence of one of these competitors from a neighbouring community, for example as a result of the faunal impoverishment of islands, allows the species in question to widen its diet. This presents the problem of interspecific equilibrium through the sharing out of niches and the regulation of populations, which are largely though not exclusively determined by food requirements.

Quantities of food and diurnal feeding rhythms

Birds as a whole are gross feeders. It has been calculated that, to eat proportionately as much as a hummingbird, a man would need to consume every day 285 pounds of meat, or 370 pounds of boiled potatoes, or 130 pounds of bread (Greenewalt). Birds' demands for food are related to their very high metabolic rate, the severe heat losses to which they are subject, and their very intense activity. In a Czechoslovakian forest, the total bird population daily consumes food amounting to 25 per cent of its own weight, whereas this percentage is only 20 for the mammals (Turcek). Hummingbirds, weighing 2 to 4g, take up to twice their own weight in syrup a day. Gannets and cormorants are especially greedy and in captivity eat up to 3kg, their own weight, of fish. In winter the Blackcap daily eats its own weight of mistletoe berries.

As with mammals, and indeed animals generally, small birds eat proportionately much more than large ones. Among terrestrial birds, those whose weights range between 100 and 1,000g eat on the average 5 to 9 per cent of their body-weight daily, while those between 10 and 90g eat 10 to 30 per cent.

The amount of food required varies greatly within the annual cycle. Migrants eat more before they depart, in a very characteristic period during which they accumulate fat. In the season of nuptial displays the males, which are then more active than the females, eat more than they do (about 15 per cent more among ducks), while this proportion is reversed during laying, when the females may eat more than twice as much as when they are sexually inactive. Food intake is also increased during the moult, and of course proportionally much more so among young birds at the height of their growth. It also varies with the conditions, especially the temperature, of the external environment, as many observations made on birds under natural conditions and in captivity have shown. Thus the Village Weaver *Ploceus cucullatus* takes food amounting to 20 per cent of its weight at 18°C, to 25 per cent at 9°C and to 28 per cent at 7°C (Schildmacher).

The search for food goes on from the earliest hours of the day, and like all other activities continues at maximum intensity until the end of the morning. It slows down in the middle of the day, and picks up again towards evening. Small birds, whatever their diet, take food repeatedly throughout the day. Lacking reserves, they must at all times be in a state to metabolize food. A House Sparrow will die after starving for sixty-seven hours at 29°C, and sooner at higher or lower ambient temperatures. The Waxbill *Estrilda angolensis* dies four or five hours after its alimentary canal has become entirely empty. Swifts survive longer, while larger birds are more resistant still. Thus a domestic fowl dies from starvation only after about ten days, a pigeon rather more quickly, and a duck after twenty-one days.

Carnivorous birds, especially raptors, are very clearly more resistant to starvation. A Snowy Owl can fast for twenty-four days in captivity, and an eagle for a month. Just as in carnivorous mammals, this is related to their much more irregular feeding rhythm, resulting from the hazards of the chase and the uncertainty of success. When the rhythm of a Red-tailed Hawk *Buteo jamaicensis* was followed in the United States, five captures were recorded in twenty-one days with an average daily intake of 100g. This very loose periodicity contrasts with the strict rhythm of small birds.

Some birds undergo exceptionally long fasts at certain periods in their annual cycles, especially during incubation. Female Eiders and Golden Pheasants do not eat for three weeks, and albatrosses not for twenty-four days, while penguins fast still longer. The male Adelie Penguin *Pygoscelis adeliae* does not eat for from one and a half to two months while incubating, during which time it loses up to 40 per cent of its weight. The male Emperor Penguin *Aptenodytes forsteri* broods the eggs for sixty-two to sixty-six days without taking any food, apart from snow which supplies its need for water. Not having fed since its arrival at the rookery, it actually starves for four and a half months, and thereby sometimes loses almost 50 per cent of its weight. This resistance to starvation, incredible at first sight, is made possible by fatty reserves much greater than those of flying birds.

It is notable that, in order to face long periods of scarcity, some birds store provisions in the autumn for their use during the winter. This is true of woodpeckers, such as the Acorn Woodpecker *Melanerpes formicivorus* of western North America which stores acorns; of corvids, such as the Jay and the Nutcracker; and of tits, especially Crested Tits in Norway, which accumulate dry fruits and seeds in hiding-places (Haftorn). Shrikes (*Lanius*) impale their prey, consisting of large insects, small rodents and lizards, on the thorns of bushes, which serve them as larders from which they draw later. Nutcrackers have gone still further, and give the impression of fore-thought on behalf of their future offspring. Thus in Sweden during the autumn they store up nuts and conifer seeds, burying them in little caches in the earth under moss and lichens. These birds nest at the very beginning of the spring, while a thick layer of snow still covers the ground. They raise their nestlings on the previous autumn's provisions, their success being strictly dependent on the amount available. Surprisingly, they find their little stores without difficulty, even when they are still covered with a layer of snow renewed by daily falls.

While the quantity of food must satisfy the birds' needs in calories, its quality is equally important. Like all animals birds have particular needs which vary from species to species. Ducks for example must have about 19 per cent of proteins if they are to reproduce normally. Vitamins and trace elements are also indispensable, in amounts depending on the species, and so it is with minerals and especially calcium. The presence of particular birds on one soil and not another is often explicable in strictly pedological terms, by the richness of the soils in calcium and phosphorus.

Communal foraging

While some birds gather food alone, others form flocks which are more or less organized for foraging, at least outside the breeding season (Rand). Every intermediate state exists, from simple accidental encounters resulting from the attraction of many individuals to the same food source (a tree in flower for nectarivorous birds such as hummingbirds, sunbirds and honey-creepers, or in fruit for frugivorous ones such as thrushes and parakeets, or a carcass for carrion-feeders) to highly organized multispecific flocks.

Several individuals of the same species, sometimes a mated pair, may occasionally hunt in a co-ordinated way, driving a prey animal which one of the birds captures for the benefit of all. Bald Eagles *Haliaeetus leucocephalus* have been seen to hunt ducks, on which they occasionally feed, in pairs, while Arctic Skuas *Stercorarius parasiticus* do the same. Collective fishing, in which large numbers of individuals take part, is clearly more common among birds, especially pelicans and cormorants. The Double-crested Cormorant *Phalacrocorax auritus* of North America gathers in flocks of up to 1,900 individuals, by the successive arrivals of small groups which settle on the water side by side. Fishing begins as soon as the flock is large enough. The cormorants then arrange themselves in a close line at right angles to their course, moving forwards by swimming and diving in turn. A quarter to a third of the birds are underwater at any moment, swimming at the same speed and thus surfacing almost in line with those already above water. In this way, they chase schools of fish which are systematically exploited by taking advantage of the disturbance set up by the attacking fleet. Flocks of pelicans fishing in the same way have been seen to encircle a shoal and force it into shallower water, where the fish could be caught more easily.

In other cases there is no such straightforward co-ordination, if only because the birds involved belong to different species and pursue different prey. This is true of flocks of land birds which may comprise only a few species, such as those formed in temperate countries by tits, sometimes associated with nuthatches and creepers. Much more important mixed flocks form in tropical forests, in the Old as well as the New World. These flocks assemble at dawn, the birds coming together as they leave their night shelters. Certain species of especially sociable disposition form the nucleus of the flock, to which less gregarious birds nevertheless congregate, drawn by the communal disturbance. Such flocks, comprising very varied species are clearly organized. They may contain several hundred individuals, and show real social structure, 'multilingual' alarm and contact calls with semantic significance between the species, and often a composition which is virtually fixed both in the number of individuals and in the various species which comprise it. In America cotingas, tanagers, ovenbirds, antbirds, tyrant-flycatchers, barbets, woodhewers and thrushes associate in this way with woodpeckers, trogons, hummingbirds and parakeets. In Africa various thrushes, flycatchers, sunbirds, warblers, bulbuls, babblers, woodpeckers, drongos, weavers and parakeets are involved. In Australia the mixed flocks are made up of flycatchers, whistlers and other insectivorous passerines belonging to various families. Thus while the specific composition of such flocks varies between different parts of the world because of their different avifaunas, the species concerned are always ecological counterparts.

These flocks spread out according to the ecological preferences of the component species; the terrestrial birds on or near the ground, the others in the various layers of vegetation overhanging the assembly point. They move ceaselessly through the forest, giving an observer at a fixed point the impression of a forest filled with birds which chatter and bustle on every side, then emptying again as the flock moves on. These movements follow fixed routes, as though the flock held a kind of territory on which its neighbours did not encroach. These apparently semi-permanent associations have a social and territorial organization, though not one which should be too rigidly construed since the composition of a flock may change from day to day.

The individuals integrated in a mixed flock clearly derive advantages from the association. The food resources of a particular habitat are systematically exploited, by birds which are each sufficiently specialized to avoid over-keen competition. Most of these birds are insectivores, and the passage of such a flock throws into confusion the whole insect fauna of the area it occupies for the moment. The disturbed and confused insects thus fall prey much more easily to the birds, which hunt them according to their several preferences and methods of capture. These flocks sometimes even follow columns of army ants, not to feed on them but to take advantage of the alarm which they excite in the small inhabitants of the forest. This special situation involves a very curious association between insects and birds.

It has been argued that mixed flocks provide better defence against predators, which are more efficiently detected by a number of birds than by a solitary individual. This is no doubt true, but the fact that a flock is much noisier than a single bird must make it more conspicuous, thus counteracting this advantage. The improvement in getting food is the one which is finally important.

The same motives have undoubtedly determined the associations of birds withmammals. One of the longest-known examples is that of the Cattle Egret *Bubulcus ibis* with large African and Asiatic ungulates and domestic cattle. Walking beside and seeming to guard them, it actually benefits from their trampling, catching insects and other small prey disturbed by their presence. Recently, by counting the number of captures per unit of time, it has been shown that egrets following cattle capture 25 to 50 per cent more prey than solitary individuals, with a one-third economy in movement (Heatwole). The same motivation causes them to follow agricultural machines in action, behaviour which is shown equally by other birds including gulls and crows. So it is with the Cowbird *Molothrus ater*, formerly dependent on bison and now on cattle across the American Great Plains. In South America the beneficiary of such an association is the Groove-billed Ani *Crotophaga sulcirostris*, a cuckoo of very specialized habits which also hunts insects near herds. The Crested Wood Partridge *Rollulus roulroul* has been observed in Borneo to follow wild boars engaged in rooting with their snouts, profiting from their efforts in its hunt for insects, larvae, worms and seeds which they uncover from the layer of dead leaves and humus (Pfeffer). Forest birds associate in this way with monkeys, the passage of these noisy restless animals disturbing insects which the birds seize as they flee. The African White-crested Hornbill *Tropicranus albocristatus* is nicknamed 'monkey bird' because of this habit. In Borneo and Celebes drongos (*Dicrurus*) have been seen to follow troops of macaques. In the Philippines the Fairy Bluebird *Irena cyanogaster* does the same, so that the natives call it 'sentinel of the monkeys' believing that it gives warning of the approach of enemies. Possibly for the same reason, in Africa the Carmine Bee-eater *Merops nubicus* rides on the backs of Great Bustards, pouncing on all the insects disturbed by their walk before settling again on its mobile perches.

Similar associations have even been observed at sea, between birds and carnivorous fish. Frigate birds have been seen to catch flying fish in association with bonitos, respectively above and below the surface. Thus the birds profited from the disturbance caused by the predatory fish, from which the flying fish tended to flee by taking to the air.

An association deserving special mention is that which exists in Africa between honey-guides and the ratel *Mellivora capensis*, and which is transferred to man. Adroitly, these birds announce the presence of a honeycomb, and lead towards it a mammal which breaks it open to eat the honey and the insect brood. The honey-guide, too weak to achieve this itself, is thus enabled to join in eating the brood, and also the wax.

Some birds also seek the company of mammals in order to eat their ectoparasites. The African oxpeckers *Buphaga erythrorhynchus* and *B. africanus* hunt insects and ticks on the hide of large game animals and domestic cattle. Their commonest hosts are Elan, Hippopotamus, Kudu, Rhinoceros, Sable Antelope, Roan Antelope and Buffalo. The preferences probably result from the high infestation and relatively scanty coats of these particular mammals, which make it easier for the birds to make their living than in the thicker pelts of other antelopes. Oxpeckers show special adaptations to their way of life: needle-sharp strongly curved claws which give them a secure hold of the hide; rather long rigid tails which support them somewhat like those of woodpeckers; and strong laterally flattened bills with sharp edges for grasping arthropods firmly. They feed primarily on ticks and the larvae of parasitic flies, but secondarily tend to feed on the blood and even the flesh of their mammalian hosts, opening wounds and keeping them open. By their behaviour, and especially their ringing whistled calls, these birds announce the approach of man. Even a sleeping rhinoceros is alerted by their cries of alarm, and when they take off the animal is at once on guard (Attwell 1966).

A similar association has been found in the Galapagos, between Darwin's finches and the marine iguanas. While the latter sleep in the sun the birds rid them of parasites, especially mites. A race of finch endemic to Wenman Islands, *Geospiza difficilis septentrionalis*, has also recently been observed to feed on the blood of Red-footed Boobies *Sula sula*. The finch hops on to the booby's tail and then its wing, which it attacks with its sharp bill where the secondaries are inserted. Having made a wound it licks the blood as it oozes out, behaviour which is unique among birds. This diet is probably secondary as in the oxpeckers, the finch having begun by taking the plumage parasites, especially hippoboscid flies, by which the boobies are heavily infested. Having tasted the blood with which the larvae of these insects are engorged, it has turned to attacking the boobies themselves. This diet gives it first-class nitrogenous food, as well as liquid which is scarce in these waterless islands. Nevertheless, blood is only a balancing component in the diet of these finches, which consists primarily of insects and cactus flowers (Bowman & Billeb). The origin of this behaviour seems to have been like that of the habit or vice of the Kea parrot *Nestor notabilis* in New Zealand, an originally frugivorous bird which has taken to attacking sheep. It perches on their backs and tears shreds of flesh and fat from their loins with its strong curved bill. The severe wounds thus caused may result in the death of the unfortunate victim. This habit has been acquired very recently, since sheep have been introduced to these islands only within the last two centuries.

Food parasitism

Despite the last few examples, the behaviour of most birds, except predators, towards other birds and mammals does them no harm and may even be beneficial. This is not true of those which frequent members of another species and molest them in order to steal the prey which they have caught. This behaviour is called food parasitism or piracy. Skuas (*Stercorarius*) with their fast agile flight, chase gulls and force them to release their prey, catching it before it falls to the ground. Frigate birds, though they can themselves catch fish basking near the surface, prefer to live by pirating gannets. As soon as one of these birds manages to catch a fish, the frigate birds chase it so that it has no time to swallow its prey, attacking it with their strong and formidably hooked bills. Thanks to their highly developed flying ability, they are able to harass it without respite, often doing it an injury if it will not give up its prey: gannets have had their legs broken by frigate birds. This stratagem is only used over the sea. Once the gannet passes over the coastline the chase is called off, and the two species live on good terms on land, even establishing their colonies immediately next to one another.

Water intake

The need of birds for water is very variable, but relatively less than that of most mammals. This is because they have no sweat glands, excrete solid urine, and have water-saving devices in their alimentary canal and kidneys. On the other hand, their respiration involves heavy water loss. While they all need water, some of them find enough in their solid food to avoid the need to drink. This is true of marine birds whose food, especially fish, contains much water – though some seabirds do drink brine, which even seems to be indispensable to the physiological equilibrium of penguins. In contrast other birds must drink more or less free water in order to meet their needs, either from streams and puddles or by swallowing the drops which trickle from leaves. Graminivorous birds, taking dry food, have the greatest need for water.

Birds find sufficient water in most surroundings except desert habitats. Many desert birds can entirely do without free water, finding enough in their diet or possessing organs for conserving water, especially in their caeca.

Renal excretion

Birds have a very intense metabolism, so that large amounts of waste products have to be rapidly eliminated by a very efficient excretory system. Its kidneys, symmetrically placed in the pelvic cavity, are relatively larger than those of mammals and reptiles, making up from 1·0 to 2·6 per cent of the body weight. Structurally they are reptilian rather than mammalian, especially in not being divided into a cortex and a medulla. Their microscopic arrangement is that of all vertebrates. However, the glomeruli of birds are smaller and more numerous than those of mammals. The total number in both kidneys varies from 30,000 in a small passerine to 274,000 to 353,000 in a pigeon and 1,989,000 in a duck. This represents four to fifteen times as many glomeruli in a given volume as in a mammalian kidney, allowing much more efficient excretion. The kidneys are richly supplied with blood by a circulatory system more complex than those of mammals. Besides the standard system of arteries and veins there is a supplementary venous system, comparable to the hepatic portal system. This arrangement allows the blood to pass through the kidney more rapidly and to be more efficiently purified.

The chief component of the urine is uric acid as in reptiles, not urea as in mammals. This nitrogenous compound is produced in the liver by reactions more complex than those which result in urea. It is a physiological necessity, arising from oviparity by means of a cleidoic egg, that nitrogen should be eliminated in the form of urates and not of urea. Since urates are insoluble they can be stored in the crystalline state in an outgrowth of the embryo, the allantois, which serves as a temporary urinary bladder (Benoit). Because of its high solubility urea cannot thus be stored, since it would poison the embryo. Uricotelic excretion is therefore imposed on the adult bird by selection acting on the egg. It is undoubtedly useful to the adult also, in allowing economy of water, since the urine can be more easily concentrated if the waste product which it contains is insoluble.

A bird's urine arrives at the cloaca as a clear liquid, through the ureters which empty into the middle section (*urodaeum*), without passing through a storage bladder. Only in the ostrich does the bursa of Fabricius perform a storage function. The urine flows back into the proximal sac or *coprodaeum*, where it mixes with the faeces. Here water is strongly absorbed from the mixture, continuing the action of the urinary tubules where water and glucides are withdrawn. This operation is very important in water metabolism, since it results in an economy of water which is especially useful to desert birds. The concentrated mixture of urine and faeces is expelled from the cloaca.

Sea birds have a special problem in the excretion of large quantities of salt which they take in with their food. Paradoxically they can also drink sea-water, which no mammal can regularly do. The excess salt is not eliminated through the kidneys, but through specialized glands which open into the nasal fossae and act as supplementary kidneys for saline excretion.

Chapter 6

Respiratory System and Circulation, Temperature and Thermoregulation

BIRDS are especially lively animals, which seem to live at a higher tempo than most others. They must make rapid movements, and their displacement in all three planes of space demands a large expenditure of energy. They work at high output and a correspondingly high metabolic rate, their food is oxidized rapidly, and their circulatory system is effective in carrying metabolites quickly round the body. Homeothermy frees birds from unduly strict dependence on the environment, allowing them to colonize habitats which poikilothermal vertebrates are unable to penetrate, by allowing them to make better adaptations to external conditions of temperature.

Some features of the organization of the respiratory and circulatory systems allow us to appreciate the physiological efficiency and mode of adaptation of birds.

Circulatory system

Though the circulatory system of birds is derived from that of reptiles, and preserve more reptilian features than that of mammals, it is remarkably efficient. The pulmonary and systemic circulations for oxygenated and deoxygenated bloods are completely separated, by the division of the heart into two isolated parts. The heart, built on the same plan as the mammalian one yet with differences, deserves special attention, since part of the secret of birds' high output is to be found in this organ.

The left-hand, systemic, part of the heart is very strongly muscled and much more developed than the right-hand, pulmonary, part. This difference is clearly shown in a cross-section, in which the thick walls of the left ventricle leave open only a narrow star-shaped lumen, whereas the right ventricle is less muscular and shows a wide cruciform opening. In mass the two parts bear on the average a relation of one to three to one another. As a result the tip of the heart is formed entirely by the left ventricle, the right ventricle being displaced upwards. This arrangement is in accordance with the work called for from the two components of the heart, the left-hand part having to pump blood round the whole body while the right-hand part sends it only to the lungs. It is commonly found among vertebrates, but attains a quite striking disproportion among birds, though among them it is highly variable as the following figures show (after various authors, and particularly from Hartmann 1955 and Johnston 1963).

Species	*Heart weight as % of body weight*
Ostrich *Struthio camelus*	0·98
Black-throated Diver *Gavia arctica*	1·12
Bonaparte's Tinamou *Nothocercus bonaparte*	0·21
Brown Pelican *Pelecanus occidentalis*	0·81
Great Blue Heron *Ardea herodias*	1·00
Domestic duck	0·95
Black Vulture *Coragyps atratus*	0·90
Broad-winged Hawk Buzzard *Buteo platypterus*	0·57
American Kestrel *Falco sparverius*	1·27
Bobwhite *Colinus virginianus*	0·39
Willow Grouse *Lagopus lagopus*	1·35
Sandhill Crane *Grus canadensis*	0·87
Killdeer *Charadrius vocifer*	1·35
Common Tern *Sterna hirundo*	1·04
Common Gull *Larus canus*	1·17
Mourning Dove *Zenaidura macroura*	1·11
Violet Sabre-Wing *Campylopterus hemileucurus*	1·96
Scintillant Hummingbird *Selasphorus scintilla*	2·40
Swallow *Hirundo rustica*	1·42
American Common Crow *Corvus brachyrhynchos*	1·20
Willow Tit *Parus atricapillus*	1·45
House Wren *Troglodytes aedon*	1·53
Siberian Tit *Parus cinctus*	1·72
Starling *Sturnus vulgaris*	1·30
Wheatear *Oenanthe oenanthe*	1·51
Arctic Warbler *Phylloscopus borealis*	1·59
Song Sparrow *Melospiza melodia*	1·18

As a general rule the relative weight of the heart bears an inverse relation to the size of the bird, in accordance with the much more intense metabolism of small birds. However, the relation is more complex than this suggests, especially since the pulse rate is involved and can compensate for the small size of the heart by high frequency of beats. Other factors are also important. The way of life of the bird may require an unusually high or low output of energy. Thus ground-feeding birds which seldom fly, divers and gliders all have relatively smaller hearts than good flyers. For example vultures and buzzards usually have hearts weighing less than 0·7 per cent of their body-weight, whereas in falcons the value is more than 1·2 per cent. Again, the heart is relatively larger in birds of cold countries, which have to generate more heat: arctic birds (for example those of Alaska studied by Johnston) have larger hearts than tropical ones. Altitude also has its effect, increasing the size of the right ventricle especially, and so ensuring the blood supply to the lungs which is so important to birds which must compensate for lower oxygen tensions. Systematic position, reflecting degree of organization, also plays a part; so that for example game birds have smaller hearts than passerines of the same weight, and the difference cannot be correlated entirely with ecological differences.

The two parts of the heart beat almost synchronously, a wave of contraction beginning in the right auricle and passing on to the left auricle and the two ventricles. Well developed nervous centres within the heart control the rhythm of contraction under the influence of the sympathetic and parasympathetic nervous systems, the former accelerating and the latter retarding the rate. The heart rates of birds are generally high in comparison with those of mammals. The following values were taken on birds at rest.

	Heart rate (beats/min)
'Raptor'	301
Domestic duck	212
Turkey	93
Domestic fowl	243–341
Pigeon	192–244
Starling	460–800
Canary	1000
Crow	342

The rate naturally varies very much for a given bird according to its output of work. In aquatic species the rate slows markedly during a dive, without producing a corresponding drop in blood pressure.

Bird blood retains reptilian characters, especially the form of the red corpuscles or *erythrocytes*, which are nucleated ovoid discs. They are more numerous in the blood of small than of large birds, as this table shows.

	Count of erythrocytes (million/mm^3)
Ostrich *Struthio camelus*	1·89
Peacock *Pavo cristatus*	2·7
Pheasant *Phasianus colchicus*	4·8
Domestic pigeon	3·0–4·0
Jackdaw *Corvus monedula*	4·5
Blackbird *Turdus merula*	6·4
Slate-coloured Junco *Junco hyemalis*	6·2
Ruby Topaz Hummingbird *Chrysolampis mosquitus*	6·69

The concentration may also vary with sex, (for example, the domestic cock has 3·24 million to the hen's 2·77 millions per mm^3) and sometimes according to the times of the year (in the Starling the number increases in winter from 3·9 to 5·3 million).

The erythrocytes vary greatly in size between species, being larger in primitive birds and smaller in those which are more highly evolved. They measure $17{\cdot}5 \times 9{\cdot}1\mu$ in a cassowary, $15{\cdot}4 \times 8{\cdot}5\mu$ in the Ostrich, and $15{\cdot}8 \times 10{\cdot}2\mu$ in the Lined Tiger Heron *Tigrisoma lineatum*; while in contrast they are only $10{\cdot}7 \times 6{\cdot}1\mu$ in the Rufous Hummingbird *Selasphorus rufus* (Hartman & Lessler 1963).

Birds have white corpuscles or *leucocytes* like those of mammals, and similarly separable into distinct classes, but they are relatively more numerous. Birds sometimes have as many as one leucocyte to every seventy erythrocytes, which is ten times the frequency in man.

Respiratory system

The respiratory system of birds is unique. Besides lungs with characteristic interbronchial connections, it consists of huge air-filled sacs throughout the body. This means that the circulation of air is much more complicated than in other vertebrates, whose lungs are dead ends. The air can pass twice across the surfaces where gas exchanges take place, which compensates for the restricted movements of the thoracic cage. In addition, the distribution of the air sacs throughout the body allows the considerable heat, resulting from the intense metabolism of a bird, to be dissipated where it is generated.

NASAL CAVITIES

The respiratory passages open in the bill at the base of the upper mandible, except in kiwis whose openings are at the tip. The nostrils, usually elongated and situated in depressions, are sometimes protected by vibrissae, frontal feathers which are highly modified into bristles. Sometimes too the nostrils are partly covered by opercula which leave only narrow slits for the passage of air. In gannets and cormorants these opercula even close the nostrils completely, depriving them of their function which in gannets is taken over by openings which act as secondary nostrils – a permanently open slit at the gape of the bill allows the bird to breathe without opening its jaws. This slit is protected by a horny lid, which probably closes when the gannet dives and may remain closed as long as it is under water. This arrangement, found also in cormorants, has been interpreted as an adaptation to diving head-first from a great height (Macdonald 1960).

The nasal cavities communicate with the mouth by elongated openings through which the air passes into the trachea.

RESPIRATORY PASSAGES

The trachea opens into the back of the buccal cavity in a larynx, braced by cartilages, which is formed on the characteristic vertebrate plan and resembles that of reptiles. It shows no anatomic differentiation for the production of sound. The trachea, stiffened by cartilaginous rings, is very flexible so as to allow free movement of the neck. It is sometimes thrown into huge loops, especially in swans and cranes where it is even folded into the specially hollowed-out breastbone.

Where the trachea branches into two bronchi there is a highly specialized organ, the *syrinx*, functionally replacing the mammalian larynx.

LUNGS

The lungs are housed in the thoracic cavity, and separated from the abdominal cavity by a diaphragm (not homologous with that of mammals). They are proportionally very small: whereas in man they represent about 5 per cent of the volume of the body, they are reduced to 2 per cent in the duck. They are firmly bound to the spine and the dorsal ends of the ribs, and their tissue is comparatively inelastic.

In structure the lungs of birds differ fundamentally from those of mammals and reptiles, which are closed systems. In these each bronchus forms a tree-like structure by repeated divisions, opening into the closed alveoli where the gaseous exchange takes place. In contrast, a bird's lung is an open system. The bronchus, known as the mesobronchus from the point at which it enters the lung, passes right through and opens into the abdominal air sacs, as we shall see. Thus the lung is an intermediate organ, placed between the surroundings and the air sacs. The branching of the bronchus within the lung is also quite different from that of mammals. As it enters the lung it gives rise to four to six ventrobronchi, and opposite to them six to ten dorsobronchi (while some birds develop further supplementary series). The normally very numerous ramifications of these ventral and dorsal bronchi intercommunicate by way of parallel *parabronchi*, which are the anatomical units of the lung. From their walls hundreds of fine bronchioles radiate at right angles, dividing many times as they leave the parabronchi. These air-filled capillaries are enclosed in a rich network of capillary blood vessels, and it is here that the gaseous exchanges take place. In birds which do not fly much each unit, made up by a parabronchus and the air capillaries arising from it, is more or less independent. In the best flyers, however, many anastomoses linking the systems to one another virtually obliterate the divisions between them. In the hummingbirds even the walls of the air passages disappear, so that gases pass even more easily from the highly developed system of interconnecting spaces, having to cross merely the walls of the blood capillaries themselves.

In birds the pulmonary ventilation is so arranged as to allow free passage to the air in every direction.

AIR SACS

The lungs are prolonged into huge air sacs which lead out of the main bronchi and are situated outside the lungs and even outside the thoracic cage, being distributed throughout the body of the bird – especially in the abdominal cavity, but also in the neck and the upper segments of the limbs. There are normally nine sacs:

two cervical sacs, extending along the neck on either side of the vertebral column, and opening into the first ventrobronchi.

one interclavicular sac, the only unpaired one, extending in front of the trachea, oesophagus and heart, and also opening into the first ventrobronchi.

two anterior thoracic sacs, extending from the base of the lungs to the tip of the sternum, and opening into the third or fourth ventrobronchi.

two posterior thoracic sacs, also in contact with the base of the lungs and opening by large passages into the mesobronchi.

two abdominal sacs, the most highly developed of all, occupying the intervisceral spaces of the abdominal cavity and opening into the posterior ends of the mesobronchi, of which they are in a way expansions.

These sacs are often divided into two groups, the first (cervical, interclavicular and anterior thoracic sacs) opening into the fore ends of the lungs, and the second (posterior thoracic and abdominal sacs) into the hind ends. The interclavicular sac also has a double origin but its two parts fuse during development, though maintaining their communications with both lungs. The other sacs sometimes fuse also, and except for the abdominal sacs, which open only into the ends of the mesobronchi, they often open through multiple bronchial passages.

Although always arranged in this general way, the air sacs vary in development from group to group. Sometimes even a single pair is asymmetrical, as in the pigeon whose left posterior thoracic sac is larger than the right while the reverse is true of its abdominal sacs. The sacs are of course not fixed in size, but vary in shape according to the fullness and movements of the organs. Their ramifications and diverticula are insinuated between organs and even between muscles: for example around the joints of the humerus and

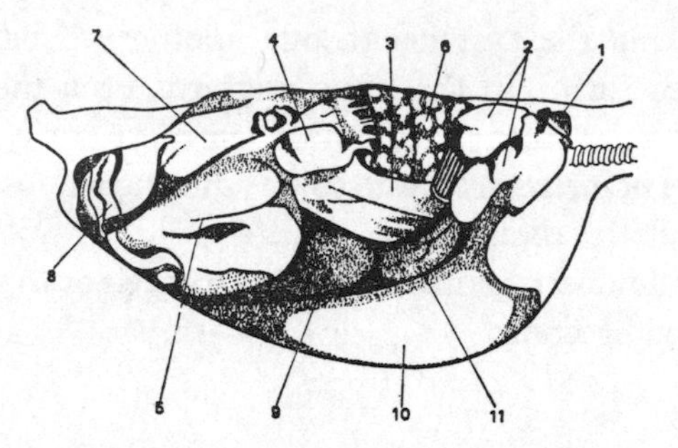

Figure 27. The internal anatomy of a pigeon. The lungs and air-sacs have been injected with latex, which has somewhat altered the shapes and sizes of the latter. 1. cervical sac. 2. interclavicular sac. 3. anterior thoracic sac. 4. posterior thoracic sac. 5. abdominal sac. 6. lung. 7. synsacrum. 8. intestine. 9. liver. 10. sternum. 11. heart.

the femur, and between the thigh muscles. Other ramifications extend into the bones, penetrating their substance and there replacing the marrow. Such bones are called *pneumatized*. Some diverticula occur as subcutaneous layers, which may form inflatable pouches used during specialized displays. The sacs as a whole occupy the dorsal rather than the ventral part of the body, an arrangement which is very important to stability in flight by lowering the centre of gravity. Despite the reduction of the lungs, the presence of the air sacs greatly increases the volume of air-filled organs within the body. It has been calculated that on the average the whole respiratory system of a bird represents 20 per cent of the total volume of the body, whereas it is only 5 per cent in man.

The air sacs may be considered as expansions of the walls of the bronchi. They are formed of a thin membrane traversed by elastic fibres, but rich in neither blood vessels (except within the bones) nor muscle fibres. They thus cannot contract by their own power, and cannot be the site of respiratory exchanges.

PHYSIOLOGY OF RESPIRATION

We still understand very poorly the physiology of birds' respiration, and especially the passage of air within the lungs and air sacs. The respiratory movements made by the lungs are much restricted, because of the comparative rigidity of the thoracic cage. Inhalation is produced by the contraction of muscles, which open the angles between the vertebral and sternal segments of the ribs, forcing the sternum forwards and downwards, and thus increasing the vertical diameter of the thorax.

The air sacs are much more important than the spongy tissue of the lungs in the variations in volume which follow the respiratory movements, although they are passive and are moved by muscles in other parts of the body. When the thoracic cage is enlarged during inhalation, the pressure in the abdominal cavity is lowered and the air sacs within it fill. The ventilation of the anterior sacs being more difficult, they seem to play a relatively minor part, at least during rest. Measurements made on domestic fowl have shown that during inhalation 6 per cent of the air passes into the anterior thoracic sacs, 8 per cent into the posterior thoracic sacs, and 80 per cent into the abdominal sacs. This is confirmed by the higher concentration of carbon dioxide found in the anterior sacs. The latter are probably involved in respiration only when the bird is active and has a greater need for oxygen. Since pulmonary ventilation causes a backwards and forwards flow of air in the lungs, it must be produced primarily by the filling and emptying of the posterior air sacs. There is no doubt that respiration continues when the air sacs have been destroyed, but that pulmonary ventilation is then reduced: the volume of air inhaled or exhaled at each respiratory movement (amounting to about 10–15 per cent of the total capacity of the lungs and air sacs) is reduced by 18 per cent when the sacs are destroyed.

Experiments appear to demonstrate a continuous circulation, during inhalation as well as exhalation, from the dorsobronchi to the ventrobronchi through the parabronchi (the 'd-p-v system' of Hazelhoff). This one-way circulation through the bronchial system is in striking contrast to the tidal flow in mammalian lungs.

RESPIRATORY RHYTHM

As in all vertebrates and especially mammals, the respiratory rhythms of birds are regulated by a nerve centre in the medulla oblongata. This centre is strongly affected by variations in the temperature, p_H and content of dissolved gases of the blood. In the pigeon an increase of internal temperature from 41·7°C to 43·6°C is followed by an acceleration of the respiratory rhythm from 46 to 510 inhalations per minute, the volume of air inspired in this time increasing from 185 to 610cm^3. In the Hour Sparrow the respiratory rate rises from 50 inhalations per minute during rest to 212 during excitement or flight, and an increase in the ambient temperature has the same effect. In the resting Willow Tit *Parus atricapillus* the rate rises from sixty-five per minute at an ambient temperature of 11°C to ninety-five at 32°C. Naturally the rate changes during flight, as a result of the repeated physical exertions involved. Some birds seem to breathe in a rhythm synchronized with that of their wings, such as the pigeon whose rate is 490 beats per minute, but others beat their wings too rapidly for synchronization of the breathing rhythm to be efficient. For example the Budgerigar beats its wings 840 times a minute, but breathes only 175 to 300 (Tucker).

The respiratory rate varies with the species and the age of the bird. Generally speaking the rate is inversely proportional to body size. The values below, obtained from resting birds, give some idea of the variation.

Species	*Respiratory rate (inhalations/min.)*
Domestic fowl	20
Pigeon	26
Red-shouldered Hawk *Buteo lineatus*	34
Baltimore Oriole *Icterus galbula*	107
American Robin *Turdus migratorius*	45
House Wren *Troglodytes aedon*	83
Willow Tit *Parus atricapillus*	64
Song Sparrow *Melospiza melodia*	63
Canary *Serinus canarius*	57
Red Cardinal *Richmondena cardinalis*	45
Starling *Sturnus vulgaris*	84

Diving presents aquatic birds with special problems. While they hold their breath they must show a special resistance to the rising concentration of carbon dioxide in their blood, and this seems to be the case especially in those such as divers which remain underwater for a long time.

Metabolism

Birds in general live at a very high metabolic level. They show intense activity, are of relatively small size (some of them very small), and must maintain a constant high temperature despite this handicap. They compensate their energy losses by the intake of a relatively large quantity of food.

The physiological state of the subject (whether ovulating, in moult, etc.) also affects the basal metabolism, which is merely the energy output which the animal must make in order to keep alive. The table below gives values in large calories for a number of species (various authors, from King & Farner 1961).

Species	*Weight (g)*	*Kcal/kg/24h*	*Kcal/24h*
Bennett's Cassowary *Casuarius bennetti*	17,600	29·3	516
Brown Pelican *Pelecanus occidentalis*	3,510	75·2	264
Great Blue Heron *Ardea herodias*	1,870	68·4	128
Trumpeter Swan *Cygnus buccinator*	8,880	47·1	418
Domestic goose	5,000	56	280
Domestic duck	1,870	84	157
Golden Eagle *Aquila chrysaetos*	3,000	34	102
Kestrel *Falco tinnunculus*	108	157	17·0
Quail *Coturnix coturnix*	97	235	23

Species	*Weight (g)*	*Kcal/kg/24h*	*Kcal/24h*
Domestic cock	2,000	47	94
Domestic hen	2,000	68·6	137
Wood Pigeon *Columba palumbus*	150	113	17·0
Domestic pigeon	311	105·9	32·9
Anna's Hummingbird *Calypte annae*	4·07	1,410	5·83
Rufous Hummingbird *Selasphorus rufus*	3·53	1,601	5·67
Raven *Corvus corax*	850	108	92
Great Tit *Parus major*	18·5	451	8·36
House Wren *Troglodytes aedon*	10·8	589	6·36
House Sparrow *Passer domesticus*	24·3	449	10·90
White-crowned Sparrow *Zonotrichia leucophrys*	26·4	324	8·55
Yellowhammer *Emberiza citrinella*	26·4	354	9·35

These figures show firstly the high metabolic rates of birds, usually higher than those of mammals of the same size. Only the insectivores show rates comparable to those of birds. The figures also show that birds follow the general rule for vertebrates, that the intensity of metabolism is inversely proportional to weight. Large birds produce less heat per unit weight than small ones, according to an exponential rather than a linear relationship. In fact an animal's basal metabolic rate varies with the 0·73 power of its weight.

Basal metabolism is of course higher in young birds, because of their smaller size and more especially because of the demands of growth. In adults it varies primarily according to a circadian rhythm, which passes through a minimum during the night, (except of course in nocturnal species), the drop amounting to 15 to 30 per cent in the domestic fowl and pigeon, to 49 per cent in small passerines (*Passer domesticus*, *Fringilla montifringilla*,) and even to 68 per cent in tits. This leads to the idea that birds, especially small ones, have two basal metabolic rates, one during waking rest and the other, amounting almost to semi-torpidity, during sleep. In some birds there is also a seasonal cycle.

The basal rate is increased by physiological activity, as for example in the breeding season and especially during moult. While the remiges are growing the rate increases by about 25 per cent in the Chaffinch *Fringilla coelebs* and nearly 50 per cent in the fowl. This has been explained as the result of more intense loss of heat following loss of part of the plumage by the moult, but the true explanation lies rather in the intense activity of the thyroid, which plays an essential part in the determination of moult. This affects the whole metabolism, including the increased synthesis of keratin. In the House Sparrow the formation of plumage keratin requires about ninety-four extra Kcal, apart from the additional expenditure of physical energy. Obviously the metabolic level changes still more according to the activity in which the bird is engaged. It has been calculated that the metabolic rate of a pigeon in flight is twenty-seven times greater than when it is at rest (Zenthen). The normal metabolism of hummingbirds is already considerably higher than that of other birds, yet it increases still more during activity. In Allen's Hummingbird *Selasphorus sasin* the mean hourly oxygen consumption per gram of body weight is 11·8–16·0cm³ at rest and 85cm³ during flight, an increase of six times (Pearson).

Finally, the basal metabolic rate changes profoundly when the ambient temperature exceeds the limits of thermal neutrality. When birds are exposed to varying temperatures it is found that their oxygen consumption, which expresses their metabolic level, decreases when the environmental temperature increases, then increases again slightly until the upper lethal temperature threshold is reached. Above this limit the bird cannot combat the excess environmental temperature, and dies.

Homeothermy

The high metabolic level of birds is expressed in high internal temperatures, of which the values in the following table give some idea.

Species	*Body temperature (°C)*
Ostrich *Struthio camelus*	40·0
Brown Kiwi *Apteryx australis*	39·0
King Penguin *Aptenodytes patagonicus*	37·7
Brown Pelican *Pelecanus occidentalis*	40·3
Black-headed Gull *Larus ridibundus*	41·4–42·1
Mute Swan *Cyngus olor*	41·0
Domestic duck	42·1
Quail *Coturnix coturnix*	42·0
Mourning Dove *Zenaidura macroura*	42·7
Sparrowhawk *Accipiter nisus*	41·2
Tawny Owl *Strix aluco*	41·0
Greater Spotted Woodpecker *Dendrocopos major*	42·1–43·3
Ruby-throated Hummingbird *Archilochus colubris*	38·9
Rufous Hummingbird *Selasphorus rufus*	39·0
Sombre Hummingbird *Aphantochroa cirrochloris*	44·6
Raven *Corvus corax*	41·9
Kinglet *Regulus* sp.	42·4–42·7
Robin *Erithacus rubecula*	40·6
Wren *Troglodytes troglodytes*	41·3
Blackbird *Turdus merula*	41·0–45·1
House Sparrow *Passer domesticus*	43·5
Starling *Sturnus vulgaris*	41·5

The internal temperature of birds as a whole are clearly higher than those of mammals. In the latter they vary on the average between 36°C and 38°C, whereas they attain 40°C to 42°C in most birds and as high as 44°C in some passerines. Relatively primitive types such as the kiwis have lower temperatures.

Homeothermy is relative rather than absolute and, even apart from the special cases considered below, body temperatures do fluctuate significantly. As in mammals there is a daily rhythm, temperature mounting to a maximum during the afternoon, and the difference in temperature between times of activity and rest may reach 2°C to 3°C. Thus in the Red Cardinal *Richmondena cardinalis* the internal temperature is 41·0°C – 42·5°C during the day and 38·5°C – 40·0°C during the night. In nocturnal birds such as owls and kiwis the cycle is reversed (38·2°C – 39·9°C at night, 36·4°C – 37·2°C by day).

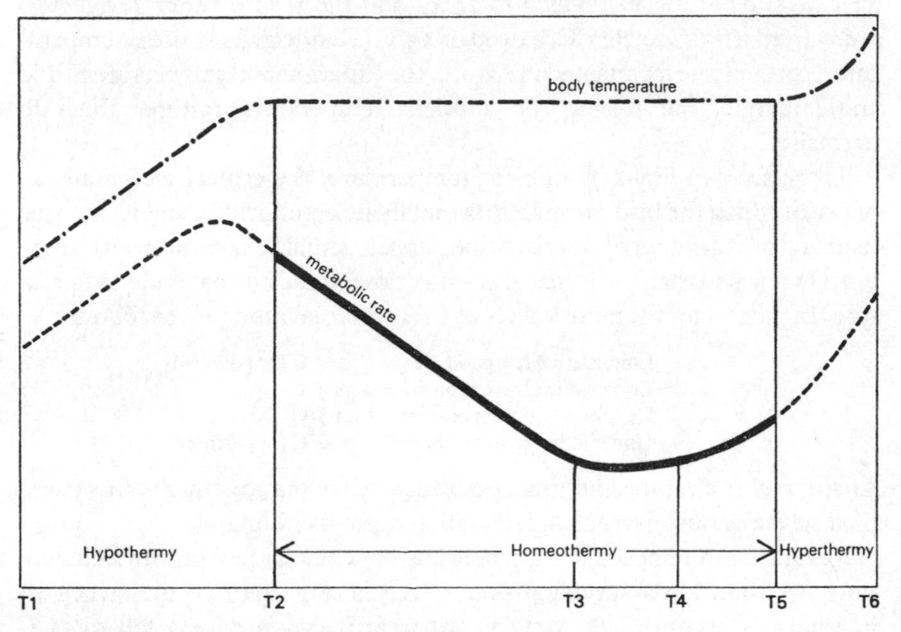

Figure 28. Thermal regulation and ambient temperature:
T1, temperature lethal through hypothermy. T2, lower critical temperature.
T3 & T4, limits of the zone of well-being. T5, upper critical temperature.
T6, temperature lethal through hyperthermy.

Some birds can even become torpid during enforced fasts, a state accompanied by a marked lowering of body temperature. This is true of Swifts *Apus apus*, which unlike any other insectivores can fast for five days. This is an adaptation to a strictly insectivorous diet, which may be unavailable for several days because of bad weather. When fasting, their temperature falls markedly to about 20°C, and their oxygen consumption decreases by 80 per cent. This lethargy, which is reversible as long as it does not go beyond a certain limit, is accompanied by a loss which may be as much as 38 per cent of the starting weight. It allows the birds to reduce the demands on their energy, economize their reserves, and thus face unfavourable periods (Koskimies 1950). A similar situation is shown by the American White-throated Swift *Aeronautes saxatilis*, whose internal temperature can fall to 20°C–25°C during fasts. Such hypothermy is really a state of poikilothermy, or dependence of the internal upon the ambient temperature.

Many other birds, even ones which appear susceptible, can experimentally withstand a considerable drop in body temperature, and afterwards resume their normal activity. The House Sparrow, whose normal temperature is 43°C, can be artificially chilled to 24°C, and the House Wren *Troglodytes aedon* from 41·3°C to 26·7°C, a drop of 15°C (Kendeigh). At these temperatures, certainly never attained in nature, the experimental subjects are torpid and lethargic, but raising the ambient temperature restores them to normality.

There are two limits of ambient temperature, the critical temperatures, between which the bird maintains its metabolic equilibrium, and its internal temperature at normal levels. The upper lethal temperature is more quickly reached than the lower, since it is close to the normal body temperature. In passerines the mean values of these thermal limits are as follows:

Lower lethal temperature	21·7°C (hypothermy)
Lower critical temperature	38·9°C
Upper critical temperature	44·8°C
Upper lethal temperature	46·8°C (hyperthermy).

All this makes clear the importance of much behaviour, by which birds guard themselves against overheating as well as against chilling.

Resistance to temperature fluctuations depends largely on physiological state, well-fed birds surviving better. A starving sparrow dies when its internal temperature falls to 32°C, whereas it can survive a fall to 21°C without serious harm if it is well nourished. The stage of the bird's annual cycle is also important, birds being generally more resistant during the winter than in summer.

Thermoregulation – defence against cold and heat

Temperature flexibility is an advantage, allowing birds to adapt to environmental conditions better and without mathematical strictness. Homeothermy itself brings undeniable advantages, guaranteeing the stability of the internal environment, and independence from the ambient environment, which is the first necessity for superior mental ability.

In birds as in mammals, we must recognize an internal hot *core* containing the vital organs, whose temperature must be held constant, enclosed by a *shell* consisting of the lining of feathers, the skin, the outer layers of the body proper, and distant parts especially the limbs. This shell serves to insulate the core, forming an enclosure whose insulating power can be varied so as to maintain the balance between thermogenesis and thermolysis. It is at a temperature lower and more variable than that of the core, so that there is a temperature gradient between the centre of the animal and its surroundings.

THERMOGENESIS

Birds like other animals derive their energy from their food. Increased demands on their energy are reflected by increased intake of food, as is seen when the ambient temperature drops. Thermogenesis increases also as a function of muscular work. We have seen how spectacularly the consumption of oxygen increases in flying birds, which generate heat in proportion.

THERMOLYSIS

Because of their rather small to very small size, birds lose heat very rapidly. The increase of body size shown by the populations occupying the coldest parts of the range of a species, known as *Bergmann's rule*, can be explained as an adaptation to climate as a result of the demands of thermoregulation. It is shown clearly by the Raven and the Wheatear.

Raven *Corvus corax*	Wing length (folded) (mm)
Greenland (*principalis*)	440–475 (455)
Iceland (*varius*)	400–465 (425)
Tibet (*tibetanus*)	470–490 (479)
Europe (*corax*)	375–442 (414)
North Africa (*tingitanus*)	380–420 (401)
Wheatear *Oenanthe oenanthe*	
Greenland (*leucorhoa*)	100–109 (105)
Sweden (*oenanthe*)	94–101 (98)
North Africa (*seebohmi*)	92– 97 (95)

The convergence between populations is so close that it is sometimes almost impossible to distinguish between them by their morphological characters. This is particularly true in the Raven because of a *cline* running parallel to the temperature variation.

Birds like mammals need to regulate their heat loss according to their own production of heat. Since birds have no sweat glands, evaporation can take place only from the respiratory system. Convection and radiation are usually difficult because of the unbroken sheathing of feathers, and the absence of external 'radiators' like the ears and tails of mammals. Accordingly birds achieve thermolysis differently from mammals, putting to use special physiological mechanisms and behaviour patterns.

The plumage constitutes the first line of defence against undue chilling, forming an effective barrier to cold which shows special adaptations in the birds of rigorous climates. These have many close-packed feathers, usually with a well developed basal layer of down. As a corollary, birds have more feathers in winter than in summer. Thus a Carolina Chickadee *Parus carolinensis* taken in February had 1,704 feathers, and another in June had only 1,140. A Louisiana Water-Thrush *Seiurus motacilla* had 2,146 feathers in April, and another taken in June only 1,525. Many similar examples show that feathers reach their maximum numbers in winter, and decrease gradually during the spring as the temperature rises. By shedding feathers the thermal insulation of the plumage is adjusted to environmental conditions. At the postnuptial moult the contour feathers do not all grow together, so that the maximum number appears only later in the season. This differential growth also agrees with the thermal conditions of the environment in autumn, since the bird does not immediately need a dense covering, which would be harmful while the weather was still warm. The full number of feathers grows only when the temperature falls and the bird is faced with more rapid heat loss. This adaptation is shown only by sedentary birds, since migratory ones grow their whole postnuptial plumage very quickly (Wetmore). Furthermore those species which flee the winter always have fewer feathers than those which are sedentary. In areas which do not suffer large annual variations of temperature the number of feathers remains more or less constant throughout the year, as in the Laughing Dove *Streptopelia senegalensis* in South Africa (Marcus).

The disposition of the feathers, which the bird can control by the play of its dermal muscles, is important in thermal regulation. When exposed to low temperatures a bird makes itself as spherical as possible (thereby exposing the minimum surface) and ruffles its feathers, thus increasing the thickness of the trapped layer of air. It also withdraws its feet into its plumage, or only rests on one at a time, and tucks its head and neck under its wing; behaviour which by itself reduces the heat loss of a fowl by 12 per cent. Besides, we have seen that in aquatic birds which remain for a long time in the water, an environment which conducts heat well, the plumage is made up of dense feathers tightly packed on one another in an especially water-repellent texture; and that the fatty secretions of the uropygial gland, which the bird spreads on the feathers with its bill, impregnates them and helps to hinder the water from penetrating their insulating layer.

Birds lose a significant amount of heat through their feet. This is especially true of web-footed birds such as ducks, geese, divers, gulls and terns, which often live in cold climates and move about in water or over ice. Nevertheless their feet very rarely freeze, and do not seem to cause them any thermal difficulties. Unlike those parts of the body covered in feathers, they are the site of a very effective temperature control related to a greatly reduced flow of blood. Their cutaneous temperature is very much depressed, following a gradient along the length of the leg, as has been shown especially by measurements on gulls (Irving & Krog 1955). The ends of the toes are almost at the ambient temperature, which prevents heat losses through them by conduction or convection. This conservation of energy is achieved by the provision of heat-exchangers between the warm arterial blood flowing towards the toes and the cold venous blood flowing back. These heat exchangers are in the forms of arterio-venous networks in gulls and ducks, or of veins and arteries running side by side in the tarsi of cranes, herons and flamingos. Similar arrangements are found in the flippers of penguins.

By contrast the feet can serve as 'radiators', when increased heat loss is necessary. Regulation of the arterio-venous networks allows a greater flow of blood towards the feet, which are then at almost body temperature and dissipate excess heat. This is shown strikingly by the American Wood Ibis *Mycteria americana*, which goes further by squirting liquid faeces over its legs so as to lose heat more rapidly by evaporation. Such arrangements and behaviour like that of the Wood Ibis are found also in other large wading birds, notably the European, Adjutant and Marabou Storks (Kahl 1963) and in seabirds such as gannets.

Defence against cold also involves a series of behavioural responses, especially the search for the most favourable microclimate. Even outside the nesting season some birds shelter during the coldest hours in special roosts or in abandoned nests: woodpeckers in hollow trees, and some Furnariidae of the high Andes in the burrows in which they nested. Sometimes group behaviour is involved, as when certain Andean passerines shelter in communal roosts. The White-winged Diuca-Finch *Diuca speculifera* has been found in flocks of 200, clustered in fissures of glaciers at an altitude of 5,500m in Bolivia (Niethammer). The short-toed Treecreeper *Certhia brachydactyla* gathers in flocks of twenty or more in tree fissures, huddled together with tails outwards so as to keep warm during the night (Löhrl). Measurements made on Starlings resting, singly and in roosting groups, show that a very important mass effect reduces by half the metabolic rate of the latter, thus leading to a considerable economy of energy (Brenner 1965). Penguins caught by antarctic blizzards, blowing at 200 km/hr at temperatures of −5° C to −30° C, form compact huddles. Only the outermost birds feel the blast of the icy winds, from which they shelter those in the middle. All these examples demonstrate a very effective social thermoregulation.

Defence against excess environmental temperature, by more rapid loss of heat, calls into play various physiological mechanisms. The respiratory system is remarkably well adapted to that purpose. It is large in volume and in area, and the flow of air within it can be greatly increased over the normal flow. At a high ambient temperature or during intense muscular effort, as during flight, the respiratory rhythm accelerates greatly until the bird is actually panting (under the control of a 'panting centre' in the midbrain). In the Budgerigar the normal respiratory rate of 72 – 100 per minute increases to 300 at an ambient temperature of 38° C. A bird exposed to high temperatures gapes widely and expels a rapid stream of air from its mouth, behaviour accompanied by a very characteristic throbbing of its throat, producing accelerated evaporation.

The air sacs are important in heat regulation. During rest or at low temperatures they form a layer of air which insulates the hot core of the bird to a remarkable degree, since there is then little circulation of air within them. In contrast, when body temperature rises because of muscular work or a rise in the ambient temperature, the air circulation increases, with more effective mixing in the anterior sacs, so that the organs which produce heat are cooled at source. Further, there is considerable evaporation from the walls of the air sacs, and the flow of blood increases so that the whole body is more effectively cooled.

These regulatory processes take place during flight. The acceleration in respiratory rhythm, and consequently increased flow of air through the lungs, are clearly greater than are necessary merely to meet the increased oxygen demand due to muscular exertion, and serve more particularly to eliminate the heat generated by the muscles. Furthermore, increased heat must be lost by convection from the moving wings (especially their undersides which are incompletely protected by feathers) and from the legs (Eliassen, Pearson).

Species which must remain immobile in full sunshine while incubating, especially seabirds, show similar responses. Experiments on gannets, albatrosses and frigate birds have shown that the loss of excess heat takes place essentially by evaporation from the mouth. The broad webs of the feet, providing additional surfaces for the loss of heat by radiation and convection, also play an important part in thermolysis (Howell & Bartholonew

1962). If the bills of these birds are tied so as to prevent them from panting, their internal temperature quickly rises to lethal levels. Demands on heat-regulation have had a profound influence on the breeding biology of these birds. Boobies can nest in full sunshine, since their chicks as well as the adults can eliminate excess heat by a marked and rapid throbbing of the throat which increases the flow of air. In contrast tropic birds have to seek the shade of bushes, since they lack this mechanism for speeding up the flow. Thus defence against high temperature calls into play patterns of behaviour, which reduce the effect of direct radiation. Other birds find shade beside cliffs or in cavities. Saharan larks arrange their nests against stones, whose shadows fall exactly over the sectors where the nests are hidden. It is obvious that some desert birds endure very high temperatures. They travel over ground heated to as much as 70° C, which has much more serious thermal effects than are produced by air at the same temperatures.

Thermal regulation in young birds

Thermoregulation is only gradually acquired by young birds, and never at hatching, while the speed with which this physiological mechanism is established depends on the rate of development of the bird. In nudifugous chicks it appears very quickly, a few hours after hatching. These young birds are still warmed and protected by their parents during the coldest hours, for during the early stages of growth thermoregulation is far from perfect. In contrast nidicolous nestlings are clearly poikilothermal, thermoregulation and homeothermy appearing only very gradually. In the Budgerigar, at ambient temperatures of 22° C to 27° C, the young behave as real poikilotherms, their temperature following that of the environment and exceeding it by only 0·5° C – 1·6° C. In the House Wren *Troglodytes aedon* the temperature drop of the young is at first equally parallel with that of the external temperature, but from the third day the drop becomes less and less discernible as thermoregulation is established. Young a fortnight old are as precisely homeothermal as adults.

The poikilothermy of young birds may be of great importance in the survival of the species, since it is noteworthy that the adults, of swifts especially, show great thermal flexibility. Young Swifts *Apus apus* can withstand prolonged fasts, of up to twelve days, during which they pass into a state of lethargy in which they sleep and their expenditure of energy falls to a minimum. In this state their internal temperature can drop to 21° C, scarcely 2° C above the ambient temperature. The respiratory rate, normally forty breaths a minute in the young, may drop to eight. Weight, first of reserves and then of the living substance itself, decreases markedly, the loss being up to 60 per cent of the total weight. This massive reduction in the metabolism allows the young Swifts to survive spells of cold or bad weather, during which the flying insects, on which they are exclusively fed, disappear. This is thus an adaptation eminently favourable to the species (Koskimies 1950).

Poikilothermy in birds

On the whole homeothermy, characteristic of birds and mammals, guarantees them a relatively stable internal environment, and great independence of their surroundings. However, far from being rigid, it shows marked flexibility. Many birds can survive a temporary drop in their internal temperature, and thus adapt better to the thermal conditions of the environment or to the possibility of meeting their need for energy by intake of food. This flexibility is in some cases so marked that we may call it true poikilothermy.

A nocturnal poikilothermy, comparable though opposite to the diurnal cycle of bats, is shown by certain hummingbirds such as Anna's Hummingbird *Calypte annae* and Allen's Hummingbird *Selasphorous sasin*, both desert birds of the American west. They can enter into a state of lethargy, which has been observed in the field and confirmed in the laboratory. During the night when the temperature falls below a certain limit, the bird seeks a hiding place and falls into a torpor. This is reflected in a fall of the internal temperature to less than 19° C, only a few degrees above the ambient temperature, whereas it is normally about 40° C. Oxygen consumption, 10–16 cm^3/g/h during waking rest (and up to eighty-five during flight) falls to 0·84. The bird is then inert, showing no reflexes, but wakes in a few minutes in the morning. An identical nocturnal lethargy has been found in the Andean Hillstar *Oreotrochilus estella*, which shelters by night in caves and rock holes from the intense cold of the high regions in which it lives.

This physiological specialization, which may be shared by the African mousebirds, allows these birds – without energy reserves and living at a high metabolic level – to face the cold of the night during which they are unable to feed. Not able to satisfy their energetic needs they thus put their physiology into low gear, spending minimum energy and making considerable savings during twenty-four hours. The *mean* metabolic rate of a hummingbird, corresponding to a mean oxygen consumption of 9·8 cm^3/g/h, is much less than that of a shrew, which corresponds to 15 cm^3/g/h. Although very high during active periods, during torpor the metabolic rate of the bird is only one twenty-eighth, whereas in the mammal it decreases by only a half

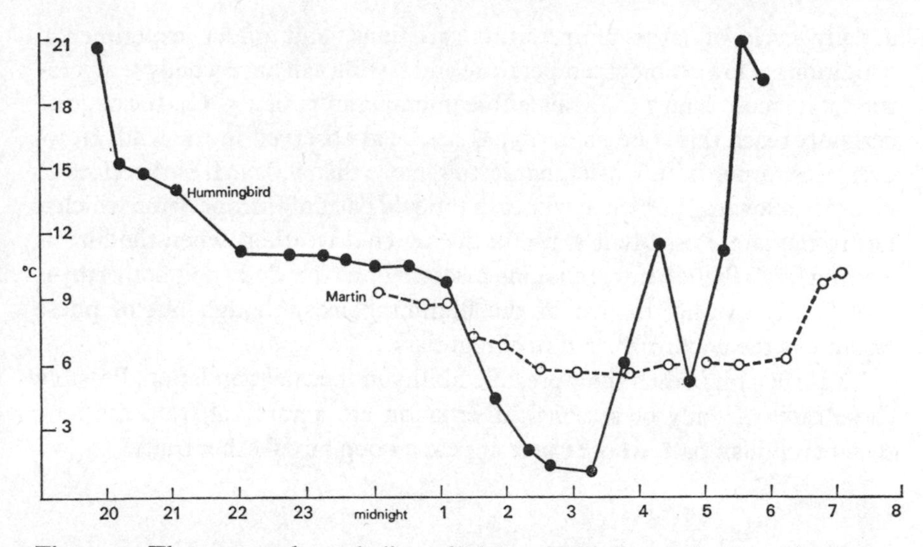

Figure 29. The nocturnal metabolism of a hummingbird, and of a Crag Martin which does not become torpid.

(Pearson 1950). This state of torpor is not obligatory, its frequency being controlled by environmental conditions and the physiological state of the individual bird. However, it is shown by several tropical species in Brazil, which appear to be sensitive to falls in temperature much smaller than those which produce torpor in North American hummingbirds (Morrison).

Seasonal poikilothermy is shown by the Poor–Whill *Phalaenoptilus nuttalli*, a nightjar of the North American west and centre. Individuals of this species, which is notable for great variability in temperature even when active (from 35·0° C to 43·5° C), have been found in midwinter crouched head-downwards in holes in the rock. They are then completely inert, no cardiac or respiratory movement being discernible. A bright light shone in their eyes elicits no reaction. Their internal temperature is about 18–20° C at an ambient temperature of 17–24°C whereas their normal temperature is 41° C. Thus this nightjar shows a hibernation comparable to that of poikilothermal mammals (Jaeger, Culberson), though less extended. It is almost certain that it does not pass the whole winter in this state, but returns to active life and hunts during favourable periods, as is further shown by its small loss of weight. Nevertheless this specialization gives it an advantage, allowing it to live a sedentary life, and to face unfavourable periods in a state of suspended animation. The European Nightjar *Caprimulgus europaeus* can also enter into a state of torpor, in which its internal temperature drops greatly and its metabolism is considerably slowed (Peiponen 1966). It shows a daily cycle of large temperature variations, and under experimental conditions of low ambient temperature and fasting can have a body temperature of no more than 7° C (at an ambient temperature of 4·5° C). It can quite certainly reach this state naturally, as has been observed in the wild. However, this torpor is only brief, not lasting more than a day. It is therefore of no use in allowing the bird to winter in the cold parts of Europe, from which it has to migrate. Possibly it is useful during cold weather, when the bird is waiting for twilight before pursuing insects. Thus this daily poikilothermy is ecologically parallel to that of the hummingbirds, though out of phase because of the nocturnal rhythm of nightjars.

These few birds thus show great flexibility in thermal regulation. Possibly these cases of daily or seasonal hibernation are a survival from the long distant reptilian past, whose traces appear among birds' other traits.

Chapter 7

The Sensory World of Birds

BECAUSE of their very active way of life, birds need to be precisely and quickly informed about their surroundings. Indeed their perception of the outside world is excellent, equal to that of mammals in quality though very different in kind. Their most fully developed senses are sight and hearing. Birds live in a world which is primarily visual, and indeed coloured. Many of the 'messages' by which information is conveyed, between mates, members of the same colony, rivals, foes and prey, consist in showing off coloured areas making up specific patterns. Their hearing too gives birds very highly evolved perception and analysis of the world of audible vibrations. Here again their auditory capacity is correlated with well-developed vocalizations, of which the songs of passerines are only the best known and most elaborate examples. These two senses are strictly related to the way of life of birds, which move rapidly about a three-dimensional world and thus need to be able to judge distances quickly and to explore their surroundings.

In contrast their sense of smell is poorly developed, being useful only to species which pick up food from the ground or from the surface of the sea.

The sense organs are built on the plan common to all vertebrates. Although they preserve some reptilian characters they are none the less as highly evolved in structure as in function.

The brain

Brains of birds are much larger and better differentiated than those of reptiles. A lizard weighing ten grams has a brain of 0·05g, whereas a bird of the same weight has a brain of 0·5g. This difference is largely due to the development of the cerebral hemispheres and the cerebellum. For birds of effectively the same weight (80-90g), the weight of the brain varies as follows:

Quail *Coturnix coturnix*	0·73g
Spotted Crake *Porzana porzana*	1·1g
Starling *Sturnus vulgaris*	1·8g
Scops Owl *Otus scops*	2·2g
Great Spotted Woodpecker *Dendrocopos major*	2·7g

The brain is housed in a very specialized, rather narrow cranium, and is compressed by the enlargement of the eyeballs – to such an extent that in snipe for example, the cerebrum is thrown to the back by flexure of the cranial axis.

The cerebral hemispheres are well developed, though in quite a different way from those of mammals. The *corpora striata* which form their floors are very massive. Being responsible for instinctive behaviour (especially the complex activities of reproduction, food gathering and locomotion), they are of cardinal importance in the life of birds, and form the co-ordinating centre of the whole brain. In contrast, the cerebral cortex, forming the roof of each hemisphere, is much reduced and never shows the convolutions seen in all mammals. The part it plays is certainly modest, all the more since it does not have the direct connections with the spinal cord. Birds do not therefore show the 'intelligent' behaviour of mammals, or at any rate not in the same way. Birds seem to act according to admirably well-regulated mechanisms, but by wholly stereotyped reactions.

The midbrain formed by the optic lobes is well developed, largely in relation to vision and its prime part in a bird's perception of the external world. The fibres from the optic nerve reach the optic lobes, retaining a

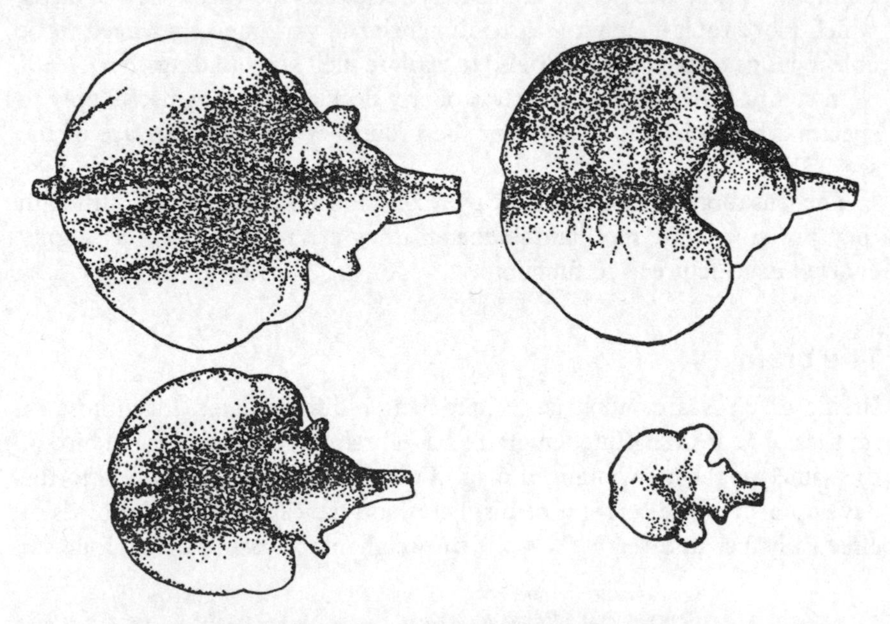

Figure 30. Brains of the galliform Scrub Fowl *Megapodius freycinet* and passerine Blackbird *Turdus merula* (right). Adults above, newly-hatched young below, to the same scales.

primitive arrangement of reptilian type. The cerebellum is remarkably well developed, since this centre for the co-ordination of movements and for the control of muscular contractions is naturally of great importance in animals whose locomotion is as complex as that of birds. Finally, the olfactory lobes are reduced, together with the sense of smell.

The proportions of the various parts of the brain vary between different kinds of bird. The cerebral hemispheres increase in volume according to degree of evolution, so that a macaw of the same weight as a fowl has hemispheres eight times as heavy. These differences have to be related to very different levels of mental process, which become more and more complex with progress along the avian evolutionary lines. Thus study of cerebral anatomy shows that game birds occupy the lowest mental level, followed by pigeons and some waders, while the highest level is occupied by crows, woodpeckers, owls and parrots. The analysis of behaviour provides entirely parallel evidence of evolutionary progress. However, though these behaviour patterns have been perfected in relation to more and more complex nervous mechanisms, and though birds have an excellent perception of the outside world, their whole biology bears the mark of their deficiency in cerebral cortex. This prevents them from adapting their behaviour to unforeseen circumstances, and to some extent of integrating their actions. The automism of birds contrasts with the flexible behaviour of mammals, and their actions lack the 'understanding' characteristic of the latter.

Eyes and vision

Since vision plays a predominant part in the life of birds, it is not at all surprising that their eyes are particularly highly developed. All birds have well-developed eyes, though those of kiwis are comparatively small, only about 8mm in diameter in a bird weighing as much as a large fowl. In other birds the eyes are large, or even enormous to the point at which they actually deform the head. The eyes make up 15 per cent of the weight of the head in a Starling compared to only 1 per cent in man, while the eyes of some eagles and owls are even larger than human eyes in absolute terms.

In most birds the eyes are placed very much to the sides. The angle between the two optical axes is of the order of 120° in most passerines, and as much as 145° in pigeons. This arrangement forces the bird to turn its head sideways in order to look at a particular point, but allows an extended field of view. In contrast the binocular field, which gives perception of depth, is reduced. In the pigeon the total field is 300°, with a blind area of only 60° behind the head, but the fields of the two eyes overlap in front by only about 30°. This tendency is reversed in raptors and other hunting birds (including insectivores which capture prey on the wing), whose eyes in contrast are turned towards the front of the skull, so as to increase the field of binocular vision. In the Barn Owl the binocular field is 60° wide, but the total field is reduced to 160°. Nocturnal birds of prey compensate for this deficiency by the extreme mobility of their necks, which allows them to turn their heads backwards. These differences are certainly adaptations of vision to different ways of life. The wide visual field which most birds enjoy is very advantageous to animals which move about in three dimensions. However, the resulting largely monocular vision is a disadvantage, since it does not allow perception of relief and especially of distances. This handicap is much reduced in fast-flying hunting birds, which need to estimate quickly the distances between themselves and obstacles or prey. Birds' eyes are of a very characteristic general shape: the front and back sections are parts of spheres, of very unequal radii with the posterior one much the larger, joined together by a tapering centre section. This shape is advantageous in reducing the

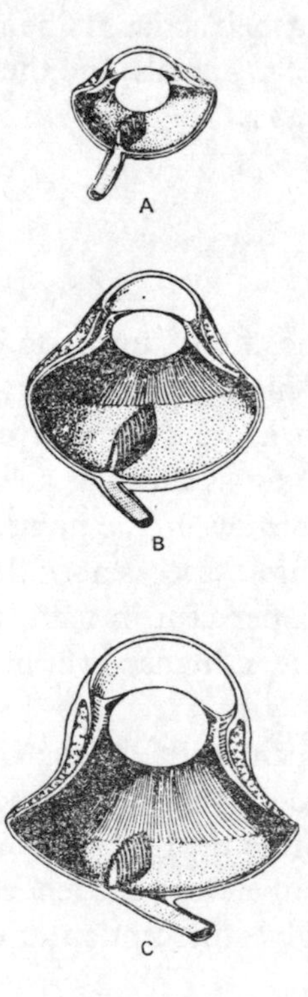

Figure 31. Birds' eyes of very different shapes: A. Mute Swan *Cygnus olor*. B. Golden Eagle *Aquila chrysaetos*. C. Eagle Owl *Bubo bubo*. Type A is the commonest among birds.

volume and weight of the eye without impairing its optical qualities. The shape of the eye varies considerably between groups of birds. In some, such as game-birds, pigeons and many passerines, it is comparatively hemispherical, the front section being flattened and the scleral ring almost a flat disc. In others, especially birds of prey, it is elongated and the scleral ring has the shape of a truncated cone, the front and back sections being more clearly distinct. In nocturnal raptors this ring becomes tubular, and the eye has a very specialized shape due to contraction of the middle and expansion of the hind sections. A bird's eye, because of its shape, can scarcely move within its orbit, and head movements must compensate for this lack of mobility. There are a few exceptions, such as the Bittern *Botaurus stellaris*, which directs its eyes under its bill when it 'freezes' with bill pointing upwards, so that it can keep watch on a potential enemy.

ANATOMICAL STRUCTURE

A bird's eye is formed like those of other vertebrates, except for some remarkable specializations. The sclerotic which maintains the shape of the eye is stiffened towards the front by a bony ring and elsewhere reinforced by a cartilaginous sheet. The cornea is thin and bounds a deep anterior chamber. The iris, coloured variously but characteristically for each species (often dark brown but sometimes yellow or orange), surrounds a pupil which is most often round, rapidly contracting, but small. The lens is made up of an elastic substance without a rigid core, and is very variable in shape. Some birds, such as parrots, have lenses which are almost flat in front and very convex behind; others, such as ducks and hawks, have both faces convex. These differences in shape correspond to different types of vision, related to different ways of life. The choroid is of the ordinary type, in two layers one of which supports the visual cells of the retina while the other nourishes them.

RETINA

The visual cells consist of cones and rods. The first are much the more abundant in diurnal birds, which may even lack rods or have them only in limited parts of the retina. Nocturnal birds in contrast have more rods than cones. Cones contain oily globules, mostly coloured a brilliant yellow, sometimes orange or ruby-red, rarely green. Though richly coloured in diurnal birds, these are almost colourless in nocturnal ones. The visual cells are of the same type as in other vertebrates, and are distinguished only by their dense packing. Up to 120,000 cones per mm^2 have been counted at the back of a wagtail's eye, whereas in the human eye a little way from the fovea the density is only 10,000. The numbers of underlying nerve cells and of efferent nerves is correspondingly very high in birds, and despite the difference in size between the two eyes there are almost as many fibres in the optic nerve of a pigeon as of a man.

AREAS AND FOVEAE

The density of nerve cells increases considerably towards special retinal zones termed *areas*, in which there are about 1,000,000 cones per mm^2 and the retina is correspondingly thicker. At the centre of the area there is usually a fovea, produced rather by a thinning of the retina, which is here composed of nothing but cones, and here visual acuity is at its highest.

In some birds, especially graminivorous ones, there is only a rounded area with a *central fovea*. However, water birds and others of open habitats have the area horizontally elongated, while in some of these the fovea itself is lengthened into a linear depression, an arrangement which probably corresponds to a particular mode of vision. The retinal pattern is more complicated in birds which catch their prey in rapid flight (such as raptors, swifts, swallows, kingfishers and hummingbirds), since these have two areas and two foveae. The central fovea, corresponding to the single fovea of other birds, lies on the optic axis of the eye, whereas the *temporal fovea* is towards the side. Although of the same structure, it is less highly differentiated than the former (except in swifts, whose central fovea is often considerably

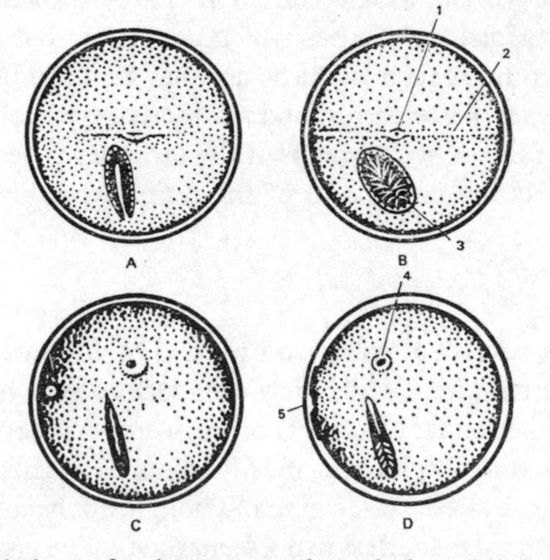

Figure 32. Internal views of retinas as seen through the ophthalmoscope, showing the positions and extents of the fovea and area. A. Herring Gull. B. Ostrich. C. Shrike. D. Hummingbird. (1. fovea. 2. area. 3. pecten. 4. central area and fovea. 5. lateral area and fovea).

reduced). Birds with two foveae use them in distinct modes of vision. Monocularly, those light rays which pass more or less along the optic axis fall on the central fovea, whereas in binocular vision the position of the eyes relative to the object causes the images to fall on the lateral foveae. These various arrangements are clearly adaptations to the ways of life of the birds concerned.

PECTEN

At the back of the eye on the optic axis is a vascular membrane, the *pecten*. It varies to some extent in shape from group to group, being conical in kiwis and made up of branching leaflets in ostriches and rheas, while its usual form is a pleated strip. Very deeply pigmented and richly vascularized, it is held in place by attachments along its crest, so that it seems to float in the vitreous humour pointing towards the centre of the lens. In the pigeon, with an optic axis 12mm long, the pecten measures 4mm.

Many hypotheses have been proposed on the function of the pecten. Some authorities consider it to be a primarily nutritive organ, its vascularization allowing metabolites to diffuse from it through the vitreous humour towards the retina. Others believe it to be a spongy organ made elastic by the filling or emptying of its tissues, which thus compensates for changes in pressure caused by the movements of accommodation – the more necessary since the eye as a whole is rigid and inextensible. Others again hold that the pecten plays a part in actual vision, increasing the sensitivity of the eye to the perception of movement. Menner made a sort of camera as a facsimile of a bird's eye, and showed that when pointed at a moving object this produced a clearer image when a cardboard 'pecten' was put between the 'retina' and the lens. Though the pecten must have nutritive functions corresponding to its rich vascularization, it seems likely that its position in the centre of the eye is also directly connected with the function of vision, for which that organ is so highly specialized.

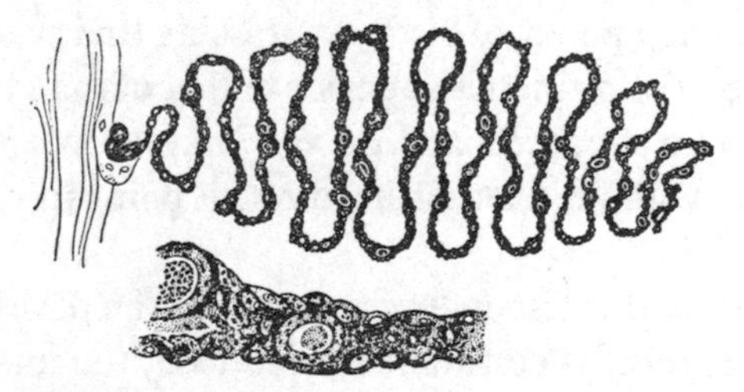

Figure 33. Cross-section of the pecten of a Gannet *Sula bassana*. Below, part of the section at higher magnification.

THE VISION OF BIRDS

(a) *Accommodation:* Diurnal birds have considerable powers of optical accommodation so fast that it is believed a bird can accommodate from infinity to 2cm while it turns its head.

This ability, indispensable for vision in the air, is no less so to aquatic birds. When these dive they find themselves in a medium with a much higher refractive index, in which their corneas suddenly lose their optical effect. They instantly need a much shorter focal length, with an increase of at least twenty diopters over their vision in air. It has been demonstrated that in cormorants a very strong *sphincter iridis* can instantly and markedly decrease the diameter and increase the thickness of the lens, thus changing its curvature. These birds can thus accommodate by forty to fifty diopters, whereas a human child is capable of only 13·5 diopters (while an adult has still less power of accommodation). Other diving birds (ducks, divers and auks) make use of a supplementary lens in the nictitating membrane. In the middle of this third eyelid, which covers the eye when the bird is underwater, is a transparent window which is convex and of high refractive index, and thus acts as a 'contact lens' or 'corneal lens' (Wall 1942). Thus in waterbirds the nictitating membrane has taken on a function additional to the cleaning of the eyes.

(b) *Visual acuity and field of vision:* The observation of raptors alleged to search for prey while gliding at great heights, and of vultures similarly detecting carcases, has led to the belief that birds have extraordinarily keen sight. However birds as a whole seem to enjoy visual acuity comparable to that of man. Some have very fine retinal structure with more numerous cones, and with many coloured globules which are important in filtering the light and so sharpening the image; yet the visual acuity of many passerines is of the order of 1·5 minutes of arc, which is about three times coarser than human acuity. In contrast the acuity of large raptors and vultures has been estimated as two or three times finer than that of man.

Though the resolving power of birds, expressing their visual acuity, is not extraordinarily high, their vision is superior to that of man in its much wider visual field. The avian eye forms a clear *simultaneous* image of much of the surrounding space, without fixating it point by point in order to establish details.

As a consequence of this last adaptation, and of the position of the eyes, a bird's perception of relief is considerably reduced, the image on the retina being 'flat' by reason of this uniform clarity. In order to perceive relief, the bird must somehow integrate the images formed as it moves through successive points in space, which is no doubt easy for it during flight. During slower movement or at rest the trick of moving the head, either along the line of march or laterally, may be a device to obtain stereoscopic vision from differences of parallax in the successive images thus cast on the retina.

Clear vision over a wide field is essential to birds, not least for their long-distance orientation, which is largely based on astronomical guiding-marks. The wide vision of a bird allows measurement of very wide angles and the perception even of seemingly slow movements.

(c) *Colour vision:* Experiments on pigeons have shown that the curve, expressing the relationship between the wavelength of light and the smallest change in this wavelength perceptible to the bird, has the same form as in man. This confirms what we know by observing the behaviour of birds, that they show excellent colour vision, possibly better even than our own. Birds' retinas contain many coloured globules, which absorb radiations in the yellow and red parts of the spectrum especially, making these exceptionally conspicuous to the bird without correspondingly reducing the perception of those colours sensed by other cells. The effect of these globules must be to heighten the colour-contrast between various objects in the field of vision, but there is no evidence that the bird's spectrum of perceived light differs from that of man. This resemblance does not mean that their impressions of colour are the same, and no doubt these vary from species to species.

Birds' development of coloured plumage is correlated with this excellent colour vision, without which their mosaic patterns of coloured patches would be useless. A bird shows this during any specialized display, which would be in vain if its partner did not receive the message. Thus interspecific and intraspecific communication has been established, by parallel evolution of colour perception and coloured sign-stimuli.

(d) *Nocturnal vision:* We have repeatedly referred to the vision of nocturnal birds, especially owls, whose eyes are of very specialized construction, apparently for maximum light-gathering power. This is seen especially in the large diameter of the pupil, while the focal length is short and almost fixed, and the retina consists almost entirely of rods whose terminal segments are elongated and packed with rhodopsin pigment. Ophthalmologists estimate that an owl's eye absorbs a hundred times as much light as a human eye adapted to vision in semi-darkness. The eye is correspondingly richer in ganglionic cells, which suggests that visual acuity increases in proportion to the perception of very feeble light. The visible spectrum for owls as a whole is the same as that of other birds, though perhaps somewhat more extended into the violet. No evidence has ever been provided for the hypothesis that these birds can see infra-red radiations, by which they might locate themselves at night and find their warm-blooded prey, and it is contradicted by physical considerations.

Hearing and balance

The ears of birds differ significantly from those of mammals. They are sensitive to faint vibrations, which play a most important part in birds' lives. The external ear lacks a pinna and is reduced to a plain and relatively short tube, opening under feathers which guard its mouth. In the eared owls the opening is closed by two folds of skin, one of which may be erected and serve to catch noises coming from behind. The duct leads to the middle ear, which is shut off by the tympanum or eardrum, and is full of air in communication with that of the buccal cavity by the Eustachian tube. Because of the inertia of this air, rapid changes in pressure produced by vibrations deform the tympanum and cause the tympanic cavity of the middle ear to resonate, and to transmit these vibrations to the inner ear. Thus the middle ear serves as a resonance chamber, the more effective since it is frequently pronged into cavities ramifying within the surrounding bone. The tympanum is linked to the fenestra ovalis by a partly bony and partly cartilaginous rod, the columella, homologous with the mammalian stapes, which transmits the sound vibrations to the inner ear. There is no chain of auditory ossicles as in mammals.

The very complex inner ear is a closed system filled with a somewhat viscous liquid, consisting primarily of a labyrinth analogous to those of other vertebrates. The labyrinth is formed principally by three semicircular canals arranged in the three planes of space and connected by the utriculus, which play an essential part in the life of birds. Beside the utriculus is the

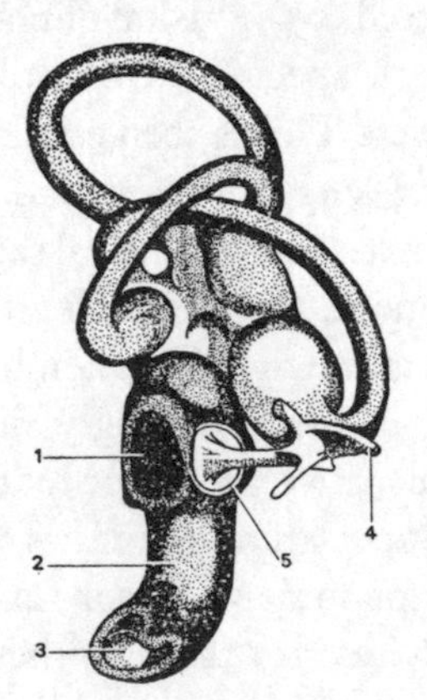

Figure 34. The labyrinth of a Bullfinch *Pyrrhula pyrrhula*. 1. fenestra rotunda. 2. cochlea. 3. lagena. 4. columella (= stapes). 5. annular ligament.

sacculus with below it an important extension the cochlea, corresponding to our own, in the form of a long curved chamber. In the middle of this essential auditory organ is a tubular sac, the *scala media*, filled with *endolymph*, which is closed on one side by the tegmentum vasculosum and on the other by the basilar membrane. The cavities on either side, both filled with *perilymph*, are known respectively as the *scala vestibuli* and the *scala tympani*. Near the basal end of the latter is another opening closed by a membrane, the *fenestra rotunda*, separating it from the middle ear. Thus the cochlea is divided into three parts, and is filled with liquid which unlike the air in the middle ear is incompressible. Deformation of the fenestra ovalis by a movement of the columella causes a corresponding deformation of the fenestra rotunda, while displacing the basilar membrane in the same direction. At the basilar membrane the sensory nerve cells end in auditory hairs, grouped in what is known as the *organ of Corti*, where the mechanical responses to changes in pressure caused by vibrations are converted into nervous input. The avian cochlea also differs from that of mammals, most obviously in that it is not coiled in a spiral. The differences of fine structure are more important and are reflected in auditory performance. The comparatively short basilar membrane suggests that birds hear a more limited range of frequencies than mammals, while in contrast their greater number of sensory cells per unit length points towards increased sensitivity to faint vibrations, and quicker reactions resulting from reduced inertia of the system.

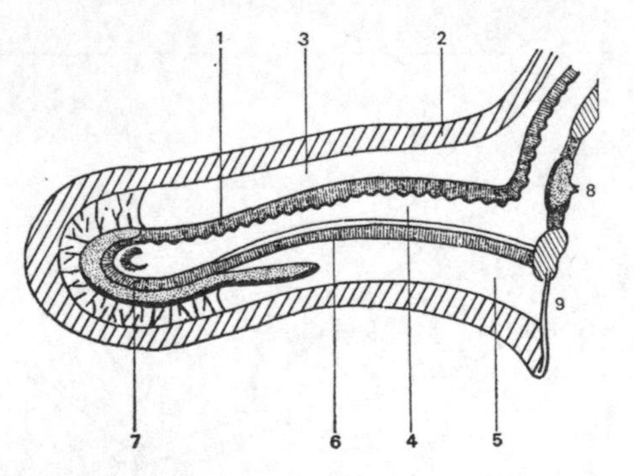

Figure 35. A bird's cochlea in diagrammatic cross-section. 1. tegumentum vasculosum. 2. bony wall. 3. scala vestibuli. 4. scala media. 5. scala tympani. 6. basilar membrane. 7. papilla of the lagena. 8. end of the columella. 9. fenestra rotunda

HEARING

It is difficult to assess levels and qualities of hearing in birds, since few objective measurements have been taken. Those made on the Bullfinch

Pyrrhula pyrrhula show that the curve expressing the auditory response of this bird is broadly comparable to that of man, but displaced, with a maximum sensitivity between 2,000 and 4,000 hertz (cycles per second), on either side of which there is a rapid decrease. The lower and upper limits of hearing have been determined for several species, either by using conditioned subjects or by physiological measurement of the cochlear potential, and values in hertz are shown below (after Schwartzkopff 1955).

	Lower limit	*Maximal sensitivity*	*Upper limit*
Mallard *Anas platyrhynchos*	< 300	2,000–3,000	8,000
Pheasant *Phasianus colchicus*	< 250		10,500
Budgerigar *Melopsittacus undulatus*	40	2,000	14,000
Long-eared Owl *Asio otus*	< 100	6,000	18,000
Eagle Owl *Bubo bubo*	60		> 8,000
Starling *Sturnus vulgaris*	< 100	2,000	15,000
Chaffinch *Fringilla coelebs*	< 200	3,200	29,000
Canary	1,100		10,000
House Sparrow *Passer domesticus*			18,000

As a whole, birds show maximal auditory acuity between 2,000 and 4,000 hertz. Apart from parrots and owls, which differ in the unusual length of their cochleae, they do not seem able to make real use of sounds higher than 10,000 hertz, and certainly not of ultrasonic vibrations.

Birds' discrimination between frequencies appears to be good, the

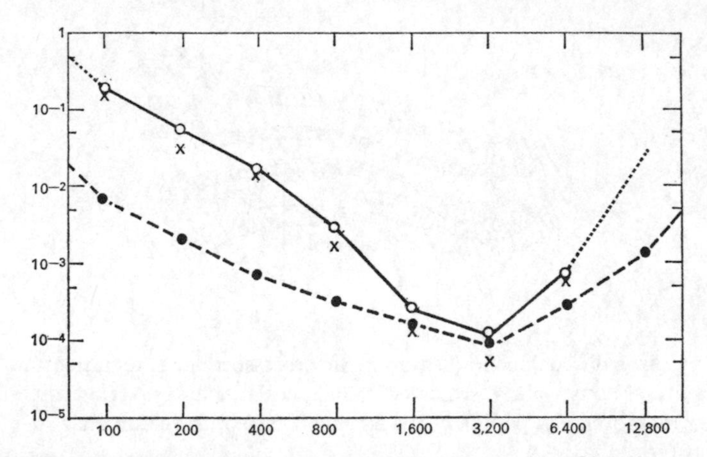

Figure 36 Curves showing auditory sensitivities – thresholds of hearing at different frequencies. Open symbols and solid curve: human sensitivity, for comparison. Solid symbols and broken curve: sensitivity of the Bullfinch *Pyrrhula pyrrhula*. XXX: thresholds of a Bullfinch with an unusually low minimum.

Budgerigar and various passerines being able to distinguish clearly between sounds differing in frequency by only 0·3 to 0·7 per cent. Their localization of sounds is equally good. This is known to depend not only on the phase of the sound waves received by each ear, but also on differences of intensity resulting from the orientation of the head relative to the source. Because of the small separation between the auditory orifices, the latter difference is the more important in birds. Measurements have shown that sound intensities at the two ears may differ by about 14 per cent, allowing localization of the source within 20° to 25°. Owls localize sound still better, thanks to their large skulls, which separate their ears more widely; to their possession of external pinnae of a sort, and further to an asymmetry between their ears, which in some species even affects the structure of the skull. They can close their ear flaps, reducing their acuteness of hearing but making it highly directional along the axes of the ear openings. This ability, in conjunction with other characteristics of their hearing, allows them to hunt rodents and other small mammals by hearing at dead of night. Experiments have shown that in total darkness Barn Owls localize their prey solely by hearing, with an error of not more than 1° (Payne). Like man, birds find more difficulty in localizing pure sounds than complex ones which are rich in harmonies, which explains the correlation between the quality and content of their sound emissions.

For each species there is a definite relationship between the pitch of its own sounds and the optimal sensitivity of its ear. This adaptation is shown as clearly in the interaction between the young and their parents, so that a hen is very sensitive to the shrill cheeping of its chicks (maximum emission at around 3,000 hertz), while the chicks hear especially their mother's deep clucking (about 400 hertz). However, hearing need not be adapted only to the calls and songs of the species itself. Many other sounds are received, especially those made by enemies and prey, and owls are highly sensitive to the shrill squeaks of the rodents on which they feed.

ORGAN OF BALANCE

As in all vertebrates, the utriculus of the inner ear carries three *semicircular canals* which constitute a statical organ concerned with balance. This complex structure lies at the back of the otic region, at the side of the foramen magnum, where it is completely enclosed by the bone. Within each bony canal, and separated from its walls by the liquid perilymph, is a membranous tube containing the differently constituted endolymph. At one end of each canal is an ampulla within which is differentiated a richly innervated area, the sensory crest (*crista acustica*), formed from epithelium folded into a ridge and carrying cells with hair-like terminations. Connected to this crest is a small one, which is swung like a door by movement of the liquid contained in the canals. The anterior semicircular canal (usually the largest) is in the sagittal plane, the posterior canal in the transverse plane, and the horizontal canal extends forwards. The largest in a pigeon measures up to 14mm, which gives an idea of the size of the organ.

This apparatus works by the inertia of the liquid in the canals, for which the utriculus provides a kind of reservoir. When for example the bird raises its head, the endolymph in the anterior canal flows towards the ampulla of the canal, in the opposite direction and at the same speed as the movement, and this causes excitation of the sensory crest. Any movement of the head, in whatever plane or direction, thus causes a flow of liquid in the corresponding canals. The resulting excitation of the nerve endings informs the bird of the movement its head has undergone, so that it knows from instant to instant its exact position in space. However, such a system relying on inertia suffers from certain functional defects. When the head ceases to move the sensory organs of the canals are no longer excited, and the bird can then keep track of the position of its head and return it to the horizontal only by using visual cues. This inconvenience is mitigated by another system depending on otoliths, provided by an organ situated in a recess in the anterior basal part of the utriculus. This *macula utriculi*, which is strictly in the bird's horizontal plane, is a highly differentiated area of epithelium, rich in sensory cells which end in hairlike processes, above which is another membrane on which are statoconies or otoliths. These calcium carbonate concretions never form large statoliths as in fishes but remain particulate, with an average diameter of 30μ. They do not roll freely upon the membrane, but are held in a gelatinous secretion, so that they excite the underlying sensory cells by acting through this relatively fluid mass to control the cell hairs. The inertia of the mass causes, and its viscosity prolongs, the movements of the otoliths, which can thus inform the bird of the position of its head even when immobile.

Birds thus have precision instruments for judging the positions of their heads, and therefore of their whole bodies, in space.

Smell

Birds have a poorly developed olfactory organ in the nasal cavity. The nasal cavities are separated by a median septum, and each is divided into three chambers by transverse thresholds, folds or turbinals. The anterior chamber or atrium hides a fold, always coiled on itself, which is marked at most by furrows and protruberances. The middle chamber, which opens widely into the buccal cavity by an internal nostril, contains a very large concha or turbinal corresponding to the maxillo-turbinal of mammals. This is very variably developed, usually making no more than one turn of a spiral, but in some birds, such as ducks, tube-noses and vultures, it may show up to five turns and be further elaborated by the development of longitudinal furrows. The nasal septum itself may be decorated by excrescences which similarly increase the surface of the nasal fossae. This surface is covered by epithelium which is partly ciliated, and richly vascularized. It serves to filter, warm and humidify the incoming air.

The third chamber is behind and above the second, into which it opens. It protects the posterior turbinal, homologous with the mammalian naso-turbinal, which is often reduced to a simple tubercule. However, in tube-noses, vultures and kiwis it is much better differentiated and coiled like the middle turbinal. It is here that olfactory receptors are to be found, with nerve fibres leading to the olfactory lobes of the brain. The sense of smell, still very poorly understood in birds, is atrophied in most of them. It is relevant in this connection that birds lack all cutaneous glands, apart from the uropygial gland whose very specialized functions are unrelated to odour. The olfactory parts of the nasal fossae receive odours coming principally from the hind part of the mouth through the internal nares and the second chamber, which allows the food to be checked by smell. This explains for example how certain raptors which feed only on fresh meat will sometimes take rotten flesh, but reject it as soon as the smell reaches their nasal fossae.

Birds' sense of smell has been studied in many experiments, with highly contradictory results. According to some authors positive results have been obtained from a variety of passerines (Robin, Blackbird, warblers, tits) and ducks, while the results from pigeons have sometimes been positive (Bajandurow & Larin) and sometimes negative (Walter, Calvin). Recent experiments on a variety of birds have shown that in fact they have no sense of smell comparable to that of mammals (Neuhaus 1963), but that geese do show a reaction, changing their respiratory rhythm as soon as they are exposed to a strong smell. This behaviour is explained as a reflex originating in a different part of the brain (the diencephalon) from that responsible for mammals' responses to smell.

Thus despite their possession of an olfactory apparatus most birds seem to be practically devoid of a sense of smell, which even if it exists is certainly not used. However, mode of life and olfactory perception are obviously compatible in those few birds which seek their prey on the ground or on the surface of the sea. This is especially true of kiwis, whose eyes are small and poorly differentiated, their exclusively nocturnal life not predisposing them to reliance on the sense of sight. They clearly make use of smell to find the prey which they dig out of the ground layer of vegetable matter. Their external nostrils open at the ends of their long curved bills, and lead the smells to a well developed olfactory region, which as mentioned above has no fewer than five coils. As in mammals the nerve fibres remain separate instead of forming a single nerve as in other birds. In tube-noses too the nasal region is well differentiated. The posterior turbinal is of relatively large surface area, covered in sensory epithelium connected with a well-developed rhinencephalon. The nerve fibres are combined in a single nerve, though this is

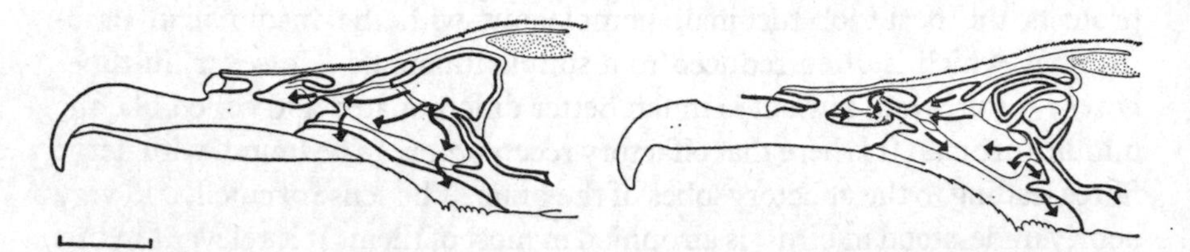

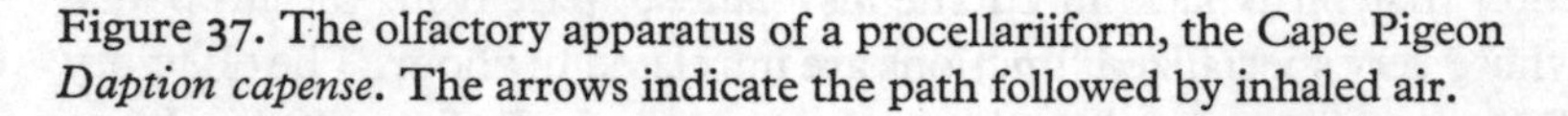

Figure 37. The olfactory apparatus of a procellariiform, the Cape Pigeon *Daption capense*. The arrows indicate the path followed by inhaled air.

divided up by bony spicules which functionally recall the cribriform plate of mammals. This apparatus, highly developed in comparison with those of other birds, is undeniably useful in the lives of petrels and albatrosses. Experiments in which smelly fish oils have been spread on the surface of the sea have shown that these birds can be attracted and guided solely by their sense of smell. Such a sense is certainly useful to birds which feed not only on marine animals but also on all sorts of refuse floating on the surface.

A sense of smell has also been postulated both for Old World and for New World vultures, to which it could be useful in detecting carcases. There is voluminous literature on this controversial subject, to which Audubon in 1826 and Darwin in 1834 contributed articles. They both concluded that these birds did not perceive odours and were guided solely by sight, and recent results confirm this opinion (Stager 1964). Experiments carried out in India have shown that White-backed Vultures (*Gyps percnopterus bengalensis*) and Egyptian Vultures (*Neophron*) cannot locate hidden carcases even if they are stinking. The same is true of most New World vultures. Thus these birds survey huge expanses by soaring at great heights, guided to carcases by watching the behaviour of predators or of small mammalian or bird scavengers – the latter also guided by sight while exploring the terrain in low level flight. In contrast the Turkey Vulture *Cathartes aura*, found throughout the New World, seems to be able to locate its prey by scent, as various

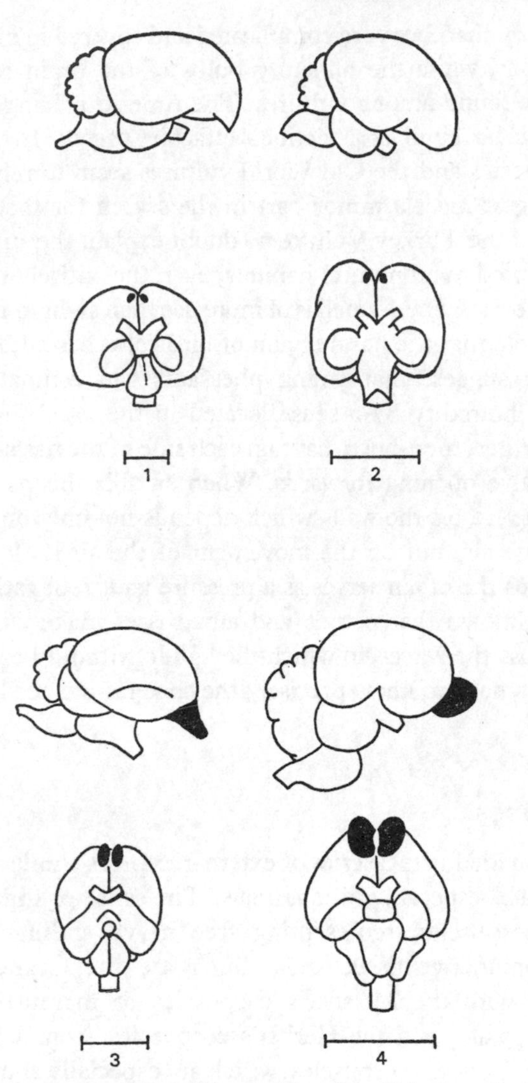

Figure 38. The appearance of the olfactory lobes (black) in four families of birds. 1. Corvidae. 2. Psittacidae. 3. Rallidae. 4. Procellariidae.

experiments have proved. It is attracted by the smell of corpses placed in the airstreams from blowers, and even by the smells of certain chemicals such as mercaptans, while, in contrast, a stuffed deer carcas did not attract one of these birds. Perception of the smell of carcases is helped by the fact that this vulture, unlike the other American species, flies low in seeking food. Study of its olfactory apparatus shows it to be well developed and functional, with the nostrils proportionally larger than those of other species and the turbinal in the olfactory chamber very complicated and covered in highly differentiated epithelium, while the olfactory bulbs of the brain attain the largest volume to be found among vultures. The American King Vulture *Sarcorhamphus papa*, has similarly functional olfactory organs. In contrast the other American species and the Old World vultures seem to rely on sight alone, smell playing at most a minor part in the search for food. The olfactory capabilities of the Turkey Vulture no doubt explain the success of a species which is adapted to all natural habitats, even the entirely closed one of the Amazonian forests where smell is of more use than sight in finding carcases.

It is possible that the nasal region of birds also has additional functions. Experiments suggest that young pheasants can estimate differences in atmospheric humidity by a sense located in the nasal fossae (Shelford & Martin). Further, tube-noses have on each side of the nasal septum a kind of triangular valve opening forwards. When air fills this as the bird flies, it exerts a pressure on the walls which depends not only on the bird's speed relative to the air, but on the movement of the air itself, that is the wind velocity. Thus the organ serves as a pressure gauge, or rather as an anemometer. It is known that petrels and albatrosses make use of air currents blowing across the waves, in which they glide without beating their wings, and that they need to know precisely the changes in speed and direction of the airflow.

Other senses

Birds are provided with a series of extero-receptors similar to those of other vertebrates and especially of mammals. The sense of touch is localized in numerous corpuscles, representing free nerve endings or enclosed in modified conjunctive tissue cells. Such are the Grandry's corpuscles, homologous with the Meisner's corpuscles of mammals, found in the tongue and palate; and the Herbst's corpuscles, homologous with mammalian Vater-Pacinian corpuscles, which are especially abundant in the bills of snipe and the tongues of woodpeckers – organs which because of these tactile corpuscles are especially sensitive and are used in foraging. Birds are also very sensitive to vibrations at particular frequencies, which has long been attributed to acute perception by the auditory apparatus. However, these vibrations are actually perceived through Herbst's corpuscles, scattered in the skin near the feather buds or concentrated in clusters in the limbs and especially the legs. This arrangement enables birds to sense very faint vibrations, and even sounds are partly received in this way. This sensibility allows birds standing on the ground to feel the approach of an enemy, and is still more useful when they are sleeping perched on a branch. Certain nuptial displays, the dances of grouse, similarly make use of such vibrations transmitted by the central nervous system. Equally, the 'prediction' of earthquakes by birds may be explained by their reception of the first movements, imperceptible to man and other animals.

The sense of taste is localized in taste buds like those of other vertebrates. However, they are never clustered on papillae like those of mammals, but are scattered over the posterior part of the tongue and the buccal cavity. There are about fifty to seventy-five in pigeons, 200 in starlings and ducks, and 300-400 in parrots. Their perception is comparable to that in mammals.

Chapter 8

Vocalizations

SOUND plays an important part in the intercommunication of many animals. Birds are certainly the best equipped among vertebrates, making use of the richest and most diverse repertoires. Sound production is very unevenly developed and follows the major evolutionary lines. The lower groups, and especially sea and water birds, have simple repertoires consisting of relatively few and structurally simple calls, whereas vocal ability reaches its finest flowering in the passerines. True song appears only in the major group among them, the oscines, which are hence known as songbirds. Their very elaborate vocalizations carry a multitude of meanings, many of which are concerned with strictly sexual activities. This development could only proceed alongside that of the corresponding receptors.

Vocal organ

The larynx is too near the mouth, and is insufficiently provided with adequately large resonating chambers, to produce loud enough sounds at medium frequencies; at most it has an effect on resonance. It is therefore replaced functionally by a unique organ, the *syrinx*, at the base of the trachea where the two major bronchi divide. This organ is very differently developed in different birds. In some, such as ostriches, storks, vultures and cormorants, it is simply formed by the modification of a few bronchial or tracheal rings, related muscles being absent or very poorly developed. In others it is a much more complex organ, which attains its highest development among the songbird. It is entirely tracheal in the most primitive of these (the ovenbirds, woodhewers and antbirds), which are hence known as Tracheophonae, and tracheo-bronchial in all the rest. Several tracheal and bronchial rings are transformed, and worked by special muscles: the extrinsic series connecting them with other structures in the neck region (the sternum and clavicles), and the intrinsic series inserted on the rings at both ends. The intrinsic muscles vary in numbers and complexity. In the Oligomyodi (the tyrant-flycatchers, manakins and cotingas) there are one to three pairs which are inserted towards the middle or all around the tracheal and bronchial half-rings. In the Acromyodi or Oscines, comprising the rest of the passerines, there are five to seven or even nine pairs, inserted solely on the sides of the half-rings. These muscles control the shape and tension of the syrinx, and thus the quality of the vocalizations. The variety of sounds produced depends broadly on the number of muscles, though many exceptions contradict this rule.

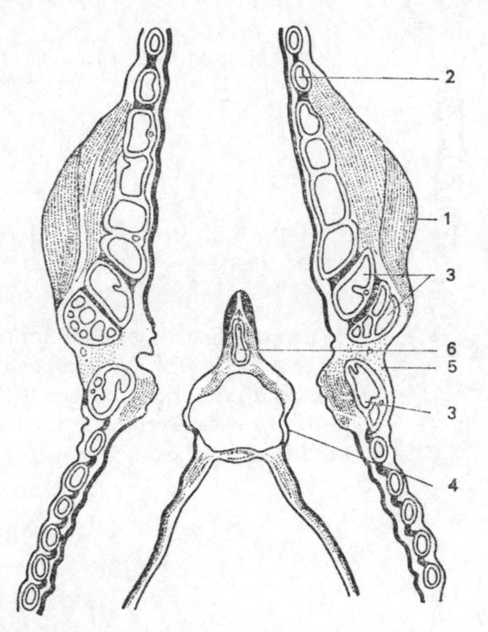

Figure 39. The syrinx of a Blackbird *Turdus merula* in section. 1. muscles. 2. tracheal ring. 3. bronchial ring. 4. internal tympanic membrane. 5. external tympanic membrane. 6. pessulus.

The vibratory organ is formed by two membranes, one stretched across the base of the trachea and the other into the base of each bronchus, known respectively as the external and internal tympanic membranes. They vibrate during the active expiratory phase of respiration, moved either passively by the rush of air or more probably actively as a result of nervous stimulation, like the vocal chords in the mammalian larynx, since the syrinx is richly innervated by branches of the vagal and hypoglossal nerves. The two membranes probably vibrate to some extent independently of one another, since sound analysis shows that several birds produce two notes simultaneously, neither being a harmonic of the other. The interclavicular sac is also indispensable to sound production, since opening it deprives a bird of its voice.

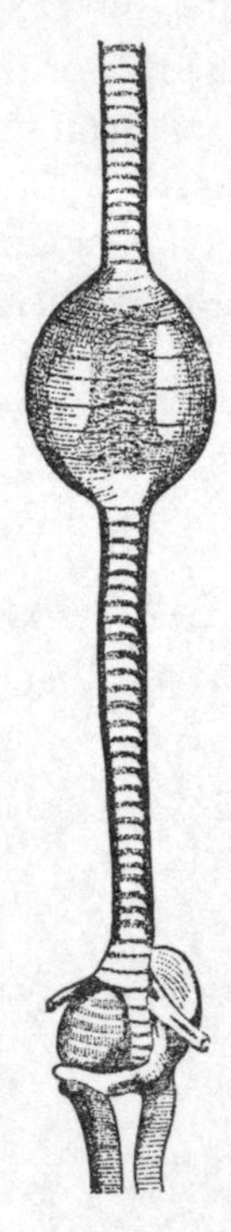

Figure 40. The trachea and bronchi of a drake Rosy-Bill *Netta peposaca* from South America (reduced to one-third). Note the swelling in the upper third of the trachea.

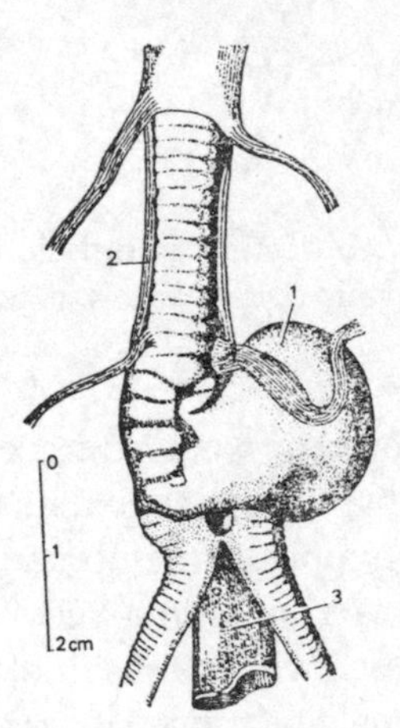

Figure 41. The lower end of the trachea in the Abyssinian Blue-winged Goose *Cyanochen cyanopterus*. The last rings are fused to form a chamber, the 'bulla ossea'. 1. bulla ossea. 2. trachea. 3. oesophagus.

The sounds produced by the vocal apparatus are amplified and altered by the trachea and the buccal cavity. The trachea is sometimes modified to this end, and in cranes and swans it traces several loops, folded upon one another and coiled within the breastbone, so that it can be as much as 1·5m long. Drakes have swellings of the trachea, supported by cartilages or even by a bony casing; but it is difficult to explain their development in terms of a possible resonating function, since the ducks which lack them sometimes have louder voices than the drakes. The sounds produced by a syrinx depend primarily on the nervous signals received by its musculature, and the organ is also hormonally influenced, especially by the sex hormones.

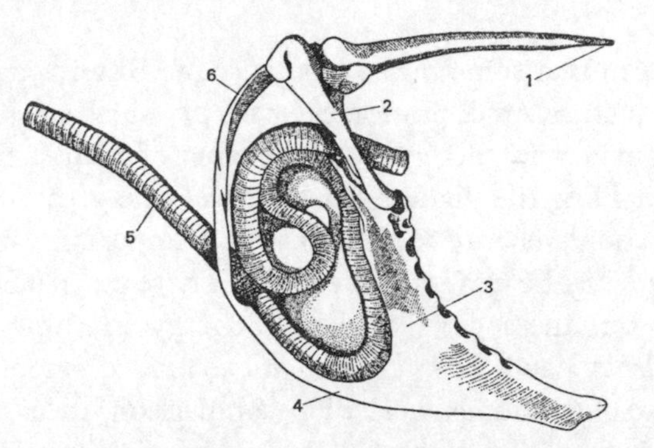

Figure 42. The breastbone and trachea of a crane *Grus*. 1. scapula. 2. coracoid. 3. sternal crest containing the convolutions of the trachea. 4. sternum. 5. trachea. 6. clavicle.

Structure of song

All sounds are characterized by their pitch (determined by the frequency of vibration), their intensity (measured as the amplitude of vibration) and their tonal quality (the result of accompanying and often harmonic vibrations).

The frequencies of vocal signals vary very much, especially from species to species, as the values in the following table show (after Brémond, Marler, Mulligan, Schubert and Thielcke).

Species	*Type of signal*	*Frequency range of signal (kHz)*	*Frequency range of maximal acoustic energy (kHz)*
Bittern *Botaurus stellaris*	Song	1	0·5
Cuckoo *Cuculus canorus*	Song	0·2–1·7	0·25–0·5
Magpie *Pica pica*	Chattering	0·8–12·0	4–6
Carrion Crow *Corvus corone*	Mobbing	0·5–11·0	1–3
Wren *Troglodytes troglodytes*	Song	2·5–10·0	
Robin *Erithacus rubecula*	Song	1·5–11·0	4–7
Nightingale *Erithacus megarhynchos*	Song	1–9	
Bonelli's Warbler *Phylloscopus bonelli*	Song	2·7–7·0	3·5–6·0
Willow Warbler *P. trochillus*	Song	2–7	
Wood Warbler *P. sibilatrix*	Song (shrill variant)	2·8–9·0	4·5–8·0

Species	*Type of signal*	*Frequency range of signal (kHz)*	*Frequency range of maximal acoustic energy (kHz)*
White-crowned Sparrow *Zonotrichia leucophrys*	Song	2·2–6·8	
Chipping Sparrow *Spizella passerina*	Song	2·4–7·0	
Song Sparrow *Melospiza melodia*	Song	2·5–6·7	

It can be seen that some birds, especially the Bittern, have deep voices, while others such as small passerines have very shrill ones. It is necessary however to consider not merely the whole range of frequencies, but also that part of it containing the highest acoustic level (the greatest intensity), and thus the most sonic energy. This is the most important part of the sound spectrum, carrying the quality of the signal. There are in addition ultrasonic components, but these carry negligible energy in comparison with the audible frequency band. Birds clearly emit sounds at much higher frequencies than those of the human voice. For comparison, speech is composed of sounds at between 80 and 400 Hz and a soprano can reach 1500 Hz (or exceptionally higher), which is the *lower* limit for many birds. In addition the frequency ranges of birds are very clearly wider than that of man.

Birds' sound signals do not of course dwell at the same frequency for any length of time, but are modulated in a very rapid rhythm. Intensity does not necessarily change in step with pitch – another essential difference from the human voice, whose intensity and frequency are more or less linked (at least in western languages). Birds emit different notes in very rapid succession. Thus the song of the Song Sparrow *Melospiza melodia* consists of an average of fifteen to seventeen notes per second, each separated from the one before by a time interval of one fiftieth of a second. The American Wood Thrush *Hylocichla mustelina* can change frequency at a record rate of 200 times per second. In this connection, remember that a bird's ear responds ten times as fast as the human ear. Birds are capable not only of emitting several notes at once, but of modulating them independently, which demands exceptionally precise nervous stimulation.

Examination of sound spectrograms shows that the sound emitted at any instant is usually at several frequencies, so that the trace appears 'smeared' and shows a number of harmonies. However, some birds do emit almost pure notes: for example the Little Tinamou *Crypturellus soui*, whose pure notes, separated by short intervals and rising steadily in pitch, are just as though the bird made use of a frequency-filter. The two notes ('titi pu . . . titi pu . . .') comprising the Great Tit's song are also relatively pure, remaining constant in pitch throughout the emission.

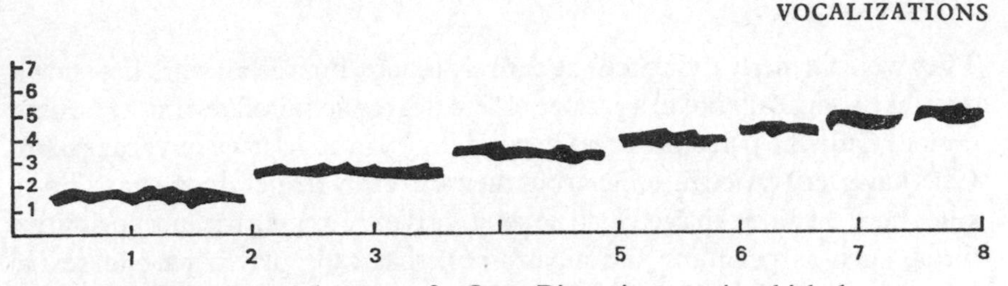

Figure 43. Sonogram of the song of a Costa Rican tinamou, in which the vocalizations become progressively shriller and shorter.

The volume of bird song, resulting from the amplitude of the vibrations and shown on sound spectrograms by the blackness of the trace, is also very variable. The power of emission of most species, and the precision with which it is controlled, are astonishing. A Wren's song carries very long distances, yet some of the vocalizations which it uses during nuptial display can only be heard at a few metres. In the first case it is advertising its holding of a territory whose boundaries it defends from a distance, while in the second its display is for the benefit of a partner perched nearby. The Cuckoo can similarly be heard over a kilometre away, the Bittern at nearly 5km like the bellbirds (*Procnias*), cotingas of South American forests whose carillon-like calls give them their name. The intensity of the sounds scarcely depends on the size of the bird responsible, but much more on the significance of the particular vocalization, while there are also strict correlations between song characteristics and the type of habitat.

The timbre of the voice, related to harmonics and complex sounds, is also very important in giving the song specific characteristics. Although it is represented in sound spectrograms it is difficult to convey, and the list of terms used in classic works to describe the songs and calls of birds shows the richness of its variation. It can be shown to be of great semantic importance by playing an artificial rendering of a bird's song with the timbre modified, to which the species fails to respond. Finally, the notes comprising a vocal signal, and especially those of the long phrases of song, are separated by short but very frequent interruptions which give a characteristic species-specific rhythm.

Call-notes

The various vocalizations of birds are difficult to classify, because every intermediate state exists between the simplest and the most complex signals. The former are known as calls, which may be defined as short sounds of simple acoustic structure, being composed of monosyllables or disyllables of not more than four to five notes, never organized in sequences or phrases.

They were formerly thought to be entirely innate, but recent work has shown that, like song, they are to a greater or less degree acquired during a learning period. Only the passerines pass beyond this primitive stage of vocalization. Calls have very varied meanings, but they are rarely sexual among passerines, since here they are replaced by true song. It is only among the most primitive birds, such as penguins and albatrosses, that calls play a part in sexual behaviour and especially in displays. The distinction between calls and songs cannot be sharp, since the former often appear as components or elementary phrases of the latter. This happens even among passerines – for example the complex and varied song of the Sky Lark is made up of integrated call-notes– but it is the rule among lower birds, and the 'songs' of waders are little more than series of such notes.

Many classifications have been proposed for dividing calls into categories. These have mostly been based on meanings, not taking acoustic characteristics into account, although in some cases there is a definite relation between the structure and meaning of a call. Following Brémond we may divide calls into three classes:

Calls to co-ordinate activities and indicate position: Many of these concern the relations between parents and their young, the former calling to bring together their nudifugous chicks, the latter when they are lost and seeking to rejoin the brood. Nudifugous young, such as chickens, give different 'satisfaction calls' when they are together and unthreatened.

Certain calls similarly serve to maintain contact between the members of a flock or colony, and most birds which take part in communal activities co-ordinate them by sound signals. Gulls draw attention in this way to a variety of occurrences in the neighbourhood of their flocks. Tits and small graminivorous passerines maintain the cohesion of their flocks by such calls, which often have semantic significance between different species, as is true of the assembling calls of the multispecific flocks characteristic of the tropics. Some birds which disperse throughout their habitat for part of their daily cycle call to one another in order to reassemble. Thus by night ducks exchange a sort of conversation, while gathering into a flock which returns to communal feeding grounds. The calls of a duck in such circumstances often brings others to settle nearby, which is why decoys are effective for hunting when ducks are on the move. Geese gathered in flocks while wintering make a continual gabbling, low-pitched but audible at a distance which keeps them together. The individual calls which go to form the murmuring of the flock become shorter when it is on the move, and can change progressively into short monosyllabic alarm calls when the birds are disturbed. In this category too are the flight calls given by birds travelling in flocks, especially on migration, which are often in rapid rhythm and serve to prevent the flocks dispersing, particularly at night.

Other calls help to maintain pair bonds, discrimination between the sexes and later individual recognition of mates making use of vocalizations. Calls play a part in pairing among species which lack true song, while others are given during search for a nesting site, nest building, or persuasion of the mate towards the nest. Much behaviour during later stages of reproduction is accompanied by calls, notably the appeasement behaviour seen when one partner approaches the other at the nest.

In contrast to all these, some calls have a negative meaning and are intended to drive away another individual, as with aggressive calls and those used in the defence of territory.

Calls used for giving information about the environment: These call the attention of other members of the species to occurrences observed by the calling bird. This is true especially of alarm calls, which are among the most important in the life of birds. There are a great variety of them, and many have precise meanings within a single species. Thus the same bird may use one call to give warning of the presence of a terrestrial predator, and another

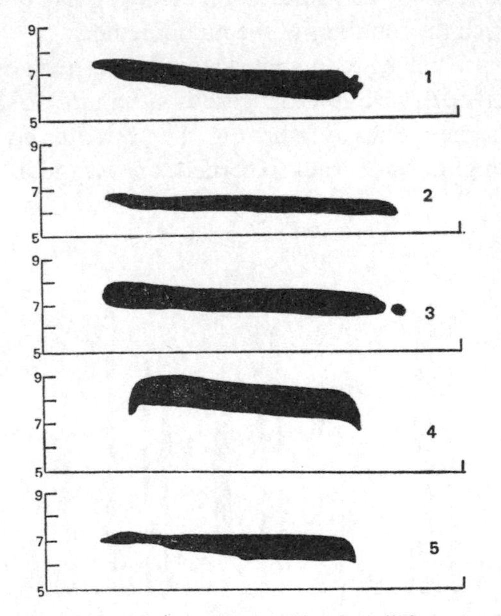

Figure 44. Sonograms of the alarm calls used by five different passerines when a raptor flies over. 1. Reed Bunting. 2. Blackbird. 3. Great Tit. 4. Blue Tit. 5. Chaffinch.

when a bird of prey flies near. Some birds announce potential and imminent dangers by different signals.

It is particularly interesting to compare the acoustic structure of alarm calls given by very different species. For example, the calls given by such diverse birds as Reed Buntings, Blackbirds, Chaffinches and various tits to call attention to an approaching raptor are all of the same structure. Thus birds, of whatever species, which are members of the same ecological community can understand sound signals which are important to all, in giving warning of approaching danger. This also applies to the calls given by many passerines when they gather round an owl to mob it. These short calls cause all birds nearby to flock together, and are all of very similar acoustic structure. Furthermore there are marked differences between alarm calls and those which cause birds to flock round a nocturnal raptor. The former are long calls consisting only of sounds of fixed pitch, whereas the latter are very short but made up of a wide range of frequencies. It is much more difficult to localize a pure or almost pure sound than a complex one containing very varied frequencies, which explains the differences in structure between these calls of different meanings. An alarm call of pure tone warns other birds without making it easy for the predator to localize the calling bird, whereas the mobbing call gives a precise indication of its origin to other small birds so that they may swell the numbers of the mobbing flock.

Distress calls also belong to this category, and are given by a bird captured by a predator or otherwise in a dangerous situation. It sometimes causes members of the same species to gather together, but then always puts them to panic-stricken flight. Such calls recorded on magnetic tape have been

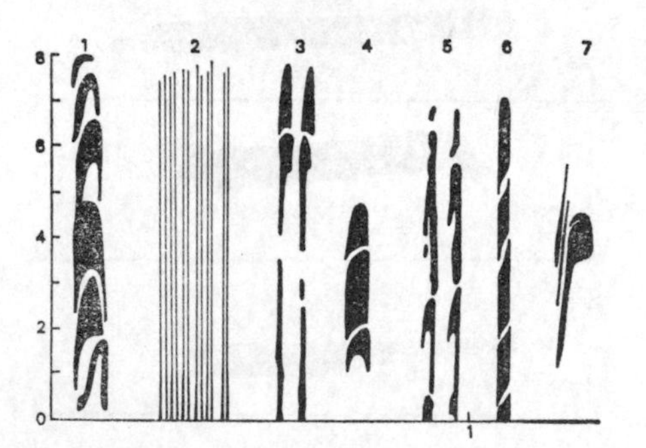

Figure 45. Sonograms of the calls of different birds when 'mobbing' an owl in daylight. 1. Blackbird. 2. Mistle Thrush. 3. Robin. 4. Whitethroat. 5. Wren. 6. Stonechat. 7. Chaffinch.

successfully used to drive away birds harmful to crops, such as Rooks and Starlings, or those which threaten human activities, such as gulls which may cause collisions near aerodromes. Unfortunately, the reaction is subject to habituation which reduces the effectiveness of the technique.

Calls used in feeding behaviour: Feeding behaviour is often accompanied by specialized calls of semantic significance, such as those by which gulls, for example, announce the presence of a source of food and attract each other to it. Many of the calls between parents and their young also fall into this category. The parents of nidifugous young, especially among the game birds, inform them of food and bring them to it. Nidicolous nestlings, on the other hand, may give soliciting calls. Together with visual stimuli, these characteristic calls emitted when the parents approach the nest elict from them the feeding response. They consist of short simple notes, rapidly repeated with rising pitch.

The calls given by honeyguides when leading a mammal towards a beehive also fall into this category.

Thus birds as a whole make use of a repertoire of call of very varied meanings. Furthermore each of their vocalizations can appear in variant forms, depending on the bird's motivation and the conditions of emission. The total number of audible signals is twenty in the domestic fowl, eight in the Blackbird *Turdus merula*, and twenty-four in the Song Sparrow *Melospiza melodia*.

Subsong

A special category of song known as subsong is a kind of elementary version of the species' definitive song. It differs from the latter by lower average frequency, wider frequency range, lower volume, different arrangements of successive notes, and extended duration of the phrases. Subsong is given by birds outside the breeding season, both before it begins and after it has finished, especially by migratory birds in their winter quarters. It is also given by young birds which have not yet mastered voice production, when they attempt the true song of the species. The Blackbird does this from its nineteenth day, the Whitethroat at thirty-eight days. Subsong thus appears as a sketch for the true song, to which it approaches progressively as the hormones increase their action, both with age and sexual maturity and seasonally with the approach of the breeding season.

Song

True song is distinguished from calls by its great richness, and by the complexity of notes emitted in phrases of distinctly longer duration. The vocalizations which go to make it have a much more complex acoustic structure. Furthermore, song differs from calls in its meaning, being concerned especially in territorial behaviour and sexual relations between mates. Among most species which produce it, song is a peculiarly male prerogative. In general it is restricted to the higher passerines or Oscines, although relatively brief songs are given by other species; and vocalizations such as those of waders, very different from the highly complicated ones of passerines, are sometimes called songs.

It is accepted that song has evolved from calls, and the various phrases of which it is formed may be considered as elements borrowed from them, whose juxtaposition has been followed by harmonization of the successive items to form a melody. The simplest songs still bear the mark of this origin, notably among pipits and in the Sky Lark. Thus the latter's song incorporates as essential elements the species' alarm call and flight call. Traces of this origin can still be noticed even in more elaborate songs, such as that of the Chaffinch which contains its familiar 'wheet' alarm call. Further, study of the origins of song in the young of a particular species shows that they first incorporate their calls into their subsong and later transform them into more elaborate and less recognizable forms.

A song is made up of a succession of notes, grouped into motifs which are themselves strung in phrases. The number of separate sounds may attain several hundreds: 300 in the Meadowlark *Sturnella magna* and up to 600 among birds which are vocal mimics. The form of the motifs and the order in which they succeed one another vary within a species, and even for the same individual, since the song as a whole transmits the message and the component parts do not have meanings of their own. Pure sounds are rare, though certain sounds of the Blackbird or Nightingale for example are sufficiently well filtered to appear pure to our ears. By contrast others have much wider sound spectra, sounds of very different frequencies being emitted simultaneously.

It is appropriate for the singer to show clearly what point in space he is occupying, in order that his mate may find him and his competitors know the limits of the territory he defends. Song is usually produced at considerable volume, in clear distinction from subsong and call notes, and as we have already noted it may carry as far as 5km. This again corresponds with the essential function of song, which is to attract the female over considerable distances, and to defend territories which are often wide. There is an obvious relation between the area of the territory and the carrying power of the song, although the song is always heard beyond the territorial boundaries. Naturally, the nature of the habitat influences the quality of the song.

The frequency with which song is produced varies considerably, mainly from species to species but also according to the phase of the breeding season. The male while unpaired sings more often than after finding a mate, when its singing rhythm tends to slow down. Thus the male Pied Flycatcher sings 3,600 times a day before forming a pair, and only 1,000 times afterwards (von Haartmann). The American Song Sparrow sings on the average 180-200 times an hour with a maximum of 310 before pair-formation, whereas the average falls to thirty to fifty and the maximum to 160 afterwards (Nice). In full song a Chaffinch emits phrases lasting two to two and a half seconds every seven to fifteen seconds, while the Red-eyed Vireo sings on the average 22,000 times a day during the maximum of its vocal activity.

Song has many functions. It is specific, capable of characterizing even a single individual thanks to the variants considered below. It has meaning as a message, which releases reactions on the part of its mate and other members of the same and of different species. Each element of its complex phrases has semantic significance: experiments in which song is artificially modified show that the sense of the message is thereby distorted and the bird no longer understands it. Thus what at first sight seems a pure luxury, like pleasant music, is really a commonplace necessity. The thresholds of excitation vary from species to species. The trumpet blasts which are penguins' sole sound signals serve to stimulate their mates to react, whereas the Nightingale's phrases, much more musical to our ears, are necessary because the female's threshold of receptivity is clearly higher. The messages vary in complexity, but maintain the same meaning within the major evolutionary lines.

Learning song

The song of young birds is at first imperfect, differing significantly from that of the adults. This can be studied by capturing young wild birds at various stages of development, and raising them in captivity in complete isolation from sounds. Such birds have been called 'Kaspar Hauser subjects', after a mysterious young German who appeared at Nuremberg in 1828, having been brought up in complete isloation. The Chaffinch has been thoroughly studied from this point of view by Thorpe (1958). Young taken from the nest at the age of five days were individually raised in soundproof enclosures, without any contact with other birds. The following spring they began to give almost normal subsong but this, far from developing into true song, only gave rise to a simplified caricature, lacking almost all the characteristics of Chaffinch song – especially the division into three phrases and the final phrase. The average pitch was deeper, and the variations in frequency greatly reduced. The experimental birds seemed to emit only a sort of outline of the species song, built up of essential elements from the sketchy subsong. All of this unquestionably represents the innate, genetically determined, component of the song.

Another experiment consisted of raising together a group of young Chaffinches taken under the same circumstances. These, living in complete isolation and never hearing other birds, formed a little community in which the species song might develop communally though sheltered from adult influences. Under these conditions all the birds belonging to a group developed identical songs, to the degree that they could not be distinguished by sound spectrographic analysis. However, there were differences between the songs of different groups. Some did not develop songs any more complex or like those of the species than those of birds brought up in complete individual isolation, while others tended towards more normal song or almost achieved this state. Thus it seems that, at least in some cases, birds deprived of any contact with adults and dependent solely on their innate component can profit from one another and so surpass the performance of solitary individuals.

This experiment may be repeated with birds not taken from the nest during their first few days of life, but caught during the autumn after fledging. These subjects have thus had contact with adults to a much more advanced stage, during which time they were receptive. By raising them until the following spring isolated from all auditory contact with adults, it can be established that their subsong develops into almost normal full song. These birds have thus *learned* during the autumn a number of characteristic elements of the song, especially the division into three phrases and the ending which are so characteristic of the species. If this experiment in turn is repeated with birds taken under the same conditions and then brought up not individually but in groups, the results are still better and entirely comparable with the normal species song of the Chaffinch. However, all birds raised together under these conditions show the same variations, and the traces of their songs can be superimposed exactly.

These experiments illuminate the situation in nature. Song certainly has an innate component: elements composed of notes at a particular pitch and in

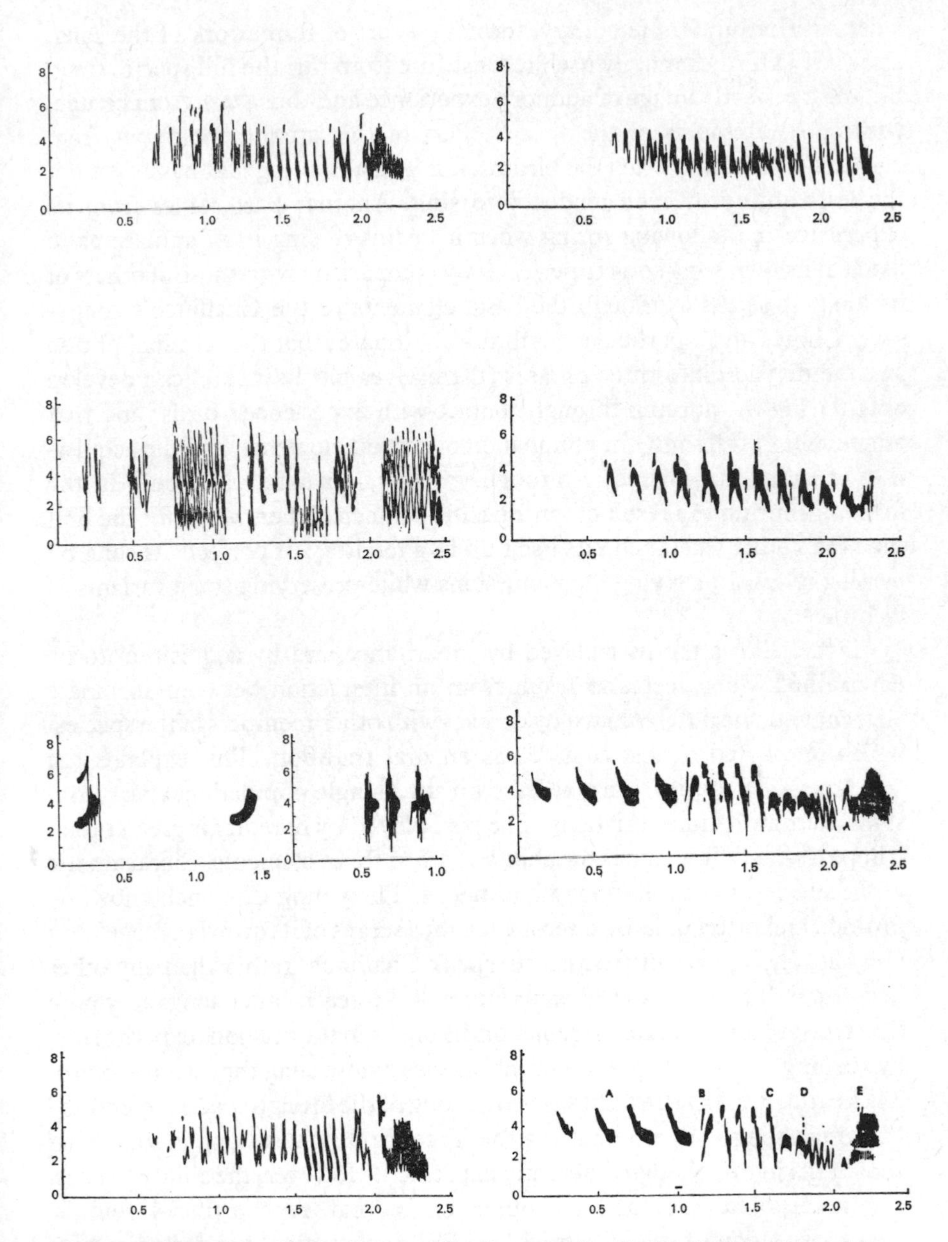

Figure 46. Sonograms of various vocalizations by the Chaffinch *Fringilla coelebs*. Song, subsong, juvenile song, and song by birds which have been reared in acoustic isolation.
Ordinate: frequency in kilocycles/second; abscissa: time in seconds.

a certain rhythm are hereditary, forming a sort of framework of the song. However, a bird cannot by itself reconstitute from this the full species song, but must take advantage of auditory experience and thus *learn* from contact with other members of the species. Part of this attainment comes from contacts with adults when the bird itself is still a nestling and has as yet not the least ability or even tendency to sing. Another part comes from its experience the following spring when it begins to sing itself and seems to listen to its own song so as to perfect it by comparison with those of others of its kind. It seems as though the basic elements of the Chaffinch's song – several notes and a particular rhythm – are innate; that the terminal phrase and the division into three phrases (themselves partly innate) can develop only during the autumn through contact with experienced birds; and that attainment of full song, the elimination of certain elements, and the acquisition of individual character through variants, can take place only in the following spring as a result of intraspecific contacts. Then, when for the first time the young Chaffinch sets itself up in a territory, it perfects its song by hearing its rivals, copying their emissions while preserving some variants of its own.

All this allows the part played by inheritance and by acquisition to be determined. Song seems to result from an interaction between an innate element and one differentiated by contact with other members of the species, which one could almost consider as an oral tradition. This explains the constancy of a species' song, at least within a single population, apart from significant individual variations. The part played by heredity is greater than it appears, since it includes the ability to learn those elements, characteristic of the species, which the young bird needs. The young Chaffinch knows by instinct, and utters of its own accord, various scraps of its own 'language', but also has a hereditary disposition to 'speak Chaffinch' rather than any other 'language'. Thorpe has tried experimentally to teach other songs to young Chaffinches, raised under the same conditions as in the previous experiments, by keeping them with an adult of the species whose song they were to learn. These attempts failed, except where the song of the foreign species resembled Chaffinch song – especially using the Tree Pipit, whose song is somewhat similar in rhythm though differently expressed. However, the ability to learn and incorporate the songs of other birds, and even artificial sounds, does exist among birds. Canaries and Bullfinches can learn melodies very different from their own song, while some have the gift of mimicry. The situation is still more complicated with birds whose songs are more complex than that of the Chaffinch, and in which learning certainly plays a still larger part.

Specificity and dialects

Certain bird calls are common to many species of an ecological community, such as those which warn of an approaching enemy. This common language is of great value to the whole community, any bird being able to sound an alarm which is understood by members of other species as well as its own. While such calls are not species specific, it is quite otherwise with song, which with a few exceptions is characterized by strict specificity. Field ornithologists justly claim that during the breeding season most birds can be identified by ear. Recognition is often easier from the voice than from morphological characters, which in a few cases are inconspicuous even when the bird is closely examined. This is true of three European *Phylloscopus*, the Chiffchaff *P. collybita*, Willow Warbler *P. trochilus* and Wood Warbler *P. sibilatrix*, which are difficult to distinguish by plumage but whose voices allow no confusion.

Two closely related species, morphologically so similar that ornithologists have difficulty in distinguishing them even as study specimens, usually have very different songs if they are sympatric. This is true of the *Phylloscopus* mentioned above, of many *Sylvia* warblers, treecreepers and tits, and also of a number of exotic birds such as the African *Cisticola* warblers. The songs of the Grasshopper Warbler *Locustella naevia* and Savi's Warbler *L. luscinioides*, resemble the stridulations of insects' but are sounded at rhythms and pitches characteristic of each species. The members of each such species pair differ from one another in their vocalizations. These differences, the more marked the more closely paired the species, are explained by the meaning of song itself as a sexual message, involved in specific recognition and the formation of bonds between mates. The message must act as a specific releaser, in the same way as nuptial displays making use of visual stimuli. The more closely related the species are the more their songs will differ, so as to prevent any hybridization. Species could probably never coexist if they were not kept apart by effective ethological barriers such as this. For species which are very similar in external appearance, song plays an indispensable part in the recognition between mates, as differences in plumage coloration can do for other species. Song as a whole indubitably has specific significance, apart from its racial and individual variations. The fundamental specific characteristics of songs are especially marked within the overlapping range of two partly sympatric species. Where the second species of such a pair is absent, the song of the first has been able to evolve rather differently since it has not been subject to competition. This has happened among Darwin's finches (Geospizinae), the vocalizations of the various species differing from island to island according to the composition of each avifauna. It is therefore easy to understand why song does not characterize higher systematic units – especially the genus, which groups together precisely such closely related and often sympatric species. A character evolved for the separation of species cannot be used to define the category which contains them. The same is true at the level of families, within which vocalizations differ greatly. At most, vocal ability is generally well developed in one family and poorly in another.

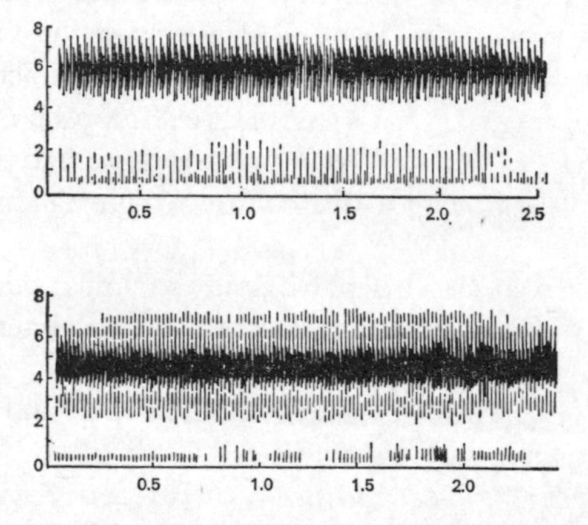

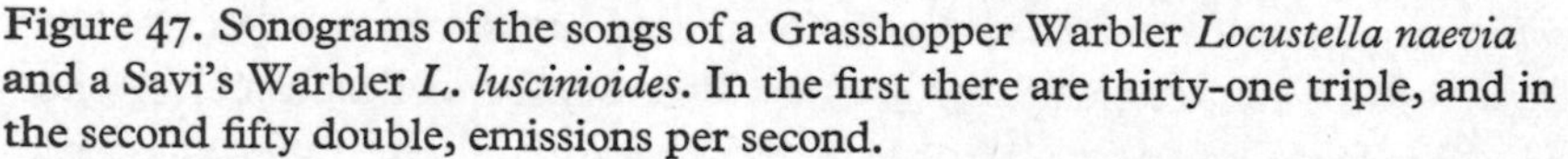

Figure 47. Sonograms of the songs of a Grasshopper Warbler *Locustella naevia* and a Savi's Warbler *L. luscinioides*. In the first there are thirty-one triple, and in the second fifty double, emissions per second.

Despite the specificity of song, it is not absolutely uniform and stereotyped within a single species, but varies to some extent from population to population. Such variation is sometimes shown between subspecies, but often between populations which systematists cannot distinguish by morphological characters – as for example between the European populations of the Chaffinch, which show racial differences in the number and sequence of notes and in the terminal phrase of the song (Marler). The differences can sometimes be detected by ear. They serve to distinguish *dialects* within the species, all sharing the essential characters of the song but showing significant local variants.

Certain populations of the same species which have long been separated no longer even understand one another. They have thus established 'linguistic' barriers, which may lead on to sexual isolating mechanisms and even to speciation. This is true of the Herring Gull *Larus argentatus* which occurs along the coasts of Europe and North America. Trials made with recorded distress calls, intended to disperse flocks endangering aircraft at airfields near the coast, have shown that some populations, although they belong to the same species, do not react to one another's calls. The same thing is found among the crows of America, especially the Common Crow *Corvus brachyrhynchos*. Populations of this species in the north-eastern United States (Pennsylvania) do not react to the calls of European crows, though those in the south (Florida) show very clear reactions. This confirms that understanding the message probably results from repeated auditory conditioning during natural association between several species in the same biotype (Busnel *et al.*)

Alongside such regional differences tending towards the formation of dialects, there are individual variations. Each bird sings the characteristic song of its species, with local variants, but interprets it somewhat originally. We have seen that during the learning period birds profit by the experience of other members of the species with which they come in contact. As a result the song of a male is influenced by those of his rivals, defending neighbouring territories. Yet he likewise has his own vocal characteristics, which the ornithologist sometimes succeeds in recognizing by ear and which his mate usually recognizes perfectly. Individual recognition between birds, by almost imperceptible differences in coloration, posture and behaviour, makes equal use of vocal differences. These are the more important the more marked the territorial instinct and the greater the population density of the species, since these necessitate free individual recognition. This is achieved by voice also among gregarious birds living in large colonies. Gull or guillemot mates recognize one another's voices, and during their first few days of life the young learn to distinguish their parents by the rhythms and intonations of their voices, and the parents are similarly conditioned to the voices of their nestlings. This allows them to recognize one another and maintain the unity of the family even in the midst of the throng. This is also true of penguins. The parents alone feed their young, although all are brought together in a crèche. At first sight the precision of this individual recognition seems surprising, but the problem is limited by a penguin or gull's ability to locate approximately, by using topographical marks, the point where it will find its nest or mate. Birds certainly recognize their mates as much by voice as by posture, thanks to differences which escape the ears of the best observers.

Mimicry

During their youth birds benefit from the experience of their elders, selectively learning the song of their own species. However, some are also receptive to other songs and incorporate them in their own, which thus become polymorphic. This is known as mimicry, defined as the vocal reproduction of sounds which are not innately determined but learned as a consequence of auditory experience. This phenomenon, known for a very long time, is rarely found outside birds. Within the British Isles alone about thirty species are known for their imitative abilities, though some use them only occasionally. The Marsh Warbler can mimic thirty-nine species, and the Wheatear about thirty. The Jay, the Sedge Warbler, the Redstart, and the Red-backed Shrike also have well-developed gifts of mimicry, while the continental *Hypolais* warblers get as far as imitating Swallows. In America the mockingbirds (*Mimus*) get their name from this ability, and the North American Mockingbird, aptly named *M. polyglottus*, has been found capable of imitating fifty-five species in less than an hour. Among the many Australian birds which show abilities as mimics the lyre-birds are indisputably the best, although as suboscines their vocal apparatus is not among th most highly developed. It has been estimated that 70 per cent of a lyre-bird's vocabulary is 'borrowed' from other birds. These imitations sound very true to our ears and this is confirmed by acoustic study, sound spectrograms showing that the mimicking vocalizations have the same structure as their models.

Not only the songs and calls of other birds are mimicked, but also artificial noises. Thus Starlings, which imitate the songs of the birds with which they share their habitat, also mimic various sounds such as the ringing of a telephone, being able to copy the intensity, timbre and even the rhythm of the tones. Crested Larks *Galerida cristata* imitated the whistles of a shepherd calling his dog, so perfectly that the dog would respond to four different commands from the birds (Tretzel). Some birds can imitate the human voice, a phenomenon long known for parrots which has therefore been termed 'psittacism'. These birds are not mimics in the wild, but in captivity they can learn words of human languages thanks to a highly developed vocal organ, and to a fleshy tongue with which they can alter the volume of the buccal cavity and thus articulate crudely. Certain starlings can also learn to speak, especially the Hill Mynah *Gracula religiosa* whose ability is even greater than that of parrots.

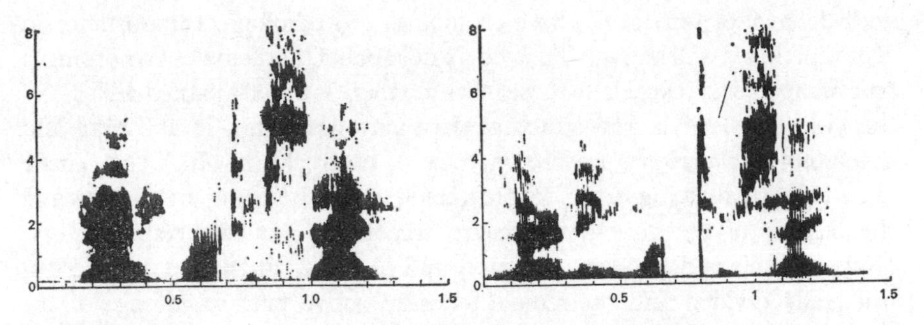

Figure 48. Sonograms of the phrase 'learning what to say', spoken on the left by a man and on the right by a Hill Mynah.

The biological significance of vocal mimicry is virtually unknown. It has been suggested that, since all bird mimics have very highly developed territorial instincts, these polyglot manifestations are advantageous in greatly increasing their repertoires; but it has been objected to this that many birds defend their territories perfectly well by using their own 'language'. One cannot any longer attribute mimicry to defence of a territory by a particular species against birds belonging to many species, and using their languages in order to make itself better understood as a man uses foreign languages, since no competition is known between the mimic and some of the species it imitates – for example between *Hippolais* warblers and swallows. Birds are not known to use this gift either in attracting their prey, nor on the other hand to be taken for those predators whose voices they imitate. No satisfactory explanation can yet be given. Possibly this tendency to assimilate the vocalizations of other species is an effective means of ensuring individuality, among species which need particularly precise recognition (Brémond, Thorpe). There is also a problem in the receptivity of the mate to this polymorphic song. This is probably very highly developed in parallel, so that despite the resemblances specificity is preserved through differences between the vocalization of the mimic and its models.

Duets and mates

In most birds song is under the control of the male sexual hormones, appearing and reaching its greatest amplitude when the testes develop, before the breeding season proper. Young birds give only the calls involved in their vegetative functions, and in their interactions with other members of the species. Song is primarily sexual in its significance, and in fact only appears in association with reproductive behaviour. It is intimately linked with defence of territory, whose ecological and ethological importance in reproduction is well known. The widely held belief that females do not sing is true of most birds, especially those in which there is a well-defined division of the tasks involved in reproduction. In many passerines establishing and defending the territory are prerogatives of the male in which the female takes no part, allowing her to devote herself to nest building, incubation and the feeding of the young, which are sometimes her sole responsibility. However, this is not always the case, and in some birds the tasks are less unequally divided and the female takes an active part in defence of the territory. Song is then no longer strictly confined to the male, as is seen in America among Robins *Turdus migratorius*, Mockingbirds *Mimus polyglottus* and White-crowned Sparrows *Zonotrichia leucophrys*. Among other species the song of the female is used not in the defence of territory but to communicate with its mate. This is true of the Bullfinch, the crossbills and some vireos, whose males and females often enjoy equal vocal abilities. In species such as phalaropes, painted snipe, buttonquails (*Turnix*) and tinamous where reproductive behaviour is inverted – the female displaying and the male performing the 'domestic' duties – the female sings (though in a rather rudimentary way) and the male is silent. Finally, some females which do not normally sing do so weakly in the autumn when the functional left ovary is quiescent, under the influence of some testicular activity from the right ovary. Old females also produce vocalizations, thus giving evidence of latent abilities, during a general masculinization which also produces symptoms in the plumage.

Thus female song is of very variable development. Where both sexes sing, their songs serve in pair formation and in maintaining the bonds between mates, taking on the aspect of a dialogue and in some cases of a true duet, either in unison or antiphonally. Sometimes such a dialogue involves only calls and not true song: the very elaborate nuptial displays of albatrosses and gannets are accompanied by calls and vocalizations of great semantic significance. Sometimes it is made up of true song, usually begun by the male with the female joining in a fraction of a second later. Songs of this kind, rather rare among birds of temperate countries, are commoner in the tropics. They are given for example by the Blue-throated Motmot *Aspatha gularis* and by various antbirds in America, and by the Blackheaded Gonolek *Laniarius erythrogaster* in Africa. In the latter the mated pair sing so perfectly in unison that the song sounds as though it came from a single bird. When the pair is newly formed this agreement is still incomplete, the mates having difficulty in co-ordinating their songs, but from the time when the bonds between them are finally established the vocalizations of the couple are amazingly well synchronized. Sometimes mates do not sing together but one after the other, with such good timing that one notices no break between their separate contributions, as in some wrens such as *Thryothorus mysticalis*. The song of the American quail *Odontophorus* can be rendered as 'corcorovado', heard as one sound, of which the male calls the first three syllables and the female the last two. Birds with this type of vocal behaviour mostly live in closed habitats, where such calls or songs help in maintaining the bond between mates.

The meaning of vocalizations

Vocal behaviour plays an important part in the lives of most birds. Except for some which are actually mute and replace vocalizations by sounds produced by other parts of their bodies, they all bring song or calls into play at various stages of their lives. These vocalizations can be placed in two categories depending on their determining factors.

Some are under external influence and in some way serve the vegetative and social life of the bird. Most calls, though of very diverse meanings, fall into this category. The Whitethroat *Sylvia communis* commands twenty-five signals, all innate; whereas some of the calls of the Bullfinch *Pyrrhula pyrrhula* are innate but are perfected with practice, while others are wholly learned.

In contrast some vocalizations and especially song are controlled by internal, hormonal, factors. The sex hormones play a fundamental part in the appearance and development of song, which in turn is significant mainly in reproduction and the defence of territory. Rather than expressing itself through the year, song follows cyclic fluctuations which are closely correlated with those of the testes, and of the endocrine system which controls its manifestations and intensity. It appears in young males as their testes develop.

A song is thus a claim to possession of a territory, a show of rivalry towards neighbouring males, and at the same time a focus of attraction for the females. During later stages these messages retain their meaning, while weakening in parallel with the territorial instinct itself, and come to play a part in reproductive behaviour following mating. The male in full song thus announces his presence, the more clearly since he usually stations himself in a conspicuous position: some birds, such as certain larks, even sing on the wing. Song is also accompanied by special behaviour patterns which provide visual stimuli.

It should be noted that external factors sometimes release internal mechanisms which play a part in producing vocalizations. The functioning of the pituitary is affected by psychological influences, so that behaviour can release hormonal mechanisms which influence the production of song. The function of nuptial displays is partly to bring the female to the same state of sexual excitement as the male. These displays, in which vocalizations often play a part, are thus set off in a complex way. It can be accepted that in certain cases the female also displays, devoting herself to optical and *acoustic* behaviour patterns, under the influence of internal hormonal controls which are themselves partly released by the visual and auditory behaviour of the male. Gregarious species often show another type of interaction, when one member of the colony starts to call and sets the others going. Thus the calls of the first bird act as an external stimulus to others which are in the same sort of physiological state as the first.

Every species has a repertoire of elementary acoustic signals, amounting to 200-300 among those with highly developed vocal gifts. These are assembled in motifs lasting three to four seconds, themselves grouped into phrases of which in turn several go to make up a song. The enormous number of theoretically possible combinations is never realized in practice since the bird eliminates most of them, while retaining only the few in which it specializes. Thus though the voice of the Nightingale *Erithacus megarhynchos* is renowned for its richness, it has only twenty-four types of song, made up of motifs containing about eleven notes. The Song Sparrow *Melospiza melodia* makes use of about sixteen types of song, which it brings together in varied combinations (Mulligan). Though this would be enough to constitute a true 'language', whose 'words' would be the melodic motifs, it is nothing of the sort as Brémond (1968) showed for the Robin *Erithacus rubecula.* Although this species' repertoire comprises only five types of signal, the song which is the most important is very diversified, making use of more than 1,300 motifs, several hundred of which are actually used by a single individual. In addition the Robin can modulate its vocalizations, in order to express its state of stimulation or aggression. But however it modifies its song the meaning remains the same, and cannot be changed by recombining its component elements. Experiments have confirmed this conclusion derived from observations, for rearrangements of a recorded song always produced the same response, unless certain fundamental structures were upset when no reaction at all could be obtained. At most the bird showed reduced responses when exposed to some of the artificial combinations, no doubt because these were less intelligible.

Influence of external factors

Vocalizations, and especially song, are under the influence of external as well as of internal factors, which change their intensity and frequency. Light is certainly the most important of these. This factor plays a large part in birds, daily cycles of activity and especially in the cycle of song. A few birds do sing at night: in Europe the best known of these is the Nightingale, which also sings during the day, but whose nocturnal song is more famous because at that time it is almost the only sound to be heard. However, the vocal behaviour of most birds is reserved exclusively for the day, and its beginning is determined by the sunrise. Although a bird under artificial conditions sings according to an internal rhythm, in nature there is a very striking agreement between dawn and birds beginning to sing. In a given region the species start singing in a well-defined order, so that one can set down the sequence of the 'dawn chorus' of European birds. The table below, based on many observations, gives the mean time for each species (Marples).

Species	*Song begins, minutes before sunrise*
Blackbird *Turdus merula*	43·76
Song Thrush *T. philomelos*	42·73
Robin *Erithacus rubecula*	34·04
Turtle dove *Streptopelia turtur*	27·00
Willow Warbler *Phylloscopus trochilus*	22·46
Wren *Troglodytes troglodytes*	21·55
Great Tit *Parus major*	17·44
Chaffinch *Fringilla coelebs*	8·69
Whitethroat *Sylvia communis*	6·66

Similar observations have been made at various places throughout the world. They show that as a whole insectivorous and worm-eating birds sing earlier than grain feeders, apart from pigeons.

The time of sunrise is an astronomical datum which varies in a regular way throughout the year away from the equator, especially at middle latitudes, and birds' rhythms show parallel fluctuations. Their hours of beginning to sing accurately follow the curve showing the time of sunrise; and still more exactly the one showing the beginning of civil twilight, when the sun comes within 6° below the horizon. Differences between species reflect different sensitivities to light intensity, each species having a threshold value at which its morning behaviour is released. This sensitivity varies a little with the bird's internal condition, and especially with its annual cycle, being lower at the beginning of the breeding season so that the bird begins to sing earlier then than it does later in the year. That light intensity determines the beginning of song is confirmed by the changes which cloudiness causes in behaviour. Overcast and rainy weather delays the beginning of activity, the birds waiting until the light reaches the intensities which are their thresholds of sensitivity.

Having begun to sing, birds enter into a very active phase. Thus the first hours of the day are the best for watching and hearing birds. They then slow down during the middle of the day, recovering towards the end of the afternoon though this second peak of activity is less marked than the first. The fall in light intensity is certainly the principal factor in determining the end of the day's activity, although the threshold of sensitivity is not the same as during the morning arousal. Though subject to great variation it is generally higher, a bird stopping its activities when the light intensity reaches a level at which, in the morning, it had already been active for some time.

This cycle is inverted for nocturnal birds. Owls begin to call at a variable interval after sunset – usually later in winter than in summer, since the length of the winter night allows them to postpone the start of their active phase. The beginning of activity, by some small waders, follows the fluctuating time of sunset, as is especially obvious from the calling of Woodcock. Some birds, particularly snipe and nightjars, are stimulated by moonlight.

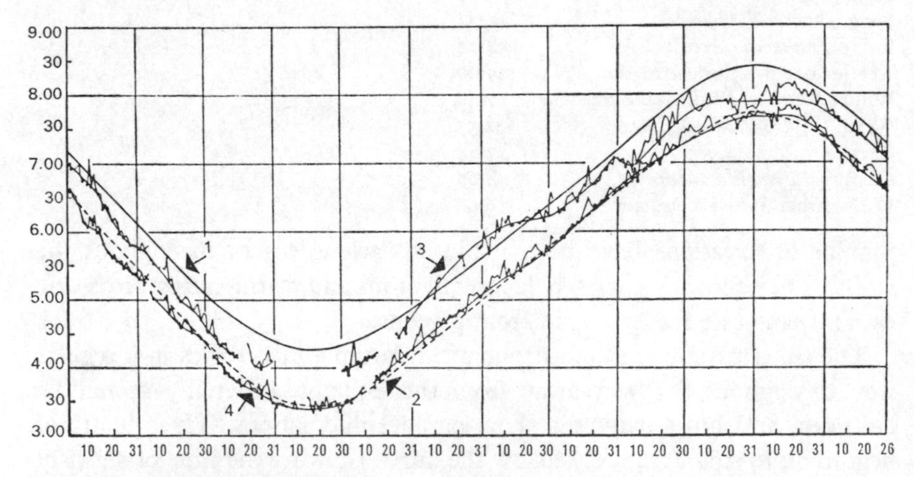

Figure 49. Graph showing the times at which the songs of two bird species, the Blackbird and Chaffinch, begin throughout the year.
1. Time of sunrise.
2. Civil twilight.
3. Times for the Chaffinch.
4. Times for the Blackbird.
Note the agreement between the curves for the songs and those for daybreak

Bird song is also controlled by temperature, rising temperature generally favouring activity up to a certain threshold, and then slowing it down. The abatement of activity in the middle of the day is probably caused by temperatures which are too high, rather than by unduly intense light. The temperature threshold varies greatly from species to species. Wind also has its effect on song as on other activities, which are slowed down by a strong wind and prevented from beginning so early in the morning. However, this influence is less clear than those of the other environmental factors.

Adaptations to the nature of the habitat

Although the peculiarities of different songs are very difficult to explain in terms of the nature of the habitat, some general characteristics can be interpreted as adaptations of this kind. Birds have at their disposal two means of expression, optical and auditory. In closed habitats specific and sexual recognition, intercommunication and localization by optical stimuli can be achieved over short distances only. It is to be expected therefore that communication by sound is much more widely distributed among the birds of closed habitats, and especially of forests, than of open ones, where visual communication at a distance is not prevented by dense cover. Most woodland passerines, such as *Sylvia* and *Phylloscopus* warblers, the Robin and many others, have loud ringing songs. Birds with powerful voices are still more abundant in tropical forests, as exemplified by the South American bellbirds (Cotingidae). However, a distinction must be made between birds of the undergrowth with resounding songs, and the quieter ones of the canopy which communicate mainly by visual means. This is equally true of birds which live in reedbeds or dense bushes, such as *Acrocephalus* and *Locustella* warblers and wrens, which are among the noisiest passerines. Not only do these birds have powerful voices, but they make frequent use of them. The Red-eyed Vireo *Vireo olivaceus* is the most prolific singer of all birds, while *Locustella* warblers will sing for more than two minutes at a stretch and for up to six hours a day. Duet singing by mates is equally appropriate to dense environments.

This does not mean that vocalizations are of no importance in open habitats, though they are often overshadowed by visual displays to which they are little more than accompaniments, serving to increase their range. In particular, singing in flight is almost entirely confined to birds of open habitats, such as the Snow Bunting, the Skylark, and pipits. Thus there is a balance between aural and visual displays, whose contributions to behaviour are complementary as is shown by the following table (simplified and modified from Armstrong 1963):

Characteristics	*Species whose signalling is essentially*	
	acoustic	*visual*
Size	small	often large
Plumage	cryptic	conspicuous
Displays	rudimentary	elaborate
Habitat	closed	open
Territorial instinct	marked	weak
Nesting	solitary	sometimes gregarious

Although few species fall clearly into one or other of these contrasting categories, the rule generally applies, and voice takes its place among the other complementary characteristics.

Echo-location

The Oilbird *Steatornis caripensis*, an intermediate between nightjars and rollers, has strange habits, for it passes a large part of its life in the total darkness of the caves in which it nests in mountains from Peru to Venezuela, a habitat which long ago captured the imaginations of the Indians who believe Oilbirds to be infernal. Despite the legends which surround them, Oilbirds are harmless, taking only pulpy fruit on which they feed during the night. This diet gives them significant fatty reserves, laid down in large quantities by their young. Sight is no answer to their problems of moving about in the utter blackness of the caves where they nest, and of raising their young on narrow ledges of the rocky walls. While these birds are notorious for their dreadful outcry, various types of call can be distinguished within it. Alongside deafening but acoustically ordinary sounds are short shrill calls, rapidly repeated but only when the bird is in flight. Measurements made on the spot have shown that these are at frequencies of between 6,000 and 10,000 hertz, and ultrasonic frequencies have never been demonstrated, whereas bats' location calls are at between 30,000 and 70,000 hertz. These high-frequency sounds, emitted by Oilbirds in trains of pulses each of which lasts about one millisecond (Griffin 1958), must surely be the means by which they steer in the dark, being reflected from obstacles and providing a measure of the distances involved. Thus the high-pitched audible sounds made by Oilbirds should have the same function as the ultrasonic emissions used by bats. To confirm this hypothesis, Griffin captured several Oilbirds and released them in a dark room where they flew about without accident, continuously emitting the characteristic calls which he regarded *a priori* as position-fixing trains of pulses. He then closed off their earholes with cottonwool and a layer of plastic, so as to block their hearing completely, whereupon the Oilbirds completely lost their steering ability, and knocked into obstacles although they were giving their characteristic calls at an increased rate. As soon as the earplugs were removed, the experimental birds recovered all their original agility. This verification makes it likely that Oilbirds do echolocate, by the perception of reflected audible sounds and not the ultrasonic vibration used by bats.

While owls do not use such a navigation system, it is found again in other cave-dwellers, the swiftlets (*Collocalia*). These birds also nest in caves, to the walls of which they stick their nests, made of saliva which hardens in air. Some of these caves are totally dark, so that the swiftlets cannot navigate visually. It has long been known that they ceaselessly utter notes, audible to man as twittering, which rebound like trains of pulses, and they probably use these to orient themselves and locate the walls of their caves.

Sounds not produced by the vocal organs

Some birds also make use of non-vocal means to produce a variety of sounds which likewise function as signals. The variety of these instrumental sounds, produced by the wings, tail, feet or bill, can be illustrated by several examples.

The wings are involved through the characteristic humming or throbbing sounds they make as they beat the air in flight. This common sound, audible from all birds, is sometimes exaggerated and plays a part in nuptial displays. Thus those of male pigeons are interrupted by strong beating of the wings which produces a characteristic sound, augmented by the intentional clapping of the wings against one another. The nuptial flights of Lapwings and some other waders, though they employ visual stimuli, similarly make characteristic sounds. The remiges of some birds are specialized so as to increase the sound produced in this way, as in manakins (Pipridae) whose males perform real aerial dances during their complicated nuptial displays. These flights make a loud whirring as the air vibrates the remiges, the outermost of which are specialized as reeds. Similarly in some hummingbirds, notably of the genus *Campylopterus*, whose males have the shafts of the outer remiges flattened and strengthened, this specialization increases the sound produced by the hovering flight characteristic of these birds. The behaviour of the Ruffed Grouse *Bonasa umbellus* is curious. During its display, the male perches on a hillock or a fallen log and beats its wings vigorously to produce a very characteristic sound like a roll of drums. African broadbills (*Smithornis*) dance on the spot while turning completely round, their wings beating very rapidly to produce a highly unusual vibration whose precise function is unknown.

The tail can be used in the same way, the outer rectrices even being modified in some birds so as to produce specialized sounds. In snipe (*Gallinago*) these feathers are very narrow and much stiffer than the central ones. During nuptial flights the bird dives at high speed while spreading its rectrices, which are thus set vibrating by the flow of air and act as a siren which can be heard a long way off. Some snipe such as *G. stenura* of Siberia even have twenty-six rectrices instead of the normal fourteen, the eight outer pairs being modified for sound production. No doubt the highly modified rectrices of the Lyre-tailed Honeyguide *Melichneutes robustus* similarly produce the characteristic sound of its aerial manoeuvres above the dense African forests, while the buzzing made by the Paradise Whydah *Steganura paradisea* during its nuptial flights also comes from its specialized tail feathers.

The bill too is used as a sound-producing instrument. Some birds, especially gannets and albatrosses, rub and clatter their bills together during display. Storks, otherwise mute, clap their mandibles together to make a sound like castanets, by throwing the neck and head sharply backwards, behaviour which takes place during nuptial displays and recognition ceremonies at the nest. Woodpeckers also use their bills to produce sounds, though they make short powerful calls. They choose as sounding boards tree-trunks within their territories which are hollow or otherwise especially resonant – sometimes using shutters or roofs On these they rap vigorously and rhythmically with their bills, thus making a hammering sound like a roll of drums which is audible at great distances. This sound, at a rhythm characteristic of the species, carries 200m from small species and up to 800m from large ones such as the American Pileated Woodpecker and European Black Woodpecker. This drumming, rarely heard outside the breeding season, is a signal used in the defence of territory and the attraction of a mate. Both sexes devote themselves to it, at a particularly high frequency at the beginning of breeding: 500-600 times a day while displaying, 100-200 afterwards in the Great Spotted Woodpecker *Dendrocopos major*.

These non-vocal sounds thus have varied meanings. They supplement and sometimes substitute for vocalizations, in support of visual stimuli. While their meanings are largely sexual, in nuptial displays and the defence of territory, they may also be involved in other aspects of communication between individual birds.

1 The take-off of a Trumpeter Swan *Cygnus buccinator*. Like other very heavy birds, it has to get up speed by running on the surface of the water before lifting off.

2 A Pied Flycatcher *Musicapa hypoleuca* in flight, seen from below. Note the widely separated outer primaries.

3 A King Penguin *Aptenodytes patagonica* in underwater 'flight'. Note the extreme reduction of the tail, the strength of the feet, the bulk of the body, and the resemblance of the wing to a whale's flipper.

4 The North Island (New Zealand) subspecies of the Common Kiwi *Apteryx australis mantelli*, 50 cm long.

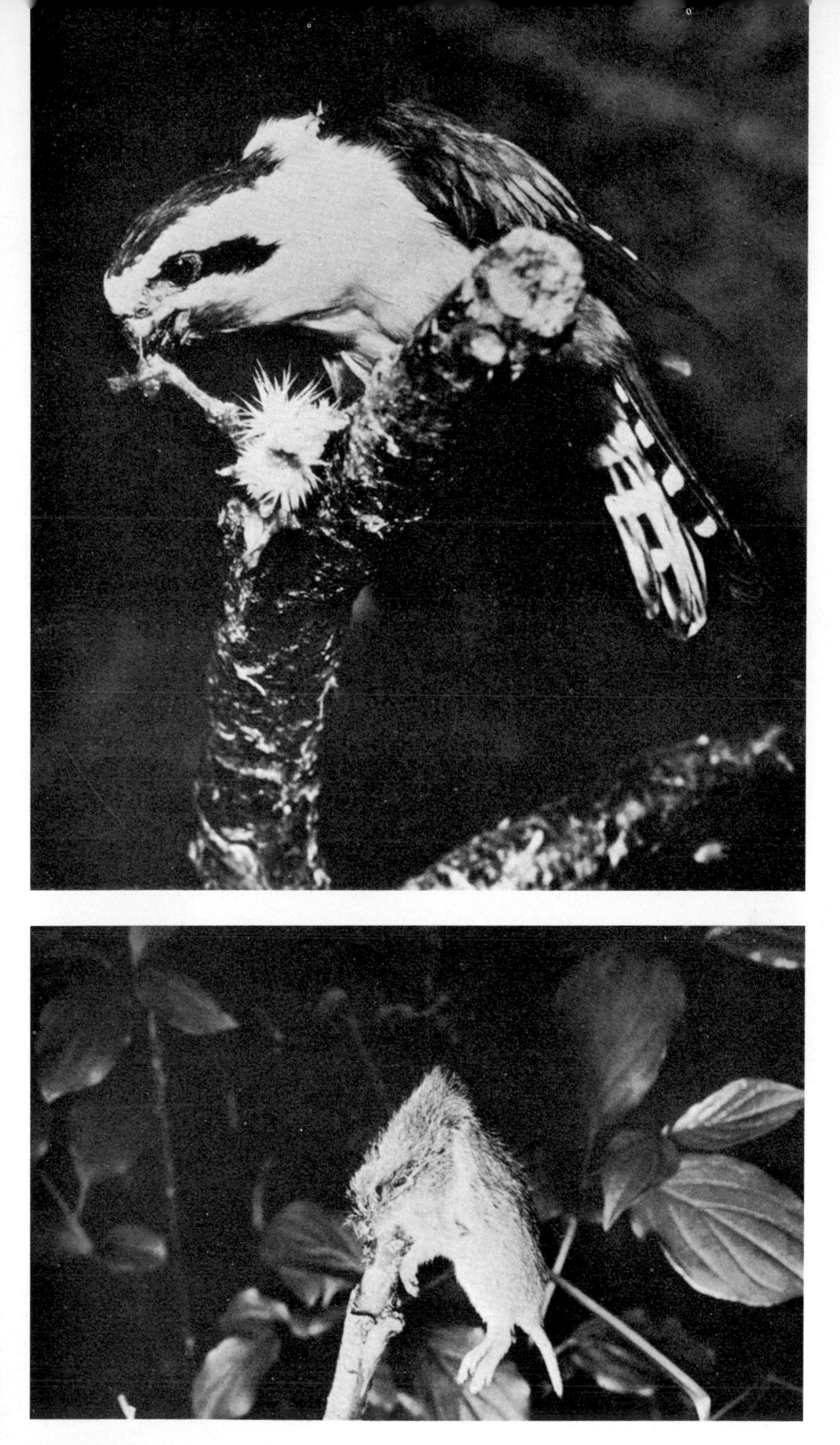

5 A falconet *Microheirax caerulescens*, 15 cm long, from India and southeast Asia. This little raptor catches both insects and birds of its own size. It nests in holes in trees.

6 A vole spiked on a twig by a Red-backed Shrike *Lanius collurio*. The habit of impaling prey is not common to all species of shrike, nor shown by all individuals.

9 An Alpine Swift *Apus melba*. In the related smaller Swift *A. apus* of our towns, the young become temporarily poikilo-thermal in bad weather.

7 Red-billed Tropic Bird on its nest.

8 A mousebird *Colius indicus*, 35 cm long, from Africa south of the Congo mouth and southern Tanzania.

10a The mutual display of Wandering Albatrosses *Diomedia exulans*.

10b In the second stage the two birds rub bills, a sign of aggression which disappears during later stages of the display.

11 Hill Mynah *Gracula religiosa*. This asiatic bird is one of the best 'talkers'.

14 Young Reed Buntings *Emberiza schoeniclus* begging to be fed. In two of them the white spots at the edge of the tongue can be seen against the background of the throat. Many nestling passerines show marks of this kind, which reinforce the stimulation to feeding produced on the adult bird by the sight of the coloured throat.

12 *Above left:* Male of the Magnificent Frigate Bird *Fregata magnificens* of the Galapagos Islands, inflating its throat pouch.

13 *Left:* A pair of Little Bitterns *Ixobrychus minutus* (the male above) with their young.

15 Male Golden Oriole *Oriolus oriolus* at its nest. 24 cm long. Europe to south-western Asia.

16 Cormorants *Phalacrocorax carbo* on their nests. This species breeds not only on the coast but also (for example in central Europe) along rivers and near lakes.

17 A young Cuckoo ejecting an egg from the Reed Warbler's nest in which it was hatched. It uses its spread wings as levers to lift eggs and nestlings with which it comes in contact.

Chapter 9

The Reproductive System and Cycle of Reproduction

THOUGH derived from reptilian ancestors, oviparous reproduction is closely adapted to the bird's way of life. In some ways it is inferior to the viviparity of mammals, which during their foetal life are better protected from predators and the other hostile factors of the environment to which eggs are subject, and which demand so much care from parent birds. Furthermore, oviparity makes birds less well adapted to marine life than mammals. Whereas whales and dolphins never leave the water and spend their whole life cycle within it, marine birds have to return to land in the breeding season in order to lay their eggs. Oviparity is highly favourable to aerial life: unlike the mammalian embryo an egg is carried for only a short while, yielding a significant economy in weight which is advantageous for flight.

Only a brief description of birds' genital anatomy is necessary, after which we can consider their secondary sexual characters, and their sexual cycle which is so closely related to the cycle of the seasons.

Reproductive system

MALE GENITAL SYSTEM

The paired testes are attached to the dorsal wall of the abdominal cavity, just anterior to the kidneys. Their inclusion within the main body cavity, and not in an external scrotum as in most mammals, is part of the streamlining of a bird's body but could be disadvantageous, since spermatogenesis takes place best at lower than body temperature. To compensate for this the testes shift considerably farther backwards in the abdominal cavity when they develop in the spring, coming to be surrounded by the abdominal air sacs and so to be kept cool. The testes are usually ovoid or rounded in shape. They change greatly in volume during the annual cycle, being 200 to 500 times as large during the breeding season as during sexual dormancy. A breeding drake's testes can weigh as much as one-tenth of its whole body, so that for economy in weight a restricted breeding season is advantageous.

The spermatozoa leave the testis through the epididymis, which continues as the vas deferens. At its distal end this forms a knot of loops, which serve as a seminal vesicle. This part of the male system also undergoes large fluctuations, increasing considerably in the breeding season – by 120 times in the American Robin *Turdus migratorius*. The vas deferens open into the cloaca, which carries an external organ only in a few species. In ratites, ducks and curassows there is a sort of penis which can be evaginated and erected, forming a tube by the juxtaposition of its edges through which the sperm is led to the female genital tract during copulation. This organ is a primitive feature which has disappeared in all other birds, which copulate simply by pressing the cloacae together.

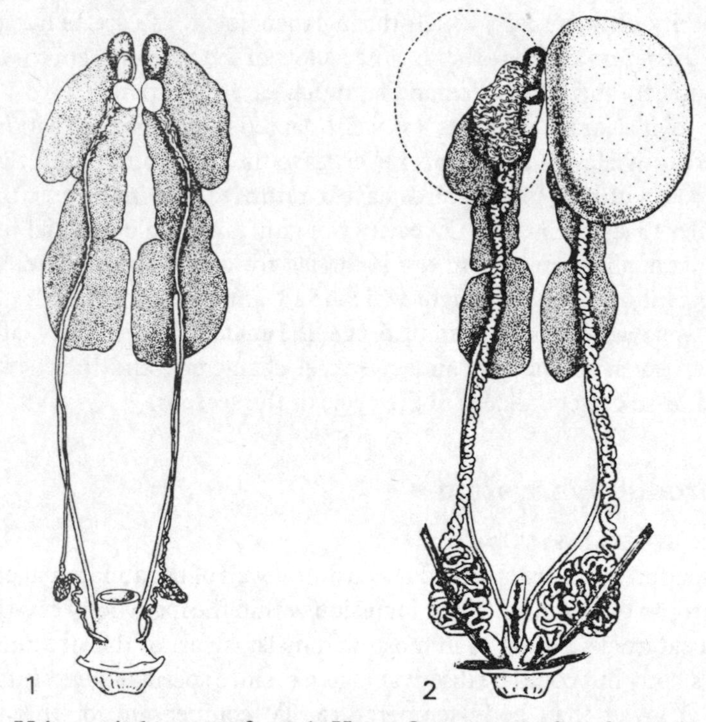

Figure 50. Urinogenital system of a male House Sparrow: 1. when dormant. 2. after seventeen daily injections of serum from a pregnant mare. In the second sketch the right-hand testis, whose position is indicated by the dotted line, has been removed to show the enlargement of the epididymis.

Copulation, representing the climax of sexual displays, is preceded by special behaviour of the female which invites the male to intercourse. He mounts his mate and clings to her back with his feet, often beating his wings.

Swifts perform the sexual act in flight. It is usually very brief, lasting not more than ten seconds in most species, and is frequently repeated.

FEMALE SYSTEM

In contrast to the male system, that of the female is remarkably asymmetrical, only the left ovary developing and becoming functional, while the right one remains rudimentary and sometimes even functions hormonally as a testis. This arrangement too probably results from adaptation for flight, serving to lighten the bird. The ovary, situated in the abdominal cavity behind the left kidney, looks like a bunch of grapes. It contains a great number of follicles (up to 26,000 in the crow), of which a very small number give rise to eggs. The follicles burst by the liberation of an ovum (ovulation) do not develop into corpora lutea as in mammals, so that the hormonal action of the avian ovary is totally different, without the permanent endocrine glands formed by the corpora lutea. Like the testes the ovary enlarges considerably in the breeding season, distended by the developing orocytes.

Once freed and immediately fertilized, the ova pass through a wide flaring opening (the infundibulum) into an unpaired oviduct, a convoluted tube in which several segments may be distinguished: a glandular part or *magnum* where the ovum is coated with albumen; a narrower part or *isthmus* where the shell membrane is laid down; and the *uterus* where the egg is completed by the formation of the shell. From here it passes through the vagina, simply a transporting channel, to the cloaca. The oviduct, which contrasts in its extreme simplicity with the complexity of the mammalian genital tract, also undergoes significant annual variation, and weighs ten to fifteen times as much during the breeding season as when it is quiescent.

Like the male gonads, those of the female act as endocrine glands by virtue of their interstitial cells, secreting hormones whose actions are complex. They in turn are under the influence of other internally-secreting glands – especially the pituitary.

Secondary sexual characters

In birds as in other animals, sex is genetically determined, all their spermatozoa being identical while they have two types of ova. In the former all the chromosomes can be paired off, whereas in the latter one of the chromosome pairs can be represented, according to some authors, either by a single chromosome (sex-chromosome) or by two unlike ones. Cytologists have still not definitely established this, because of the large number of chromosomes, some of them extremely small, which birds possess. These vary in number from sixty to eighty-four, with ten to twelve large and well-characterized pairs. As far as one can judge from the small number of species yet studied, the chromosomal situation is similar throughout the birds. Despite the difficulties, the original hypothesis seems the most probable. If the sex chromosomes are designated by the letter Z, then according to this hypothesis the formula for the male would be ZZ, that of the female ZO. These chromosomes would be the bearers of genes determining male sex when in double dose, but overruled when alone by the feminizing genes carried on other chromosomes (Witschi).

The primary sex ratio is probably 1:1, but various factors intervene to alter it, often profoundly. At hatching the proportion is still within 5-10 per cent (for example 49·4 per cent males in 12,000 hatchings of turkeys, and 51·5 per cent in 2,216,051 of chickens). Among passerines rather more males than females seem to hatch, though there are many exceptions – for example the Boat-tailed Grackle *Cassidix mexicanus* which produces twice as many females as males. The proportions of the sexes at maturity vary from group to group. In most birds they are nearly equal, though ducks and many passerines show a more or less marked predominance of males. Censuses taken on many thousand individuals of each species showed 60 per cent males among North American ducks, 73 per cent in the Pintail, 53 per cent in the Bobwhite, 55 per cent in the House Sparrow, 68 per cent in the Starling, and 84 per cent in the Red-winged Blackbird. By contrast in whydahs and some raptors and gamebirds which are naturally polygamous, females markedly outnumber males.

Although the embryo is already sexually differentiated (from the fifth day of incubation in gamebirds and passerines), young birds of both sexes in their first plumages are very much alike. In the course of subsequent moults, into a sequence of plumages which is sometimes highly complex, external differences often appear which make the sexes distinguishable at first glance, though the magnitude of such differences varies very much between species and higher groups. The male and female adults of some species are exactly alike, a condition known as sexual monomorphism. These plumages may be dull, as in many passerines such as warblers, or brilliantly coloured as in kingfishers, toucans, bee-eaters and parrots, the female being only very slightly duller than the male. By contrast the sexes of other species differ (sexual dimorphism). In some groups the differences concern only a few characters of the plumage: in woodpeckers such as the Great Spotted Woodpecker the crown is partly red in the male, and entirely black in the female. Eumelanic pigment in the male is often represented by phaeomelanin in the female, as in our Blackbird where the former is black and the latter reddish. In other species sexual dimorphism is extreme, sometimes to the extent that ornithologists described the two sexes as different species when collections without data were received from distant countries. The birds of paradise are certainly the most striking example of this: the male highly colourful with shining ornaments and highly specialized plumes, while the female is dully and meekly dressed in brownish or blackish. Hummingbirds and pheasants show similar large differences between the sexes.

It is not only the contour plumage which differs, for the distribution of supplementary ornaments is still more uneven. Ornamental feathers such as plumes and streamers are usually confined to the males, and it is the same with crests, wattles and other naked ornaments like those of cocks and turkeys. There is even one case where the shape of the bill was totally different in the two sexes. In the Huia *Heterolocha acutirostris*, a member of the family Callaeidae or wattled crows which was confined to New Zealand but is now extinct, the male had a thick, relatively short and strongly curved bill, but the female one was very long, narrow and curved like a scimitar. The sexes differed greatly in their methods of feeding on fruits, and more especially on wood-boring insects. The male attacked rotten wood, splitting the soft tissues to extract larvae, while the female probed with her bill in tunnels bored in sound wood. A sort of co-operation between mates was occasionally observed. Thus the sexes occupied distinct ecological niches – as is also true to some extent of woodpeckers in which the females, less strongly armed than the males, attack only the less-resistant woods and take a higher proportion of insects from bark.

Besides sexual dimorphism, and sometimes additional to it, is the *seasonal dimorphism* of many birds in regions where the climate is strongly seasonal. It is characterized by seasonal changes in plumage according to an annual cycle, which sometimes affect only one sex and sometimes both. A breeding plumage, which tends to be colourful and to incorporate ornamental feathers used in nuptial displays, contrasts with a dully coloured *non-breeding* or *eclipse plumage*. The breeding plumage expresses, in one or both sexes, the state of the gonads which controls its appearance. In some birds the males and females undergo these cyclic changes simultaneously, while remaining similar to one another in plumage. This is true of the Black-headed Gull which bears a dark cap during the summer, while in winter the head becomes white with mere traces of black. Similarly egrets during the breeding season carry long decorative plumes, which fall at the beginning of winter, and small waders of the family Charadriidae have two different plumages during their annual cycle. In other species only one of the sexes shows this seasonal alternation in plumage. In most it is the females which stay the same throughout the year, while the males resemble them during sexual dormancy but are very different in the breeding season, as in many of the Ploceidae–notably the weavers whose males are then brilliantly coloured, and the whydas whose males sport a long tail. In a few birds such as the phalaropes and the button-quails (*Turnix*) the situation is reversed, together with the behaviour characteristic of the sexes. The non-breeding plumages of some birds seem to be related to their annual variation in behaviour. The breeding plumage serves to advertise the aggressiveness of a male towards its rivals; so that its replacement by a less specific plumage, devoid of the patches of colour which serves to signal its presence and its identity and to release aggressive behaviour, facilitates the gregarious behaviour often shown outside the breeding season, especially by migratory species (Hamilton & Barth 1962).

Alternation of plumage most often requires two annual moults, but sometimes a change in coloration is produced by wear of the feathers altering their external appearance.

As has been shown by various experiments involving the removal of the gonads or the injection of hormones, characters of the plumage are produced by interaction between genetic and hormonal factors, the part played by different factors varying from species to species. Since the effects are expressed so spectacularly and conspicuously in the external appearance, birds are excellent experimental material for the study of the physiology of reproduction and the appearance of secondary sexual characters. In birds such as pigeons, guineafowl, sparrows and starlings, the plumage is not affected in appearance by the sex hormones, and no doubt depends solely on genetic control. By contrast in many others, especially the wild and domesticated game birds, the hormones modify the differentiating effect of the genetic factors. There is a kind of equilibrium between the hereditary and hormonal factors. Removal of a cock's testes scarcely produces any change in its plumage, whereas a hen without ovaries becomes cock-feathered. Furthermore, injection of female hormones into a cock causes the appearance of hen-feathering. It follows that the ovary must control the characters of the female plumage, by inhibiting the appearance of the latent male characters and by exerting a positively feminizing influence. Thus the male plumage is characteristic of the species and is said to be of *neutral* type, expressing itself in the absence of hormonal action. This action seems to be confirmed by the observation that under natural conditions old females of many species undergo a masculinization of plumage.

Other glands besides the gonads play a parallel part in controlling plumage differences. Both pituitary and thyroid can release the appearance of distinct male and female plumages, as has been proved for the Ploceidae. In the Red Bishop *Euplectes oryx franciscana* the female is dull throughout the year while the male, resembling her during sexual dormancy, bears a breeding plumage of vermilion and glossy black. This finery can be produced at will in castrated birds by the injection of pituitary gonadotrophic hormones. Thus here it must be the female plumage which represents the neutral type, while the male plumage is produced by the positive action of the pituitary, inhibited in the female by the action of ovarian secretions. From this arises the hypothesis that, at least in some birds, the appearance of the plumage is the resultant of action (whether direct or not) by the pituitary tending towards cock-feathering, and by the ovary tending towards hen-feathering. This is the situation among ducks. The drake Mallard is notable for its rich plumage, which it loses after breeding and regains at the beginning of winter, assuming for the brief intervening period an eclipse plumage like that which the duck wears throughout the year. Castrated drakes and spayed ducks take on the male breeding plumage, which is thus the neutral plumage of the species. The appearance of eclipse plumage in the male remains unexplained. Male hormone can in some species have a positive action, directly stimulating the appearance of certain characters such as the Ruff's ruff and ear-tufts, so characteristic of its breeding plumage. Castration abolishes these secondary sexual characters and returns the plumage to the condition of the female, which is thus the neutral type.

Some crests and wattles are similarly under the direct control of the sex hormones. This is true of the cock's comb, which is shown in a rudimentary state by the chicks and develops with sexual maturity in the cock, or can be stimulated by the injection of male hormones. In contrast it disappears rapidly after castration, but can be restored by the injection of these hormones. This response is so precisely proportional to the dose that the development of the comb is a test frequently used in assaying the androgenous activity of various hormonal preparations. The bill is also subject to very clear influence by sexual hormones. The bill of the male House Sparrow, which is pale brown like that of the female in autumn, blackens at the beginning of the breeding season under the influence of the male hormones. The pigment disappears on castration, and is re-established after injection with hormones. The Starling's bill, blackish grey in winter, becomes brilliant yellow during the breeding season in both sexes, though more slowly in the female. Castration or spaying abolishes the seasonal change, which is restored by the injection of male hormones. These are thus responsible for the appearance of this character in the breeding season, in both sexes.

In particular, the ovary secretes a masculinizing hormone, probably from its interstitial cells, which explains certain male characters shown by females. Furthermore the right ovary which remains rudimentary, and even parts of the functional left ovary, can develop as testes and secrete male hormones. Even bilateral gynandromorphism is known (in the Chaffinch, Bullfinch and Pheasant, for example), resulting from the development of a right testis and a left ovary, with the plumages of the two sides of the body precisely reflecting the glandular situation. The chromosomal constitution of such a bird is male on one side and female on the other, as a result of irregularities in cell-division during the first stages of embryonic development. Natural intersexuality occurs in many forms, which explains the mosaic of characters in some abnormal individuals, while illuminating cases have been experimentally produced. All are explicable in terms of the bisexuality of embryonic gonads.

The sex hormones permanently or temporarily determine, not only the physical characters but the behaviour of birds of both sexes. Song, territorial instinct, aggressiveness, nuptial displays, and all the other male behaviour related to breeding, are controlled by male hormones, and disappear after castration. Similarly the characteristic behaviour of females is determined by female hormones, though here male hormones which are normally present in small quantities also play some part. The result is a delicate equilibrium which varies from species to species. For example the submission which allows copulation in the female is controlled in some birds such as gulls by female hormone, and in others such as herons by a minimal dose of male hormone.

We have seen that hormones control sexual activity; but the behaviour which they release in its turn affects the physiological balance, controlling further secretion which in turn allows new behaviour. Thus stimulation of external and internal origin combines in a kind of escalation, culminating in sexual maturation, copulation and the laying of eggs. The strongest and most active males, motivated by the most intense internal stimuli, are the best fitted for the competition which precedes the breeding season and accompanies its early stages. This competition, intense as it is among birds which display individually, is still more rigorous among those which display communally. The most vigorous males secure the females at the expense of the rest, so that a few 'despots' may fertilize up to 76 per cent of the females. Gregarious birds of a single group are organized in a social hierarchy, primarily according to sexual activity (and thus finally according to the quantity of male hormone present in each individual) which is responsible for their internal motivation towards fighting and intimidatory displays.

Sexual dimorphism allows recognition between the sexes for mating, but also plays a part in territorial behaviour and social relations. Even the smallest differences in plumage or in voice may be important, since as we have seen birds enjoy excellent colour vision and very sensitive hearing. Thus for example in the Yellow-shafted Flicker *Colaptes auratus* the 'moustache' is black in the male and red in the female. By artificially colouring this ornament black in a paired female, the male was led to attack his mate when she returned to the nest. Thus this single marking is the character which indicates sex in this species. The part which a particular part of the plumage may play as a sexual signal was investigated by David Lack in classic experiments on the Robin. The vivid brick-red breast of an adult male provoked attack by another male, the more vicious as the colour was more intense. This reaction was provoked by a stuffed specimen, and even by an isolated tuft of breast feathers or indeed by a scrap of red fabric. In birds such as penguins whose plumage shows no trace of sexual dimorphism, it is differences in vocalizations which allow sexual recognition. During nuptial displays, which are audio-visual dialogues between the members of a pair, either while it is being formed or when it is already established, coloured patches which distinguish between the sexes act as optical sign-stimuli. Controlled by the sex hormones which determine their appearance, they play an essential part in the reproductive process by stimulating further hormonal secretions. Thus they are involved in a psycho-physiological recycling which is of the greatest importance in reproduction. In birds without sexual dimorphism, the sexes may be recognized by certain postures characteristic of each mate. The very complex behaviour patterns of penguins, gannets and herons, to mention only a few, show the semantic significance of minor differences in posture.

A group effect certainly acts as a stimulus in the breeding of gregarious species. Social behaviour stimulates a kind of rivalry between the pairs, which can only nest and raise young successfully when surrounded by other members of the colony. Isolated pairs or those in very small groups do not succeed in nesting, while breeding is earlier and more successful in large colonies than in small ones, and in the middle rather than at the edges of colonies. This group effect has been observed in all seabirds, especially among terns and in the Kittiwake in which breeding dates depend directly upon the density of the colony (Coulson & White 1960). However, this is not due solely to the group effect, but also to the fact that, because of the social hierarchy, experienced birds in general occupy the centre of the colony, while young ones whose reproductive success is in every way inferior have to establish themselves at the edges.

Once breeding behaviour has been released, it proceeds as a succession of well defined behaviour patterns, indispensable to breeding success, leading from the establishment of a territory and nuptial displays up to the departure of the young. The various phases are all under hormonal control, prolactin being responsible for much maternal behaviour and for the development of brood patches. External stimuli also play their part. Adults while displaying or building the nest are not yet in a state to feed their young, as can be shown by moving nestlings from one nest to another. Physiological and psychological preparation by a well established progressive process is necessary, from nesting till the breakdown of family ties. These behaviour patterns are under the control of an unstable equilibrium between external factors and the neurophysiological state of the bird, as has been shown for corvids (Gramet) and doves (Lehrman 1964).

The beginning of nesting sometimes seems to be at odds with the weather, which may at first sight still seem to be unfavourable. This is true especially when there is a long delay between the beginning of breeding and the hatching of the young. The extreme case of this is certainly the Emperor Penguin which lays its eggs during July, in the depths of the southern winter and polar night and at temperatures of $-50°$ C, so that its young have finished their juvenile growth and are fit to go to sea when the ice breaks up. In northern temperate countries birds breed at a well defined time of year in the spring, usually between April and July for most species. They thus concentrate reproductive activity into a relatively short time, leaving themselves the rest of the year for purely vegetative activities. During this well defined breeding season all their energy is directed towards reproduction and especially the raising of young. In Cambridgeshire, the Wood Pigeon *Columba palumbus* attains the maximum development of its gonads in July and remains in reproductive condition until the beginning of October; the Stock Dove *C. oenas* comes into the breeding state two months earlier and its gonads do not regress until a month later; while the feral domestic pigeon *C. livia* has a still longer breeding season, with gametogenesis beginning in December and reaching completion the following October. This timing is correlated with the diet of each species, since the food of the two last is available for longer and so allows them more extended breeding seasons (Lofts, Murton & Westwood 1966).

Annual cycle

The reproductive cycle of most birds is annual, with two or even three layings in a single breeding season. In extreme habitats a cycle may be missed out, the birds failing to breed during a frankly unfavourable year.

This happens with some arctic birds, notably predators such as Snowy Owls, Rough-legged Buzzards and skuas, which do not nest when lemmings are scarce. So too Saharan birds fail to reproduce in years of great drought: the male quail do not even show any development of their testes.

Some large birds, such as the Wandering Albatross *Diomedea exulans* and other large antarctic albatrosses, only nest every two years. A few other birds do the opposite, regularly nesting at intervals of about nine months. The

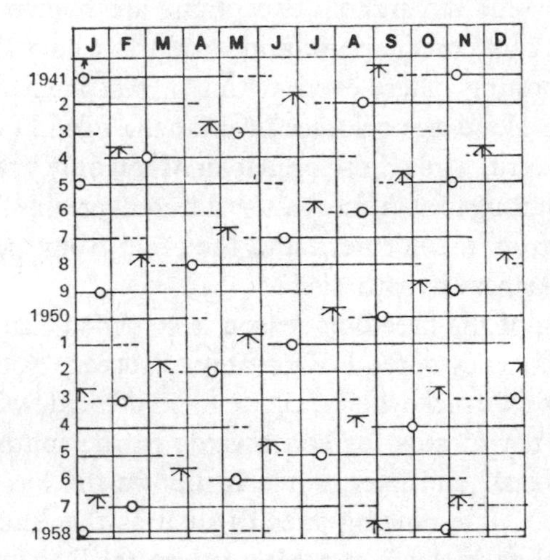

Figure 51. Breeding calendar of the Sooty Tern *Sterna fuscata* on Ascension Island between 1941 and 1958. The horizontal lines show when the birds were on their breeding grounds. The circles show the laying dates of the first eggs.

classic example is the Sooty Tern *Sterna fuscata* of Ascension Island in the South Atlantic at a latitude of 7°57′S. These birds regularly nest every nine-and-a-half months, in contrast to the other populations of the species whose nidification is always annual though very variable in date from locality to locality. Oceanographic conditions round Ascension Island seem to be constant, which may free these terns from an annual cycle but does not explain the rhythm they have adopted. It has been thought to be related to the lunar cycle – since the species is largely nocturnal, every tenth full moon might have a determining influence on their breeding. However, the interval between two nestings is more probably related to an internal cycle, allowing for the time necessary for reproduction and the raising of young, followed by a complete moult and a period for pairing, during which the birds must be exposed to social stimuli which end the maturation of their gonads (Chapin & Wing 1959, Ashmole 1963). The same seems to be true of other tropical sea birds, as has recently been observed in the Galapagos. In these islands, just below the equator, many birds reproduce throughout the year. Individuals of Audubon's Shearwater *Puffinus lherminieri* show a regular cycle of nine months, and the same is probably true of many species whose populations seem to nest continuously (Snow 1965).

In any one area all the bird species normally nest together, with differences in phase arising from their ecological needs, the peak demands of each species being adjusted to coincide with its ecological optimum. While this is true in general, some remarkable exceptions are known – even within a single species. Thus certain populations of the Sooty Tern show great irregularity in nesting dates, even within a restricted area. Within the Hawaiian Islands alone, the colonies on Manana Island (an islet off Oahu) begin to nest in April, while those on Moko Manu only 17km away begin in October. It is tempting to attribute this to different origins for the two stocks, the one coming from the northern and the other from the southern hemisphere, but this is mere hypothesis.

The beginning of the breeding season is so timed that on hatching the young have a sufficiency of food. Everything proceeds as though reproduction is adjusted so as to make their peak demand coincide with the maximum supply. Passing northwards (or southwards in the southern hemisphere) breeding begins later and later, since it follows the local spring. This is especially clear in wide-ranging species such as the Bluethroat *Erithacus svecicus*, which nests in April and May in central Europe, and in June in Lapland. In general, the breeding season is displaced by about twenty-five days for every 10° of latitude towards the pole. The timing in the southern hemisphere shows the same characteristics, though it is out of phase by six months.

In tropical regions with contrasting seasons, the breeding of most birds coincides with the rains, when the available biomass of vegetation as of insects and other animal prey attains its maximum. However, graminivorous birds nest later when the seeds ripen at the end of the wet season, as do raptors which benefit from the less dense vegetation. In drier or even desert regions the spring is the most favourable season, because it is less harshly arid, and so the birds in general nest early before the excessive heat of summer.

In equatorial regions, and especially in rainforest, breeding seasons are much less well defined. Here there cannot be said to be a breeding season for the whole ecological community, since birds breed throughout the year, the absence of periodicity being explained by the constancy of this environment. This does not however mean that an individual reproduces all year long. Although our knowledge on this subject is very limited, it is probable that every individual follows an annual rhythm with a period of sexual dormancy. Thus in Colombia at a latitude of 3°30′N the Rufous-collared Sparrow *Zonotrichia capensis* has two breeding seasons a year, with dormant periods for both sexes (Miller 1959). Generally speaking the breeding season is more extended the nearer one approaches the equator, in close correlation with the longer duration of favourable conditions. In contrast to the constancy of the equatorial environment which allows breeding to be spread throughout the year, is the rhythm characteristic of the dry tropics where an unsuitable dry season is followed on by a wet one in which breeding is concentrated. In the warmer parts of the temperate and cold zones the breeding season is slightly extended, lasting from March to July in the northern hemisphere; but it becomes shorter with approach towards the poles, as the period of ecological optimum shortens.

One consequence of the shortening breeding season is that the number of broods raised in a season by a pair is generally less at high than at low latitudes. Thus the Bluethroat raises only one brood in Lapland, but two in central Europe, and perhaps even three in western France. In contrast the size of the broods increases, as though to compensate, in the colder parts of the range, in accordance with what has been called Hesse's rule. Thus among chats of the genus *Oenanthe* the normal clutch is six eggs in *O. oenanthe* of cold and temperate zones, five in *O. hispanica* of the Mediterranean zone, and only four in *O. deserti* and *O. leucura* of the desert zone (though the desert environment also contributes to this low fertility).

Control of the annual cycle – photoperiodicity

The primary mechanism is an intrinsic physiological rhythm controlling the whole annual cycle, with its regulatory centre in the pituitary. This gland shows a seasonal cyclic activity, on which depends the functioning of the gonads and endocrine glands, and which also stimulates moult and migration. This is strikingly shown by castrated drakes, in which cyclical fluctuation in the secretion of thyrotropic hormone synchronizes with the sexual activity of control birds. This independence of the pituitary is confirmed by observations on birds transported from one hemisphere to the other. These may retain for several years the reproductive rhythm appropriate to the seasons in their place of origin, though this is in conflict with the conditions they meet in their new habitat. Even their descendants may preserve the original rhythm, but this is not always the case – many birds acclimatized to the opposite hemisphere, such as the Starling in New Zealand, have adapted themselves to the seasons of their new country. Similarly there is in spring a lack of synchronization between the sexual maturation of migrants which have come to the same tropical area from different directions, such as populations of various subspecies of the Yellow Wagtail in the Congo (Curry-Lindahl). Although subject to the same environmental conditions, these birds react differently, the timing of their response to the approach of spring being adapted to the seasonal changes in their areas of origin. Sexual maturation of the migrants must necessarily take place at least partly according to an innate rhythm, so as to coincide with the seasonal rhythm of breeding areas which are often thousands of kilometres from their wintering grounds. In the course of evolution selection has acted to adjust the rhythm of these birds, which thus seem as though controlled from a distance.

This cyclical change, though effected by an internal rhythm, is nevertheless simultaneously under the control of external factors, of which illumination and especially the daily duration of light, or photoperiod, are the most important. It has long been empirically known that illumination favours the appearance of song in birds. For centuries the Japanese have practised the art of *yogai*, which consists of getting cage birds to sing in the middle of winter by exposing them to days prolonged by artificial light. The Dutch apply the technique of *muit*, by keeping birds in the dark during the summer and subjecting them to illumination in the autumn. These birds then begin to sing at the wrong time and are used as decoys during the migration of wild birds, in order to trap them. The appearance of song is clearly released by maturation of the testes, as is confirmed by experiments which Rowan (1938) was the first to undertake. In the intense cold of midwinter at Edmonton in Canada, he was able to make the gonads, and particularly the testes, of Slate-coloured Juncos *Junco hyemalis* develop, by increasing the period of daily illumination with progressive use of artificial light. By alternating phases of short and long photoperiods, three maturations in succession were produced in a single year. These experiments, repeated many times on a variety of species, have all shown that the birds react to a real or simulated lengthening of the day by development of the testes, and to shortening by their regression. Temperature plays little or no part in this response. Ovulation and the hormonal release of nest-building seem to be more particularly under the control of temperature and the amount of food available, an important ecological adaptation. These facts explain how birds of temperate regions react in spring to increasing natural photoperiod, which is thus the stimulus for breeding. Those of the tropics, where differences in photoperiod are reduced, are also susceptible to experimental variations, though less clearly. Except near the equator the annual differences are not negligible, and probably play a part within a range of very diverse external factors.

However, an objection to this determination might be made. Northern migrants leave their breeding ranges in autumn when the days are shortening and resort to the tropics, where they suddenly find themselves exposed to longer days than they have just been experiencing. They should show a revival of sexual activity and even of nesting. That they do not is due to a refractory period following the breeding season, during which many birds show themselves to be insensitive to increasing daily length of illumination. Thus the Golden-crowned Sparrow *Zonotrichia atricapilla* and the White-crowned Sparrow *Z. leucophrys* show no response of the gonads to artificially increased length of day, from the end of the breeding season until the beginning of November. This has been found to be true of other temperate species also, but not of some tropical ones such as the Baya Weaver *Ploceus philippinus*.

The migratory instinct too is under control by variations in photoperiod. Two phases can be distinguished: a preparatory phase determined by short days – or rather by the length of the periods of darkness – and a progressive phase accelerated by lengthening days but not inhibited by short ones. This succession is of great importance, not only for migration but also in the control of breeding.

Light acts by way of vision, in an optico-pituitary reflex. Its passage through the tissues of the orbital region also produces a reaction (an encephalo-pituitary-sexual reflex), as has been experimentally shown by removing the eyes and replacing them by tubes of quartz which allowed light to pass towards the pituitary and the deep regions of the brain. In many birds this pathway of stimulation even seems to be more important than the pathway of vision. The hypothalamic region of the brain seems to play a much more important part in these actions than had been believed, as recent discoveries have shown.

Other environmental factors also play less important parts in the control of reproduction. Even muscular work can have some effect in a few birds. Juncos kept in revolving cages, and thus subject to forced exercise, showed development of their testes. The effect of ambient temperature varies according to the stage of sexual maturation. In the Wood Pigeon it has little or no effect in January, but is more important in March during active spermatogenesis, when the latter is accelerated by an increase in temperature and markedly slowed by a drop. Rising temperature and insolation, which occur together in the spring, seem to act synergistically. The effect of temperature on the females is less marked, maturation of the ovaries depending rather upon the internal cycle (Lofts & Murton 1966). In contrast, factors in the diet play an important part. The example of the Red-billed Dioch *Quelea quelea* which nests in huge colonies is especially striking, for the birds all begin to nest at the first rains. Observation on the Smooth-billed Ani *Crotophaga ani* have shown that its breeding is stimulated by a change in diet, from mainly vegetable to mainly animal, with the arrival of the rains.

The Silverbill *Euodice cantans* nests from February to April in the eastern part of Eritrea and from August to November in the western part, in accordance with the different incidence of the rains. Many birds in the semi-desert areas of Australia show similar phenomena, all the more interesting since they are accompanied by movements resembling migrations. Because of the climatic irregularities of this area, where rain falls infrequently and according to no definite rhythm, the birds have had to adapt in such a way as to take immediate advantage of the most favourable conditions. This is true of the Budgerigar *Melopsittacus undulatus* and some ducks, especially the Grey Teal *Anas gibberifrons*. (Frith, Marshall & Serventy). Populations of this duck contain two distinct components, a sedentary one whose stocks are controlled by the permanent conditions of the environment, and in contrast a fluctuating one which is very strictly influenced by the occasional rains. The individuals which make up the fluctuating component travel under the influence of the rains, with much variation in the directions and distances of their journeys, and can immediately respond to favourable ecological conditions by breeding. These ducks nest wherever they find a site which is temporarily suitable after local rains. Brooding and the growth of the young are shortened, so that breeding is completed before the habitat can become unfavourable again. This flexibility of reproductive rhythm, and the immediate release of activities which are much more rigidly determined in other birds, allow these ducks to exploit fleeting resources which are irregular in time and space.

Thus the factors which release breeding behaviour in birds as a whole are of three clearly distinct types. The first is an internal rhythm controlled by the pituitary, which is very strictly determined and not susceptible to any external influence. The second is illumination acting through variations in the length of daylight, an astronomical factor which is mathematically definable for a given place.

If in nature reproduction depended only on such rigidly determined factors its predetermined release would be seriously disadvantageous, since meteorological conditions and the biotic factors of the environment have marked effects on the success of hatching. In order that the reproductive mechanism should be assured of success, the animals must retain a certain flexibility and the timing of reproduction must be adjusted to present conditions, in the short term and from year to year. It is the role of the third type of releasing factor to take account of ecological conditions, acting especially through the quantity and quality of food available and so through rainfall, temperature and insolation. This correction allows the rigidity of a reproductive cycle controlled by an unvarying rhythm to be mitigated, giving the flexibility necessary for the onset and development of breeding to be adjusted to the most favourable environmental conditions.

After the breeding season and the post-nuptial moult (which takes place in temperate regions towards the end of summer and during the autumn), many birds show a resurgence of sexual activity. The males sing and defend territories in accordance with development of their testes, while renewed breeding behaviour is seen in the females also. Usually these activities rapidly die away with the arrival of cold weather and unfavourable environmental conditions, or with the birds' departure on migration. Occasionally when circumstances are exceptionally favourable they have led to reproduction and successful raising of the young, while still more rarely migrants have nested in their winter quarters.

Chapter 10

Territory and Territorial Behaviour

EACH breeding male establishes itself in an area whose limits gradually become more clearly defined. It defends the zone around itself, within which living space it will tolerate no other member of the species. The study of such defensive and offensive behaviour patterns has led to the definition of what is called bird *territory*. This fundamental concept was foreshadowed by the earliest ornithologists, who noticed the way in which birds divide up the habitat when breeding. However, it was not until 1868 that Altum in *Der Vogel und sein Leben* defined the concept scientifically, and it was precisely stated in 1920 by Eliot Howard in his classic *Territory in Bird Life*.

It is not easy to define territory, which explains the somewhat divergent views of various authors. It varies in space and in time and from species to species. Some authors have sought to give the term a purely sexual connotation, and have restricted it to reproductive activities. Certainly the territorial instinct then shows most clearly. However, the concept is much wider than this and it is better to give it a more general definition, agreeing with ornithologists such as Noble and Nice that *the territory is the whole area defended by a bird* at a given moment in its life. Thus defence is the most important criterion: a negative reaction to other members of the species; and a positive reaction to the site of part or all of the individual's activities, where it advertises its presence by very conspicuous behaviour patterns. In contrast an area which is regularly used by a bird, but which it does not defend, is not its territory. Examples of this situation are the hunting grounds of some raptors and the fishing grounds of some herons and other waterbirds, which may be termed their *living space*.

Types of territory

Almost all birds show territoriality, though in different ways depending on the species involved and the nature of their habitats. Of the various classifications of territory which have been proposed, Nice's (1941) seems the best and is followed here. Seven categories can be distinguished.

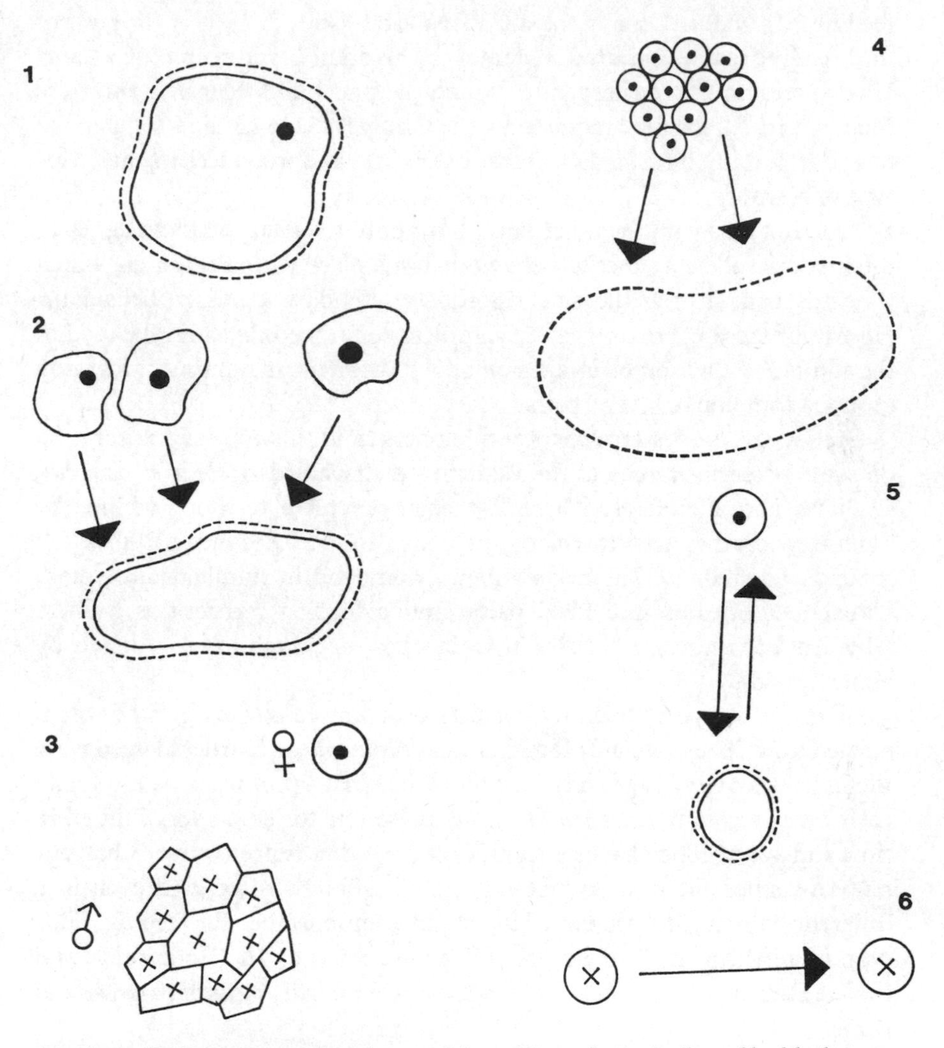

Figure 52. Different types of territory, with positions of nests indicated by black spots.
1. A territory sufficient for all breeding activities.
2. Nest territories, from which several proprietors feed in a communal area.
3. Territory used only for communal displays.
4. Territories of colonial birds, which feed together outside the colony.
5. Nesting and feeding territories.
6. Spring and winter territories.

1 The first very widely distributed type of territory is the scene of all reproductive behaviour, including pairing and sexual displays, and all feeding behaviour. Usually the male first establishes himself at the start of the breeding season and attracts a female by his audible and visible behaviour. He displays on the territory, after which the pair build their nest and raise their young. All the food required by the family is taken within the defended boundary. Many birds, notably most passerines and woodpeckers, hold this type of territory.

2 Territory may on the other hand have only a sexual significance. As in the first case all reproductive behaviour takes place within this area, which the birds, or at any rate the male, vigorously defend by aggressive behaviour. Here however the territory has no significance for feeding, since the food of the adults and their brood is taken outside its borders, in hunting or foraging grounds common to several pairs.

Grebes, swans, oystercatchers and harriers hold this type of territory. So do some passerines, such as the Redwinged Blackbird *Agelaius phoeniceus*, swallows and Fieldfares. These last each defend a territory within the boundary of the nesting colony, which also encloses neutral foraging grounds common to all the members. Some birds simultaneously hold sexual territories as individual pairs, and a feeding territory as a flock. This has been noted especially in Galapagos mockingbirds *Nesomimus* by Hatch (1966).

3 The third type of territory consists of a simple display ground, often termed a *lek*. These are only found among birds whose females alone devote themselves to the young, while the males are sexually promiscuous and spend their time in communal displays. Thus these territories are very limited in time and space. They have no significance in other reproductive behaviour patterns, since the nest is built later, at a different place, and even in a different habitat; and are not of the slightest importance for feeding. This type of territory is characteristic of grouse such as the Capercailie and Prairie Hen, manakins, cocks-of-the-rock, bowerbirds, birds of paradise and Ruffs.

4 Another type of territory differs from the second only by extreme reduction, since it encloses nothing but the nest and its immediate surroundings. Sometimes it is no larger than the area which the bird can defend with its bill while on the nest. This type is found almost exclusively among gregarious birds which nest in dense colonies, such as penguins, pelicans, cormorants, gulls, terns, herons, swallows, Jackdaws, many weavers, and some birds of prey. In Social Weavers *Philetairus socius* and Monk Parakeets *Myiopsitta monachus* a group of perhaps a hundred birds nest within a

unique and enormous structure, in which each pair defends only its own chamber against the others. Thus there are two nests to the square metre in colonies of the Brown Pelican *Pelecanus occidentalis*, and three in those of the Guanay Cormorant *Phalacrocorax bougainvillei*.

Outside the immediate boundary of the nest, individuals of the same colony seek their food in a shared living area which may be very large. The total area around the colony is often defended by all its members against intrusion by strange members of the species. Thus there is a distinction between individual territories defended by mated pairs against all others, and group territories defended by communities against strangers. Such collective defence of the colony is seen especially among terns, but also in Social Weavers and Jackdaws, and in Australian Magpies *Gymnorhina tibicen* which live in parties of up to ten individuals (Carrick).

5 Some birds have two territories, spatially separated but defended equally vigorously, one devoted to nesting and the other to food-gathering. This specialization, in the simultaneous use of separate areas, is shown notably by the Seaside Sparrow *Ammospiza maritima* of the Atlantic coast of North America, which defends both the approaches to its nest and a strip of beach where it finds its food. So too the Rock Thrush *Monticola saxatilis* holds breeding and feeding territories, which may be several kilometres apart. Very few birds have territories of this type, since the nest is usually built in habitats most favourable to the birds, providing both a suitable site and a hunting ground.

6 Non-migratory birds generally defend territories only during the breeding season, outside which they scarcely show any territorial instincts. However, there are a few exceptions to this, including the Great Grey Shrike *Lanius excubitor* which defends its boundaries throughout the year, the territory being used only for feeding during the winter.

The territorial instincts of migrants mostly disappear during migration, when birds which have been intolerant of others during the breeding season become gregarious. Certain exceptions to this rule defend territories throughout their migratory journeys – notably the shrikes, whose behaviour is unusually intolerant for passerines. Each individual defends a boundary round the perch on which it has established itself and from which it hunts, so that the territory itself is mobile, being an area surrounding the travelling bird. The same is true of flycatchers and kestrels. Some small waders, especially plovers, defend sections of the beaches on which they rest during migration.

In their winters quarters migrants often establish territories much like their breeding ones. Thus in Maryland wintering Redheaded Woodpeckers

Melanerpes erythrocephalus segregate themselves rigorously in territories, which provide food reserves in the form of hoards of acorns, and in the middle of which each bird selects a roost which it defends as vigorously as it does its nest in summer quarters (Kilham 1958). Such mobile or temporary territories are evidently of importance only for feeding, giving the birds the isolation they need. To some extent at least, they are occupied year after year by the same wintering individuals.

7 Finally, the last category comprises the roosts and resting places used by birds outside the breeding season. Many birds return nightly to the same place and use the same perch, like Starlings which assemble in flocks – often huge ones – in certain places, each individual appearing to have its own exact spot on which to pass the night. Here again the territory, strictly ethological in significance, is of greatly reduced size.

Particulars of territory

Sedentary birds and returning migrants establish themselves at the start of the breeding season, each in a sector of their preferred habitat. Every male begins by visual and auditory behaviour to declare his occupancy of a particular area, which is large in comparison with the definitive territory and has vague and fluctuating boundaries. New arrivals attempt to settle amongst those which are already established, so that all the birds are forced to accept limits to their own territories set by those of their neighbours'. While territories are not always contiguous and may be separated by neutral zones – at least in certain species, or where the population is not very dense – in other situations they form a continuous mosaic with clearly marked lines of division.

Where the environment is uniform over large areas and the birds do not have specialized ecological needs, territories are more or less circular or polygonal as the result of interactions between neighbouring landowners. This is true for instance of forest birds distributed through a homogenous block, or of savannah birds whose habitat is uniform over huge areas. More often however the boundaries follow much more complex lines, being established in relation to the nature and topography of the terrain. They often follow natural frontiers such as hedgerows or ravines, while a mere fence or line of stakes may delimit a territory, by providing favoured singing posts. The territories of birds which are restricted to a particular habitat follow the boundaries of the latter, sometimes to the point of becoming linear like those of kingfishers which closely follow streams. Their territories are thus arranged serially along a watercourse and are of characteristic elongated shape, like those of many other water birds which follow the line of the bank. The Song Sparrow *Melospiza melodia*, whose many races are confined to the western coast of the United States, exactly follows the shoreline and thus occupies a series of territories arranged linearly, each being from 40m to 52m long but not more than 9m in width (Johnston).

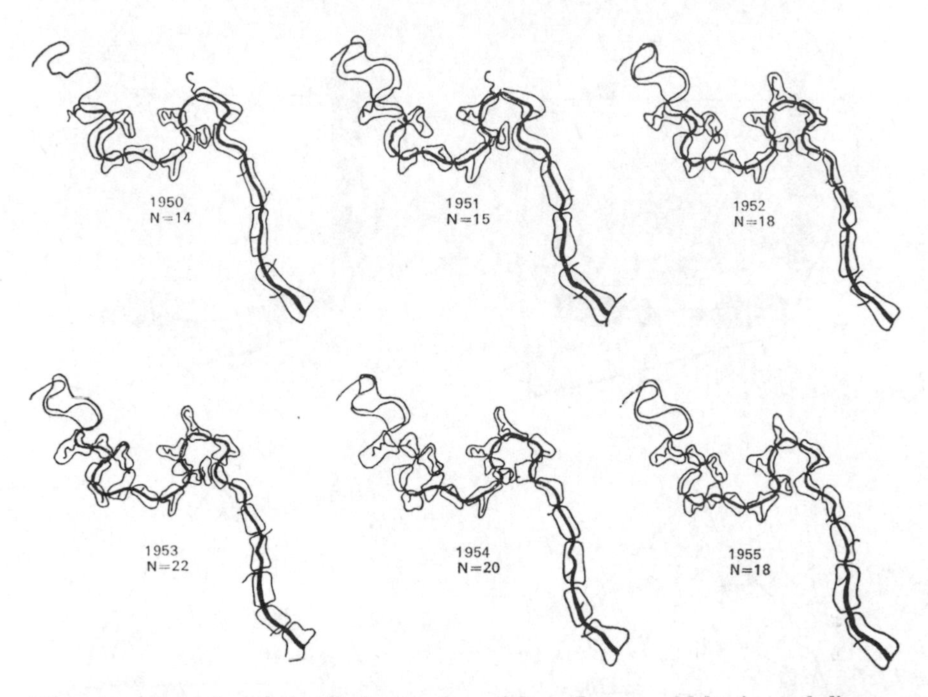

Figure 53. Annual variation in the number of Song Sparrow *Melospiza melodia* territories along a swampy stretch of North American coast.

There is a rough correlation between the size of bird and the mean area of territory; and especially between the diet and this area in relation to the amount of specific foodstuffs available. Thus birds of prey in general need very wide territories, because they have to hunt prey which is necessarily less abundant while avoiding the depletion of the populations in a small area. Graminivorous birds can occupy smaller territories than insectivorous ones of equivalent size. The kind of habitat also has an important effect, territories being generally larger in open environments than in closed ones. This is no doubt due to their reduced productivity, as well as to the behavioural differences between birds of different habitats. Ethological factors also affect the size of the territory through differences in the

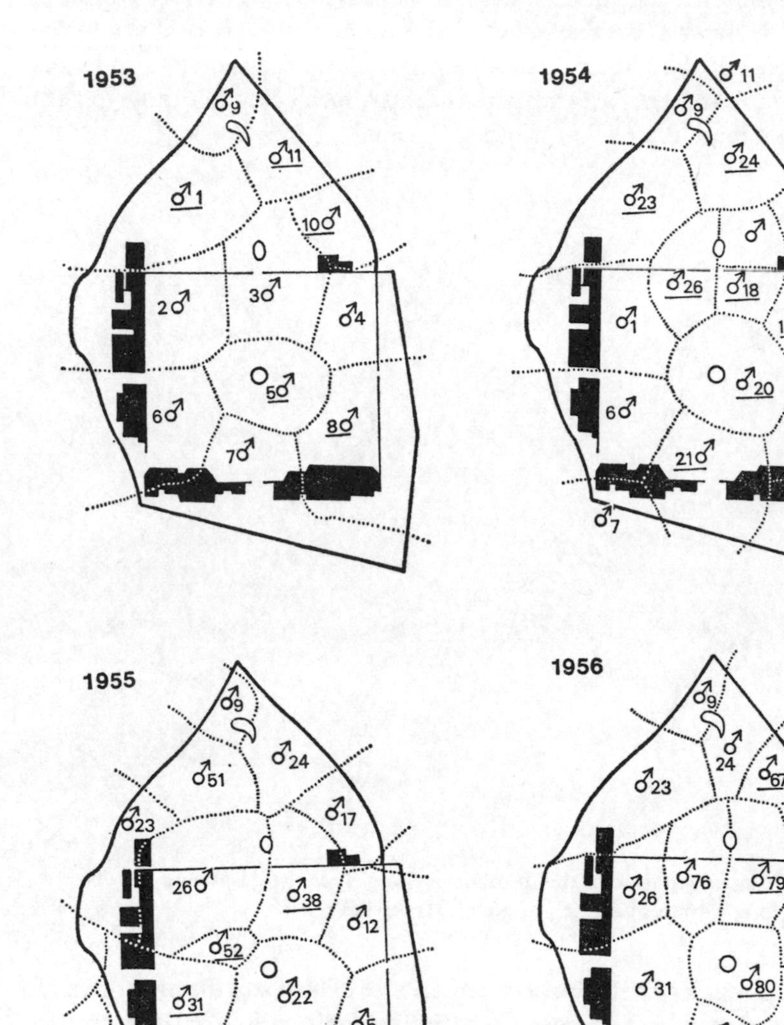

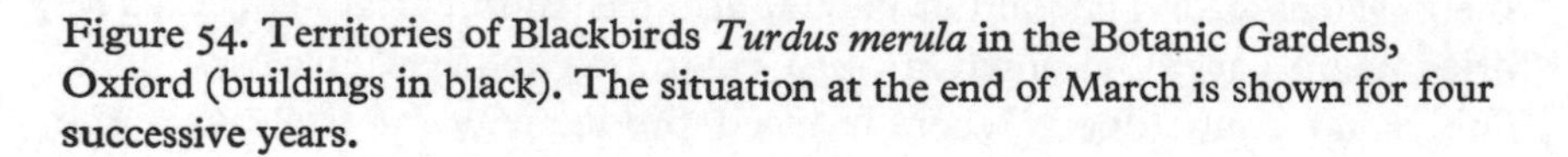
Figure 54. Territories of Blackbirds *Turdus merula* in the Botanic Gardens, Oxford (buildings in black). The situation at the end of March is shown for four successive years.

competitiveness and aggressive behaviour of the birds. Thus the Blackbird *Turdus merula* holds territories averaging 1200m² in area, while the minimum for the much more aggressive Mistle Thrush *T. viscivorus* is 150,000m².

The territories of highly gregarious seabirds include only the immediate approaches to their nests. The availability of food in the sea is very different from what it is in terrestrial habitats, and the fishing grounds are rich enough to be used communally without special competition, whereas it is important for reproductive success that the pair should hold the space around their nest. By way of example the *mean* areas of the territor ies certain species (from various authors, in Welty) are listed below:

Species	*Area* (m^2)
King Penguin *Aptenodytes patagonicus*	0·5
Black-headed Gull *Larus ridibundus*	0·3
Coot *Fulica atra*	4,000
Hazel Hen *Tetrastes bonasia*	40,000
Red-tailed Hawk *Buteo jamaicensis*	1,300,000
Bald Eagle *Haliaetus leucocephalus*	2,500,000
Golden Eagle *Aquila chrysaetos*	93,000,000
Least Flycatcher *Empidonax minimus*	700
House Wren *Troglodytes aedon*	4,000
Ovenbird *Seiurus auricapillus*	10,000
Blackbird *Turdus merula*	1,200
American Robin *T. migratorius*	1,200
Song Thrush *T. philomelos*	40,000
Willow Tit *Parus atricapillus*	53,000
Willow Warbler *Phylloscopus trochilus*	1,500
Chaffinch *Fringilla coelebs*	4,000
Song Sparrow *Melospiza melodia*	4,000
Western Meadowlark *Sturnella neglecta*	90,000
Red-winged Blackbird *Agelaius phoeniceus*	3,000

However, territories vary widely in area within species. They vary in extent from 0·1 to 1 hectare in the House Wren *Troglodytes aedon*, from 0·4 to 6·5ha in the Marsh Tit *Parus palustris*, and from 0·2 to 0·6ha in the Song Sparrow *Melospiza melodia*. The greatest part of this variation is seasonal. Territories tend to be larger at the beginning of the breeding season and to be successively reduced as further males arrive. The mean territorial area of the Pheasant *Phasianus colchicus* in Winsconsin is reduced from 5ha in April and May to 2·5ha in June (Taber), while the same has been recorded for many small passerines. These changes are fundamentally related to population density. However, this is not universal, and in some cases the mean size of territories does not change when population density decreases. Wide spaces remain unoccupied and act as neutral ground for the holders of nearby territories.

When territories are important in food-gathering, their size is as much influenced by the quantity of available food per unit area, representing the trophic value of the environment to a particular species. The territory defended by a pair will be larger the sparser the food supply, so as to assure a sufficiency to themselves and their brood. This has been demonstrated for example by the Pomarine Skua *Stercorcarius pomarinus* in Alaska by Pitelka. In years when lemmings – the prey which make up a major part of the skuas' diet – are abundant, territories are about 6 to 9ha in extent; whereas when the rodents are thin on the ground the territories expand to an average area of 45ha, thus adjusting to the amount of food available.

Ethological factors similarly influence the area of territory, through the competitiveness and vigour of the males. The territories of young males are generally smaller than those of males which have already nested (by a factor of up to one to two in passerines), and the young birds are often displaced into marginal areas which do not fully meet the ecological needs of the species. This competition is under the control of the male hormones whose quantity in a sense determines the sizes of the territories, as well as position in the social hierarchy.

Claiming and defending territory is generally a male prerogative, which to some extent forms part of the breeding behaviour and displays. At later stages of breeding the female may take an active part in the defence of the territory, which thus becomes truly the property of the pair, with the female often defending especially the approaches to the nest. However, there are variations in this behaviour. Thus among ducks, which pair in their winter quarters, mates jointly choose a breeding territory. In birds such as hummingbirds, in which all the tasks of building the nest and raising the young fall upon the female, it is she who shows territorial behaviour. Most species in which both sexes defend the territory (such as gulls and some passerines like the European Robin), show no sexual dimorphism. In others whose dimorphism is pronounced, such as the American Goldfinch *Spinus tristis* and the Red Cardinal *Richmondena cardinalis*, each of the mates may defend the territory only against intruders of its own sex. However, in most species in which the sexes are conspicuously different it is the male alone which undertakes the defence of the territory.

Defence of territory

The essence of territorial behaviour, defence of a particular space, shows itself in a series of aggressive activities: threats and pursuits leading up to real fighting. The objects of this defence are the nest, the mate and the young, and a number of localities such as the song posts which are scattered throughout the territory and mark out its limits.

Among passerines song plays an important part in defence, the male advertising his rights to a territory and defending it from rivals mainly by audible display. Real vocal duels in which two males take part are of primarily territorial significance. Birds of closed habitats scarcely have any other way of threatening, since visual contact takes place only subsequently and at close range. The importance of song is reflected in its frequency and intensity, especially at the beginning of the breeding season when territories are claimed. Display behaviour providing optical stimuli also plays a part in the defence of territory. These stereotyped species-specific movements act primarily by showing off specific coloured patches which serve as releasers, as when a Robin ruffles up the feathers of his throat and displays his red bib. Similar displays have been described for example among wagtails, starlings and buntings. In other birds threatening postures involve erection of the tail and alignment of the body, accompanied by characteristic calls. Specialized display flights, by hummingbirds and the Lapwing for example, are also meaningful in the defence of territory.

Threats and advertisement are the most important part of territorial defence, vocal and visual threat acting as combat at a distance which allows the birds to conserve their energies and to avoid actual fighting. The latter occurs only when threats have failed to achieve the flight or withdrawal of the rival, and occurs as a general rule almost entirely among colonial species whose territories are limited to the immediate surroundings of their nests. Here the density of the birds, the closeness of the territories, and the necessity of encroaching on other territories in reaching their own nests, makes frequent scuffles inevitable. This is especially true among penguins, which have to reach their nests on foot.

When two holders of territory come into actual conflict, the one on whose ground the fight takes place usually wins. It is effectively dominant within its own territory, and the intruder is submissive as though showing an 'inferiority complex'. This has been studied experimentally by trapping a territorial male Robin and caging him within his own territory. Intrusion by a neighbouring male provokes the captive bird to threatening behaviour and vocalizations which make the free bird retreat, although the territory-holder can obviously neither compel it by force nor pursue it. In contrast, if a caged bird is placed in a neighbour's territory, it behaves like an intruder and shows submission towards the territory holder, which attacks the cage (Lack). These differences in behaviour are evidence of a hierarchy which becomes established between conspecific birds, as the result of dominance which is not absolute but determined by their spatial positions.

Territorial encounters vary greatly in degree of aggression, depending on circumstances and especially on the species concerned. Tyrant flycatchers are exceptionally aggressive, and many other insectivorous passerines are scarcely less so, while other species are milder and show little hostility. Aggression is generally shown only between individuals of the same species, so that the two pairs belonging to different species may nest in the same site. However, where the territorial instinct is very strong the hostility extends to quite different species. The Fieldfare chases away jays, woodpeckers, crows and other thrushes, while the Song Sparrow is so aggressive that it shows hostility towards sixteen species of birds (Nice). Pugnacity also varies individually and seasonally, being at a maximum at the beginning of the breeding season and then diminishing until it is scarcely apparent towards the end, when the birds become social. Aggression is obviously under the influence of the sex hormones, varying with the quantity secreted from beginning to end of the breeding season. It varies also with distance from the centre of the territory where the nest is usually to be found, the approach to the nest being vigorously defended. While this is not always true of the periphery, the topography of the environment can have a great effect on the intensity of behaviour. Where the boundary is established along a clear ecological or topographical barrier this is often stubbornly defended, as is shown by the density of song posts along it.

Attachment to territory

The attachment of a bird to its breeding territory often lasts longer than a single season. Large-scale marking of birds has shown that a very high proportion return from year to year, not merely to the same locality but also to the same territories. Nutcrackers *Nucifraga caryocatactes* have been found on the same territories for ten years in succession (Swanberg). Even migrants return to the sites where they nested the previous year, as has been shown recently by large ringing programmes. While swallows are the classic example of this, they are not by any means the only ones. Raptors divide up the same area from year to year, and many passerines return to the same place. So do penguins, whose pairs occupy each year the same relative positions within the colony. This loyalty to territory varies between species (being stronger in the Blackbird *Turdus merula* than in the Song Thrush *T. philomelos*) and between the sexes, males showing greater attachment in accordance with their more marked territoriality. Thus 37 per cent of male

Pied Flycatchers in Finland return to the same place, against only 10·7 per cent of the females (von Haartman). The young show still less fidelity, showing little tendency to return to their birth places but rather dispersing widely. Not more than 5 per cent of young swallows seem to return. Other species, especially Waxwings, do not show territorial attachment at any stage and freely change their habitats, so that the populations of these nomadic birds are continually mixed.

Some species at least show a similar loyalty to their winter quarters. Experiments on Black-headed Gulls, caught on their wintering grounds in Switzerland and Denmark and transported, have shown that about 22 per cent of these birds return. Some passerine individuals, ringed during the winter, have been recovered at exactly the same places in following years. Thus a Grey Wagtail *Motacilla cinerea* was found to spend five consecutive winters in the same Garden in Bombay where it had been ringed (Salim Ali), while a Myrtle Warbler *Dendroica coronata* returned to the same winter quarters in Georgia, USA, three and four years after it had been marked. This constancy is certainly related to the possession of true winter territories.

Importance of territories

A bird travels its territory along definite pathways, invisible to our eyes yet none the less real, returning to fixed points and especially its singing posts. It is better able to escape its enemies in known terrain, and predation is markedly more intense on birds, especially game birds, when they show non-territorial behaviour. At first sight this is rather surprising, since territorial birds make their presence very conspicuously known to predators as much as to members of their own species. However, this disadvantage is amply compensated by knowledge of the terrain, which allows them to escape more effectively, and by wide dispersal of those species (such as waders) most subject to predation. Both from the point of view of feeding and of escaping predators, economy of effort favours the bird restricted within a definite boundary.

Territory and associated behaviour are essential to the males of many species for certain psychological factors in the maturation of the gonads because of the dominance which they show on their own ground. Observations and experiments (especially on the Robin by Lack) have shown that males which are subordinate or lack territory cannot reproduce successfully.

Furthermore, territory is of indisputable ecological importance. Situated as it is within the habitat best adapted for the species, it must provide everything necessary to the bird to remain alive and ensure food for the pair and its young. Differences between species in the extent of territories are explicable in terms of their diets. As we have already seen, the size of a territory on which a pair and its brood depend for their food varies to some extent within a species according to the amount of food available. A bird has no cause to establish too large a territory, which would be difficult to defend economically, if it can find its food within a more limited area. On the other hand, in the poorer parts of its preferred habitat it will on the average hold a larger territory, so as to increase the food supply at its disposal.

Other factors act jointly. The availability of nesting sites markedly influences the density of birds, and consequently the size of their territories. This is how the provision of nesting boxes may increase the number of those birds which can use them, such as the Pied Flycatcher. The population of an orchard in Germany increased in six years from two to forty-five pairs as a result of providing 100 boxes (Creutz).

The division of the habitat into territories brings about the optimal use of its natural resources. It avoids local overpopulation which would otherwise arise – experiments involving constant reduction of the population within a certain boundary demonstrate an influx, exceeding by three or four times the stock of territorial birds, originating in surplus populations which normally provide a safety reserve for the species. Furthermore, by dispersing the population and avoiding local concentrations, territory hinders the spread of diseases and parasites harmful to the species.

These diverse ecological relations pose the hotly debated question of the possible part played by territory as a factor in limiting populations. It certainly has an influence on the potential of a particular region for maintaining a certain number of pairs. In most cases the area of each territory cannot fall below a miminum, so that even if food is abundant to the point of surplus many birds (especially among the passerines) cannot tolerate the close neighbourhood of their own species. This necessarily involves a mutual withdrawal, and hence keeps the population below a definite level. In contrast, other territories are adjusted in size to the available food supply, but do not cover the whole habitat in a continuous network. Undivided zones remain as a kind of 'no man's land' which supports a fluctuating population of unpaired birds, and here of course territory is not a limiting factor.

Chapter 11

Courtship and Sexual Displays

Courtship and sexual displays

AFTER establishing his territory the male must find a mate, form a lasting pair, and begin nest building. This is done through varied and usually complex ceremonies known as *sexual displays*, which serve to attract and stimulate his mate. In the early stages the territorial male chases out all conspecific birds which intrude into his area, by threat displays, by song and calls, and if necessary by actual fighting. Females attracted by the males suffer only very mild attacks, wandering about without settling down until soon each is admitted to the territory of a male – who now tries to keep her there by exchanging his former aggression for attractive behaviour. At this stage and throughout breeding, characteristics by which the sexes may be recognized are important. These are both auditory and visual, the plumage especially serving as a complex of visual sign-stimuli. However, in species lacking in sexual dimorphism sexual recognition depends upon characteristic behaviour and attitudes. There is thus an imperfect bond between two birds of the same species and opposite sexes, which develop a sort of dialogue (both visual and auditory) tending towards the strengthening of the bond, copulation and later phases of breeding. This behaviour is especially frequent at the beginning of the breeding season after which it tends to diminish, though it recurs throughout the breeding season since it is essential to breeding success.

Sexual displays are species specific, although common characters can be detected in the displays of related species. Thus like morphological characters they serve to characterize now a group and now the species, and are therefore used by biologists in studying the evolutionary relationships between different types of bird. Warblers have very plain patterns which play only a secondary part in sexual displays, but their voices are highly developed and serve both to 'seduce' their mates and to defend their territories. In contrast brilliantly patterned birds seldom show much vocal talent, and their displays are essentially visual.

It should be noted that sexual displays are essentially dynamic, developing in successive phases during which the bird demonstrates its high mobility.

Varieties of sexual display

Sexual displays are most often characteristic of the male; though in some cases the female does take part in mutual displays, the initiative remains with him and she responds only after a latent period. In all species with marked sexual dimorphism (including the dimorphism in vocalizations shown by songbirds) sexual displays are a male prerogative, and serve mainly to show off his ornaments of plumage to the female. The plainly coloured House Sparrow merely displays its black bib, taking up postures which make this more conspicuous. The Blackcap *Sylvia atricapilla* ruffles the feathers of its head and puffs up those of its back, while alternately spreading and closing its tail. The Redstart *Phoenicurus phoenicurus* spreads its tail and gapes to show the vivid yellow lining of its mouth. The Wheatear *Oenanthe oenanthe* hops round the female fanning its black and white tail, takes off and climbs steeply, then falls to the ground and begins again as soon as it touches down.

Among the pheasants, a family characterized by very marked sexual dimorphism, the classic display is that of the Peacock *Pavo cristatus*. His special ornament is his train, which consists of enormously developed upper tail-coverts, under which are quite ordinary tail feathers which serve in display merely as a support. He shows off the magnificent feathers of the train by fanning it, strutting round the Peahen and then stopping to vibrate the fan with a characteristic whirring. In comparison the display of the common Pheasant is quite simple, the cock merely calling while walking through a covert, pausing to spread his tail and noisily beat his wings. When a female joins and follows him he invites her to forage beside him, redoubles his calls and displays, and swells up his wattles. Copulation occurs when the pair find themselves in harmony, after which the cock continues to display for a while until they separate (Kozlowa). Display is more complicated in other pheasants, especially Bulwer's Wattled Pheasant *Lophura bulweri* of Borneo. The tail plumes, which are very wide and fine, like sails, are spread during display into a pure white vertical disc in front of which the dark body is compactly hunched. The head is stretched forwards, the vivid blue wattles swelling at the height of the excitement until they are rigid and so large that they completely hide the bill. The cock sweeps dead leaves on the ground with the tips of his outer rectrices, which end in stiffened naked shafts, and accompanies the noise with shrill calls (Heinroth).

Drakes too show very complex displays. These gregarious birds use a wide variety of behaviour patterns, many of which are involved in sexual displays. In many species the pairs form during the winter and the commonest displays are only seen later, when they serve primarily to strengthen the pair bonds and to allow the maturation of the gonads, especially the ovaries. Since the displays take place when the birds are still in flocks, they are naturally group activities. Within the duck family sixteen different actions can be distinguished, each species using some of them in display. Thus the Mallard *Anas platyrhynchos* rounds its head by puffing out the feathers and curving its neck, shakes its head and tail, then dips its bill in the water and throws its head and neck up, all the while whistling and grunting. In other displays it raises its head, tail and wings, accompanying these movements too by whistling. All these displays take place on the water near the ducks, which merely swim about with necks stretched out.

Sexual displays are still more complex among the birds of paradise, a family in which sexual dimorphism is exceptionally marked. The displays differ from species to species in relation to the strange ornaments developed by the male birds, whose significance is made apparent by this behaviour. The Arfak Six-Wired Bird of Paradise *Parotia sefilata* – which gets its name from six long plumes carried behind its head, each of which consists of a bare shaft and a small terminal racket – has a bib of bronzy feathers in sharp contrast to the deep black of the rest of the plumage. During its displays, which take place on dancing grounds first cleared of dead leaves, the male widely spreads this pectoral ornament to form a shield in front of the body, while the rackets of the six head wires bob about oddly in front of the bill. The head, drawn back on the breast, swings from side to side, while the feathers of the flanks are spread round the body. The bird dances on the spot in this attitude, before flying off to perch. The Black Sickle-billed Bird of Paradise *Epimachus fastuosus* at first glance seems more restrained in its ornament, made up only of long tufts of dark feathers edged with a metallic halo, which are entirely hidden under the wings when the bird is at rest. The male looks entirely different when displaying, perched on a branch. It rises on its feet, spreads its flank ornaments, stretches its body upwards, and widely spreads its long tail feathers (the middle two of which form a train), while making scissor-movements with its shorter and stiffer lateral tail feathers. The Blue Bird of Paradise *Paradisea rudolphi* also has long flank-tufts, which are silky and of a magnificent blue. In display, this bird perches on a branch and slowly topples backwards until it hangs upside down, swinging in this position with its long hanging ornaments waving on each side.

While the plumage and its ornaments play the main part in displays, the bill and legs are also sometimes involved where they are brilliantly coloured. The Blue-footed Booby *Sula nebouxi* of the Pacific coasts of South America has intense azure-blue feet. In display it walks with slow steps and separates its toes so as to spread out the webs, picking up its feet in turn and so showing them off to its mate. Similarly the lining of the bill and throat is sometimes very vividly coloured, especially in seabirds which open their bills widely during gaping displays. For example, the Gannet separates its mandibles widely, so as to show the female its deep black mouth lining.

In many birds the inconspicuous females merely assist in the display of the male by responding with special postures, without really taking part in the ritual. However, in birds with little sexual dimorphism the mates take equal parts in the ceremony, by more complicated behaviour known as *mutual display*. Such activities are shown especially by waterbirds like the Gannet, which indulges in a very complicated ceremonial during which the mates adopt a series of stereotyped postures. Sometimes they come face to face, stretching their necks, raising their bills vertically, and spreading their wings in a characteristic ecstatic attitude. Then they lower their heads and clash beaks, seeming to whet them against each other. Such a display may appear as a dance in which the pair moves on the spot in concert. This sort of display is very widely distributed among penguins, though the details vary between the species. In the Adelie Penguin *Pygoscelis adeliae* the two birds stand upright face to face, necks stretched and bills pointed skywards, each swinging its head to right and left in rhythm counter to its mate's, while rocking gently back and forth. This movement, which is accompanied by a resonating 'ka-ka-ka', thus gives the impression of repeated embraces. It continues from five to twenty times in succession, and begins again after a rest (Sapin-Jaloustre). This mutual display, in which both birds take an equal part, acts as a means of individual recognition, a greeting ceremony, and a ritual of appeasement counteracting territorial aggression, while the mutual stimulation serves to tighten the pair bond. Penguins also use bowing displays: standing upright the mates bow deeply with bills to the ground and then straighten again with bills pointing skywards, flippers widely spread forwards in the ecstatic posture. At other times a bird will stand up and open its bill widely to emit calls like the honking of a motor horn.

Albatrosses also take part in mutual parades which look like dances. The mates face each other and bow deeply, bending their necks repeatedly. Next, spreading their wings as though to show off their span, they turn on the spot, or one revolves round the other. Then they fold their wings, stretch their necks out towards each other, and clatter their bills together with a characteristic noise. Bowing begins again, with ecstatic postures and resounding calls, so that mutual recognition can be both visual and auditory.

This is true of mutual displays as a whole, which naturally make use of vocalizations together with the visual behaviour, often in a sexually dimorphic way. Thus for example the love song of the Emperor Penguin is a kind of muted cackle, given with the head bent towards the ground. Analysis shows that this song (which lasts two seconds) is of similar mean frequency in both sexes, but that the fundamental harmonies of the male voice seem to cover a slightly wider range of frequencies than those of the female. The pulses into which this signal is broken differ between the sexes, those of the females being shorter (Prevost). Similar differences are found among ducks, geese and swans, in which the syrinx and nearby parts of the trachea show remarkable sexual differences. The voice of a female swan is deeper by a semitone than that of the male, so that when they call alternately they produce a real duet. Mallard ducks and drakes do the same. So do Tawny Owls during their mutual vocalizations, in which the part played by each of the mates may be distinguished by ear. True duets between mates, so well synchronized that they seem to come from a single bird, are also given during mutual displays.

Some of these displays take place on the water. In grebes, which are so clumsy on land, the mates swim towards each other and when they meet rise almost vertically from the water face to face. Suddenly they rush off together in all directions, thrashing the water with their feet. Travelling side by side, they really seem to be walking on the water in the midst of a great splashing, and each quite commonly holds in its beak a piece of water weed (Storer).

Birds, as the supreme aerial animals, have also taken to the air for their sexual displays. During nuptial flights the males of some species perform evolutions which serve to lay claim to territory and to attract a female. These aerobatics may be accompanied by songs, or by sounds produced by the wings or the tail. The flight of the Skylark, which sings in mid-air, is the best known. Snipe engage in display flights during which their tail feathers vibrate characteristically. Lapwings gather on a water-meadow above which the males make their rolling flights, beating the air with their wide black and white wings. The Snow Bunting *Plectrophenax nivalis* of the arctic tundras displays its strongly contrasting black and white wings to the female as it performs aerobatics above her, as though to seek her admiration for its pied plumage. Raptors also indulge in this type of display. Buzzards perform aerial games with a variety of figures, during which one bird will climb above its mate and fall upon it with wings raised, whereupon the second bird turns over in mid-air and stretches out its feet. Their talons seem to clash for an instant, before the birds separate for further manoeuvres. Such displays are still more highly developed among the hummingbirds, in which pair bonds last only a short while. Males and females have different territories, though these often overlap to some extent, the larger territory of a male extending onto those of several females. The brilliantly coloured male attracts the duller female by characteristic flights, often accompanied by noises from the wings and tail. The male lifts, hovers briefly, flies off in another direction, and traces out curves and sinuous lines broken by several pauses, before he settles on his perch. He usually orients himself so that his metallic ornaments shine in the sunlight towards his partner (Hamilton). The female, soon attracted by this exhibition, perches nearby while the male, which at first shows some aggression, immediately increases his nuptial flights. After a moment the female follows him, and the two fly closely side by side, their complicated flight-paths being absolutely parallel. Thus they show a very close though brief bond, parting after copulation.

During these mutual displays, synchronization of behaviour improves as the pair bonds strengthen. Both song and visual display are clumsy at first while co-ordination and synchronization are still lacking, but improve progressively towards their final perfection. These displays reach their climax with copulation, which is marked by submissive behaviour by the female. However, they continue for a longer or shorter time, during incubation and the raising of the young, having a much wider importance than preparation for the sexual act.

Symbolic gifts and actions

Nuptial displays sometimes involve a rubbing together of bills, which may be considered as the first step towards courtship feeding. The latter plays a part in the nuptial displays of birds belonging to very diverse groups, being found among fourteen orders and notably in seventeen passerine families (Lack). Thus the females of the Robin and many other small passerines frequently beg for food from the male, taking up the begging attitudes of the young by spreading and trembling their wings like a nestling demanding a beakful. It is also significant that begging females show infantile behaviour, with postures and calls exactly like those of their young.

Symbolic gifts of this kind are used by gulls and terns, in which the female, adopting a characteristic posture, begs a fish from her mate when he returns from fishing. Holding the fish in his bill he circles around her, half spreading his wings and extending his head and tail, sometimes for a long time before he gives it to her. Instead of eventually doing so, he may remain unwilling, and then the display degenerates into a scuffle with both birds gripping the fish, unprepared to relinquish it to their mate.

Nest-building materials too are sometimes used as symbolic gifts. Grebes in display often carry aquatic plants in their bills, like those which they use to build their nest platforms. Other waterbirds actually pass a twig from one bird to its mate, while herons give branches with many bows and erection of the nuptial plumes.

Thus sexual displays are complicated by the use of symbolically significant accessories. Such behaviour recalls activities which are biologically necessary at other states of breeding: building the nest and feeding the female while she broods. Thus they make use of motivations which will become functional later but are as yet only symbolic. This demonstrates the unity of motivations, which develop gradually without a break from one to the next.

Displays of bowerbirds

Some male birds prepare the ground to which they attract females by their sexual displays. For example, the male Argus Pheasant *Argusianus argus* of Borneo and Sumatra returns year after year to a flat space in the forest, and clears it to the bare soil to a diameter of five metres. Here he remains most of the year, attracting females by his resounding calls. When one is present he displays, calling loudly while he postures – sweeping the ground with his tail, stamping both feet, displaying his tail and spreading his wings into a disc – before bowing before her. The polygamous male has several display grounds scattered about his territory, and only unites with each female for a few days. The Australian Superb Lyrebird *Menura novaehollandiae* prepares similar platforms in its territory, visiting up to eight during a single morning. With his tail turned right forwards over himself, the male vibrates it and dances and sings most melodiously, imitating the songs and calls of many other species.

While such birds sweep their display grounds clear, more elaborate theatres are prepared by the bowerbirds – relatives of the birds of paradise which like them live in New Guinea and Australia. Bowerbirds are not generally brightly coloured or glossy, and their most distinctive ornament is an iridescent violet-rose patch on the hind-neck of some species of *Chlamydera*. During the mating season a male of this otherwise dull-coloured buff and brown genus choses a clear space in the dense scrub in which it lives, and builds a structure of twigs planted in the ground. This consists of two parallel walls, six inches apart, a foot high and a yard or more long, which arch inwards so as to form a roofless tunnel which opens at each end on to a platform. This is what is misleadingly called the *bower*. It is always aligned remarkably accurately north and south, probably so as to ensure correct illumination. All around this structure the ground is scattered with white or brightly coloured objects, such as fragments of beetle elytra, shells, feathers, berries of brilliant hues, flowers, and young green leaves. The Satin Bowerbird *Ptilonorhynchus* favours blue objects. Near houses fragments of pottery, paper and card are also used in this decoration. Some males paint the twigs of their bowers with the coloured juice of berries which they hold in their bills, a use of 'tools' which is rare among birds. The paint is repeatedly renewed when necessary, for example after it has been washed away by rain. This structure is the scene of the sexual displays, to which the male attracts a femalc before he devotes himself to showing off within and around his bower. He takes up varied attitudes and dashes hither and thither, suddenly raising and lowering the tuft of brilliant pink feathers, which in the Great Bower Bird *C. nuchalis* is normally concealed under the dull feathers of the hind-neck, like a will-o'-the-wisp darting about the gloom of the undergrowth. Every now and then he stops and picks up in his bill one of the decorative objects scattered round the bower, as though showing it to the female for her admiration.

The displays of the gardener bowerbirds (*Amblyornis*) are of the same type, though their bowers are much more complex. Around a young tree a male piles interlaced branches to form a central pillar, on which he arranges long sticks so as to form a 'roof', with a 'doorway' to one side. This structure may be 1m high and 1·5m across at its base. The floor and the surrounding platform are cleared of all rubbish, and then decorated with flowers and brightly coloured fruits which are replaced when they fade, pebbles and a variety of other objects. The shape of the hut varies from species to species, that of the Vogelkop Gardener Bowerbird *A. inornatus* being conical whereas MacGregor's Gardener Bowerbird *A. macgregoriae* builds one which is saucer-shaped with a central column supported by a tree-trunk. These constructions belong to males which tend them daily and keep near them for at least six months in the year, using them season after season (Marshall).

The displays of Archbold's Bowerbird *Archboldia papuensis* are very different. The males prepares a kind of cushion of ferns and lianes decorated with shells and coloured pieces of plants, and by calling attract a female to perch in a nearby bush. On becoming aware of her he adopts an infantile posture, flattening himself on the bed of ferns, holding a twig in his bill, and flapping his wings like a nestling waiting to be fed. He crawls towards the female, and follows her in this attitude wherever she moves around the bower. This submissive posture is all the more curious, since the infantilism which it expresses is usually a female characteristic during display (Gilliard 1959).

The displays of bowerbirds are the most elaborate in the avian world, involving as they do the building of specialized structures which serve no function but display, and are entirely distinct from the nest. The latter is built afterwards in a tree by the female – who cares for the young alone, and later leads them to the bower, where the male again displays before her (behaviour which no doubt serves in imprinting and learning). The management of the display ground shows an aesthetic sense since the effect is not produced by chance, the male choosing the objects to be arranged on the platform and rejecting those which fail to please him. The Satin Bowerbird *Ptilonorhynchus violaceus* seems to have a strong preference for the colour blue. The care with which males continue to maintain their decorations is also most remarkable.

Arena displays

Another way in which sexual displays may be complicated is by becoming communal. At the beginning of the breeding season birds which follow this pattern gather in flocks, usually at the samc place year after year even if the habitat is altered by man. This arena, a group territory which all members of the party defend against intruders, contains the individual display grounds of all the males, known as *leks* – a word used more particularly for the arenas of Black Grouse. These are separate from the mating territories, to which the pairs formed at the leks repair. This collective behaviour must have evolved from a primitive stage, in which a single female is courted by several males who gather round and display in order to attract her. Such behaviour is complicated by the fact that, besides the sexual bonds forged during pair-formation, the males also interact and a hierarchy becomes established between the participants, which mutually stimulate one another. Two successive stages can be distinguished in communal displays. In the first individual display territories simply become contiguous when the males gather in a particular area. In the second, individual territories are abolished and the displays become truly communal, taking place on a single site without spatial rivalries between members of the flock. Such behaviour has usually developed in species whose males are not involved in nesting, and thus have time for display. Under these conditions it is favourable to the species, a gathering of males being more attractive to the females than scattered individual displays would be (Snow 1963).

Cranes have long been noted for the dances they perform during courtship. They gather in flocks often a hundred strong and individually or together, circle round one another with resounding calls. Then they bow deeply, heads touching the ground, and leap into the air. Often they dance more or less on the spot, in a series of bounds described by those who have seen it as like a grotesque minuet, until the dancing individuals become still, and others begin in their turn. These dances are apparently of several types: a group dance with partners; a male dance which may be regarded as a sort of tournament; and a dance of a male with a female which has the character of individual sexual display.

Ruffs *Philomachus pugnax* return year after year to the same flat places on water-meadows, where each bird's place is marked by a circular area or 'hill' one or two feet in diameter. There the males gather in spring when they develop their breeding ornaments – a sort of breastplate of long plumes, coloured differently in each individual, from white to purple-black through every shade of russet. These shields, above which their long bills jut like lances, give them the look of mediaeval knights, and their behaviour suits this appearance. The combatants run back and forth in great excitement, stop short, subside on to the ground with bent legs, and rise again to jump on the spot and dance around each other. Though they seem to throw themselves at each other and confront one another in a joust, real fighting is rare and the strife is only pretence. The females (Reeves), attracted by these proceedings, settle nearby. The sexual behaviour of Ruffs clearly demonstrates the characteristics of communal display. At their communal grounds, on which each male has his own territory, they can devote themselves to very complex individual displays without entering into direct conflict. They attack by means of postures, and sometimes of vocalizations, in which each strives to outdo his rivals without trying to eliminate them by force. Finally, the females are attracted by the males' displays and attitudes of invitation, but it is they who choose their own mates.

This arena type of display occurs in other birds, with the same general characteristics. Among grouse the Capercaillie *Tetrao urogallus* displays solitarily in special places to which it attracts the females. Blackcock *Lyrurus tetrix* gather on a small lek where each male holds an individual territory. Here they stage mock battles, which sometimes degenerate into real fighting when an intruder crosses a territorial boundary. Each female, having been attracted by the males and subjected to their advances, enters the territory of one of them. Here the male revolves round her, flattening himself while spreading his tail like a lyre, with swollen wattles and drooping wings, all the while sounding his characteristic croaking and gobbling. American Sage Grouse *Centrocercus urophasianus* gather in flocks which may exceed 400 males, on leks about 800 metres long. There each male, about a dozen metres from his nearest rivals, defends a territory up to 6m across. In display he takes up various postures, spreading his tail, ruffling up his feathers, and inflating his external air-sacs. These take on the appearance of two brilliant yellow bags the size of oranges on either side of his throat, which emit a hollow booming as they deflate. The females appear in the area two or three weeks after the males and are attracted to the display arenas where mating takes place. The Greater Prairie Chicken *Tympanuchus cupido*, also a North American species, displays in the same way. Established on its individual territory, it inflates its air-sacs and dances on the spot, trampling the ground vigorously and very rapidly with a characteristic drumming. Here again the females gather round the lek before intruding on it and mating on the territory of their chosen males.

During communal display the males arrange themselves in a hierarchical order. One of them is clearly dominant, others rank near him, while the rest of the flock is clearly subordinate. In some species such as the Greater Prairie Chicken, the hierarchical structure seems fairly loose and all the males mate at similar frequencies, each on his own territory. Other species in contrast show a rigid hierarchy within each group, such that only the dominant males mate. This is especially truc of the Sage Grouse, whose matings take place in the special arena. Up to 74 per cent of the matings, in each group centred round an arena, is achieved by the master-cock. Those which follow him in the social scale are able to mate in their turn, while the clearly subordinate males have scarcely any opportunity except by chance or at the end of the breeding season. This conduct of the communal displays limits competition, so that arena display controls aggression while maintaining severe natural selection.

Cocks of the rock (*Rupicola*) of South American forests devote themselves to this kind of display, gathering at places which they have cleared of all litter. There each male performs acrobatic dances, strutting with fanned tail and wings, then frenziedly leaping in the air and influencing other males to do the same. Whydahs also have arenas made up of series of individual territories. Each male flattens the grass leaving a central tuft, and then dances by jumping on the spot, his long tail floating in the breeze, rising and falling in the air like a yo-yo. When a female wishes to join him she settles on the central tuft, and the male then dances round facing her, throat inflated and tail fanned. Mating takes place on the territory.

Manakins (Pipridae) of the neotropical forests gather in flocks during their sexual displays (Sick, Snow), which show every gradation between solitary and communal display. In many species (e.g. of *Manacus*) the display ground is common to all the participating males, who may number as many as seventy, but each keeps to its own territory with several feet between them.

Other species (especially of *Chiroxiphia*) have lost individual territories, and the displays are communal on an arena which has become common property. These displays sometimes take place on a branch, close to the ground or higher in a tree (as in the White-crowned and Golden-headed Manakins *Pipra pipra* and *P. erythrocephala*), while in some species (such as the White-bearded Manakin *Manacus manacus*) the scene is a patch of ground cleared of all the litter of a tropical forest. The displays make use of very varied postures, the throat feathers exhibited towards the front and the tail extended. The bird jumps on the spot, spreads its wings, and flies aerobatically, accompanying its movements with loud raucous calls, sometimes of a metallic creaking quality. Furthermore the wings, whose outer remiges often have strengthened quills, also produce vibrations and humming noises, serving to attract females in a closed habitat where sounds are particularly important. These displays have long been noted by travellers, who have called the manakins 'bailadores' or 'dancers'. The males occupy the arenas at dawn and spend 80 to 90 per cent of the day there, leaving only for hasty feeds.

Figure 55. Display of a Golden-headed Manakin *Pipra erythrocephala* in the presence of a female (A & B), and of several male Swallow-tailed Manakins *Chiroxiphia caudata* before a female or an immature male (C and D).

Birds of paradise also indulge freely in communal displays. The Lesser and Greater Birds of Paradise *Paradisea minor* and *P. apoda* gather in flocks of up to a score of males, always in the same partly leafless tree. There they display by trembling the long silky feathers of their flanks and drawing back their wings vertically above their backs. Then they jump vigorously with raucous calls, become calm, stand frozen, and begin the performance again.

Even the oddest of such displays are merely the culmination of certain tendencies, latent in many birds (such as Magpies) which gather together at the beginning of the springtime breeding season, which may be considered as forerunners of true communal display. Almost all birds which indulge in such displays show very marked sexual dimorphism. Furthermore, the gorgeously bedecked males take no part in building the nest and raising the young. The pair has only the briefest existence, scarcely persisting beyond the act of copulation. Among these birds polygamy and even complete sexual promiscuity are the rule. It is to be expected that such social behaviour should be related to a sex-ratio in which males clearly predominate (Armstrong). Indeed, it seems that males are both predominant and highly-sexed, in all birds with communal displays. If these bchaved like the majority of birds, the intense competition between them would be harmful to the species, and could even paradoxically result in some females remaining unmated. In contrast, competition on arenas for communal display is to some extent organized. Leks facilitate encounters between the sexes while, by selection acting through the hierarchical organization of the males, the females are sure of being fertilized by the most vigorous among them. However, this theory does not seem to be applicable to all instances of communal display.

The meaning of courtship displays

No-one can fail to be struck by the strangeness of sexual displays. Bizarre ornaments, which at first sight seem useless, play important parts in these behaviour patterns. Frills, collars, gorgets, topknots, trains, wings with remiges specialized as vibratory organs, are all brought to the forefront in display. Strangely modified feathers – those reduced to shafts, or altered into paddles, spoons, filaments, scales, or the arms of a lyre – all seem to play a part in these prenuptial ballets; as do inflatable and vividly coloured sacs like those of frigate birds and grouse, brightly coloured feet, mouth-linings, and all other naked areas.

The attitudes adopted in display are also unexpected. All at once every part of the bird seems put to some use other than its main function. Instead of flying, wings may be converted into vibrators or used as banners. Instead of steering and balancing the bird in flight, the tail may be exhibited cocked vertically or even laid over the back. Ornamental plumes may be realigned in positions opposite to their resting state, and the bill gaped towards the mate. There is a very clear relation between these postures and the kind of ornament involved, the bird adopting just such an attitude as will do justice to the specializations of its plumage, thus using its ornaments to the full in attracting attention. Vocalizations support the visual impressions: the song of passerines reaches its greatest volume during display, while other birds emit strange noises from hollow calls to creakings and shrill whistles.

Birds do not display statically but in perpetual movement, wriggling, bounding, stamping, pirouetting and rolling in the air. These frenzied movements are interrupted by rests, slow movements, trance and silence, in sharp contrast to the periods of excitement with which they alternate. It is sometimes possible to come very near a displaying bird, so that a Capercaillie in the climax of its excitement does not hear a hunter who approaches while it calls.

This excitement implies a considerable expenditure of energy, especially by the males, whose sexual activity is more intense than that of the females. During the period of sexual displays all their energies are turned towards defence of territory, threat displays, fighting with potential competitors, and especially actual sexual display. The latter may be very frequent at first, before pairing while the male is trying to attract the female, and then while the pair bonds are being strengthened. Copulation itself is very frequent, a Blackcock being able to copulate fifty-six times in forty-five minutes, while a Sage Grouse has been seen to mate with six females in seven minutes. Copulation is repeated very frequently in birds which form lasting pairs – from ten to twenty times at intervals of a few seconds in the House Sparrow – and in many species continues throughout laying. This expenditure of energy involves the males in appreciable loss of weight and in fatigue, sometimes forcing them to rest. This is why – in species whose sexual displays are most complex, as in fishes, reptiles and other vertebrates – very intense sexual activity is followed by the males ignoring their young.

Despite their apparent uselessness, sexual displays are of the greatest importance in the reproduction of birds. As Konrad Lorenz and many other ethologists have shown, sexual displays act as releasers. Every kind of behaviour which enters into their ritual plays its part as a stimulus, the sum of which controls the sexual cycle. At the beginning of the breeding season, the sexual drive of the male is generally stronger and more advanced than that of the females. He sets himself up in a particular place, defines his territory and defends it against rivals, and then attracts a female by his song and posturing. At this stage his sexual displays are aimed at forming a pair, and each succeeding stage of reproduction similarly requires its appropriate displays. Although the male is already in reproductive condition this is not true of the female, which is neither psychologically nor physiologically ready to mate. To any animal, a member of its own species is in the first instance an enemy, so that the female must be led progressively to accept the male. Her ovaries must also mature, and in some birds an isolated female cannot ovulate. Thus a solitary hen pigeon will not lay, but the sight of a displaying male kept in a nearby cage is enough to release laying, even though direct contact between them is prevented. Thus sexual displays act to co-ordinate the sexual cycles of the pair, establishing a kind of dialogue between the mated birds. Except in a few birds such as painted snipe with reversed sexual roles, the initiative remains with the male, whose postures and vocalizations release specific responses by the female – without which the male cannot continue his displays and reach the next stage. Thanks to this sequence of alternating stimuli and responses the female gradually attains the same state as her mate in a crescendo which is seen in the whole mating behaviour of every species. Sometimes the union of a pair is brief, pairing lasting only a short while after copulation. In other species it persists for the whole breeding season, especially while the eggs are incubated and the brood raised. Then displays continue in various ways, serving to maintain the pair bond, and take their place alongside the other kinds of behaviour which begin at the hatching of the young.

Sexual displays, consisting of the whole combination of attitudes and actions, are characteristic, since they serve as specific releasers. Only under exceptional circumstances will a male form a lasting pair with a female of a different species, since only a conspecific mate understands the auditory and visual signals which he transmits and can respond to them. Thus this ethological specificity is as effective as chromosomal incompatibility in preventing hybridization, and in birds as in other animals ethological species barriers are of great importance. However, behaviour patterns may be characteristic of a group of species (as has been shown for ducks), and have been successfully used by biologists in their attempts to establish phylogenies.

It is difficult to estimate the part played by natural selection, but it is clear that the most richly ornamented males with the most highly developed displays have the best chance of mating and reproducing themselves. This type of selection can lead to a progressive exaggeration of characters known as hypertely.

Types of pair bond

Every kind of bond between male and female is to be found among birds. Real sexual promiscuity is rare, being shown only by certain grouse (prairie chickens and Ruffed Grouse), birds of paradise and cuckoos, and by Ruffs. In these the union of the sexes lasts only during mating, which takes place close to the display ground, while except for this brief time the sexes live completely apart and only the female concerns herself with the young.

Other birds are polygamous – mainly polygynous, with one male attended by a harem of females. Among the ostriches and rheas the male gathers together several females which all lay in one 'nest', the male alone brooding and caring for the young. Polygamy is also characteristic of pheasants and of some icterids (such as the Red-winged Blackbird *Agelaius phoeniceus*) and wrens. This kind of polygamy is explicable in terms of better exploitation of the environment, especially where favourable areas are small. Then a female has a better chance of raising her brood within the territory of a dominant male established in favourable habitat, even though she has to share this with other females, than she would as the sole mate of a subordinate male in a marginal territory (Verner). In the Wren polygyny seems to depend on the nature of the habitat and its richness in foodstuffs: on St Kilda, a small island off the Hebrides where food is scarce, both parents are needed to forage for the brood; whereas in England where conditions are more favourable the Wren is polygynous. The condition of polyandry is rarer, being found only where the sexual roles are reversed. This is true of the button-quails (*Turnix*) – in which the female displays, sings and mates with several males – and of jacanas, the painted snipe, phalaropes and exceptionally of some cuckoos and cowbirds, birds whose reproductive behaviour is profoundly modified.

However, most birds are monogamous. In many species, especially among the passerines, the pair bond only lasts one season. The males return to the same territories year after year, whereas the females show this tendency to a lesser degree – so that they may mate with the same individual for several successive years, but only irregularly and because of the attachment to a territory. Sometimes mates may even change between the first and second broods in a single season, because one of them (usually the male) continues to care for the young after they fledge, while the other is immediately freed from responsibility. In contrast, the pairs of other birds last for much longer than a year. In the case of the Yellow-eyed Penguin *Megadyptes antipodes* 55 per cent of matings are with the same partner from year to year, and such a union can continue for eleven years (Richdale). The same is true of many other seabirds, especially petrels and albatrosses, and no doubt also of many sedentary passerines. Some birds no doubt mate for life, as has been suggested (though without formal proof) for large raptors, cranes, geese, petrels and albatrosses, and may also be true of parrots and jackdaws. It has even been said that if one of the pair should die its mate does not pair again; but this seems to be untrue, since the disappearance of a mate is usually followed by its immediate replacement.

Chapter 12

Nest Building

NESTS are clearly the product of long evolution and improvement, and almost all birds devote themselves very actively to the care of their young.

For most birds the construction of a nest for their eggs and young is absolutely necessary, since oviparity exposes the egg and embryo to hostile environmental factors throughout incubation. Furthermore, nidicolous birds (and especially most passerines) hatch naked and defenceless, in an almost embryonic condition, so that effective thermal insulation is indispensable to them. A nest is the only type of construction which could support eggs and young in the treetops, and so allow birds to use their arboreal tendencies in protecting their broods from terrestrial predators. During evolution birds' nests have been profoundly modified and highly perfected, so that they are very closely adapted to the biology of the species and especially to its habitat and the needs of the young. Like other biological characteristics, the nest expresses the precise adaptation of various species to their own environments.

The evolution of nests

Nest building has indisputably evolved under the pressure of ecological factors. During their early differentiation, birds must still have deposited their eggs like their reptilian ancestors, merely burying them in the soil or among decomposing matter. The megapodes or incubator birds have retained or secondarily reverted to this obviously primitive mode of nesting – as has the Egyptian Plover *Pluvianus aegyptius*, which covers its eggs with sand by day and incubates them only at night. However, attainment of accurate thermoregulation must have freed birds from their dependance on external sources of heat such as the warmth of the soil or the fermentation of vegetable rubbish. True incubation, in which the eggs are warmed by contact with the parent's body, was certainly a very beneficial development. Apart from its other advantages, the fact that the parents remain on the nest enables them to defend the eggs from predators. Some ornithologists even believe that incubation originated as an adaptation by which birds began to cover the eggs with their bodies, in order to hide them from carnivores. This habit has certainly contributed to the development of incubatory behaviour, but cannot have been the sole determinating agent, since regulated warming is too obviously an advantage not to have played the decisive part in the evolution of this behaviour. Within the present avifauna, the most highly developed nests belong to the most delicate species, or at any rate to those of small size which have the most rigorous thermal requirements. Such birds need more effective thermal insulation, and thus nest walls which are thicker and of finer weave, than hardier and larger species.

Many birds are content with a sketchy nest, often laid on the ground or on a rough platform among rocks. This is true of nidifugous species, whose young hatch in an advanced stage of development and leave the nest at or soon after hatching. Some birds seek shelter in cavities, either natural or specially hollowed out, while others build complicated nests by piling up and interweaving material to provide thermal insulation. The long stay in the nest seriously exposes the young to predators and birds therefore have to position their nests where it will be difficult or impossible for carnivores to get at them. They hide them in holes or perched on the flanks of cliffs, but above all in trees because of their own arboreal tendencies. The evolution of nest building among birds can be explained by this double selection pressure, towards thermal insulation and protection against enemies. However, such evolution need not always be in the direction of greater complexity. As with all biological phenomena, regressions towards a simpler state are sometimes produced by ecological conditions and inter-specific competition, which force the birds into new adaptations.

The building of highly elaborate nests involves birds in considerable work, and thus in expenditure of energy. Since animals in general economize their energy as much as possible, the evolution of complex behaviour involved in the nesting of many birds proves that the effort of building these structures must have been highly favoured by selection (Makatsch 1950, Collias 1964). This implies that the mode of nesting is inherited. In fact this is so among existing birds – and within such narrow limits for most species that the external appearance of the nest usually reveals the identity of the owner. Nevertheless, birds with relatively wide ecological tolerances retain some flexibility in this respect. For example the South American Bigua Cormorant *Phalacrocorax olivaceus*, which ranges from the seashore to the high Andes, builds its simple platform nest according to circumstances, either on the ground or in a tree. Variation in the form and location of the nest depends essentially on the biological needs of each species. Choice of site and type of nest are among the characteristics of a species, and contribute as much to the definition of its ecological niche as do its diet and other biological properties.

Different types of nest

The major features of the evolution of nest building, discussed above, allow types of nest to be classified more or less by their characteristics, though the variety of nests (reflecting birds' adaptation to very diverse habitats and ways of life) makes this a difficult undertaking. A quick survey of some types of nest will show this.

1 *Rudimentary nests:* We can assume that early birds merely laid their eggs on the ground, where they brooded them in order to hide them from predators and to provide the warmth necessary for the embryos to grow. Some birds have retained or reverted to this type of nesting, such as ratites, penguins, some gannets and cormorants, many game-birds and waders, most gulls, terns, bustards and nightjars. Some merely gather a few stones or some plant remains, while slightly hollowing the ground into a cup, while others do not take even this trouble. Guillemots lay on narrow rocky ledges without any trace of a nest. Their eggs are prevented from rolling off and falling down the cliff by their pyriform shape which causes them to roll in a tight circle, apparently an adaptation to this way of nesting. The Guanay Cormorant *Phalacrocorax bougainvillei* and the Peruvian Booby *Sula variegata* of the Peruvian and Chilean coasts lay their eggs on the bare soil of the desert islets where their colonies are established, and defaecate as they sit on the clutches. The droppings, accumulating in a circle, build up little by little until they form a kind of nest of sun-dried excrement, which is the guano produced mainly by these species. The Emperor Penguin incubates without any nest at all, balancing the egg on its feet and covering it with a fold of abdominal skin, which is a true incubatorium unique among birds.

Such nominal nests are none the less fiercely defended and are treated as real territories, some penguins especially indulging in violent fights for possession of the few stones which comprise their 'nests'.

2 *Nests in natural holes:* These are still rudimentary nests, merely situated in a fissure in soil or rock or between large stones, like those of some petrels, the Razorbill, tropic-birds and a few raptors. Humboldt Penguins freely nest in sea caves, but when the layers of guano used to become thick they were said to make their nests in cavities hollowed out of this material, a situation unique among birds.

3 *Burrow nests:* A more highly perfected type of nest has been developed from the preceding one, the bird using a natural hole but improving it for

habitation. Kiwis for example take advantage of cavities in the soil or between roots, and excavate the earth to make them larger. This leads on to a much more highly evolved type, the bird digging directly in loose ground to make a shaft or burrow which may be several metres long. Such burrows usually open on to soft steep slopes such as river banks, though some slant into level ground. This type of nesting has been adopted by many petrels and shearwaters, puffins, the Burrowing Owl *Speotyto cunicularia*, todies, bee-eaters, kingfishers, some parrots, and a number of passerines such as Sand Martins and a group of ovenbirds (*Geositta* and *Upucerthia*). Using its bill as a pick, the bird digs into loose earth and shovels it out with its feet. The burrow is usually a narrow slanting or winding shaft, ending in a more or less enlarged incubation chamber which shelters the brood. The eggs are laid on the bare soil, or on a mat of vegetable matter which forms an insulating layer. Such a nest is a very effective shelter against predators and environmental extremes. The temperature within remains remarkably constant despite outside fluctuations, and it is not surprising that many birds of high mountains (notably the Andes) have adopted this way of nesting.

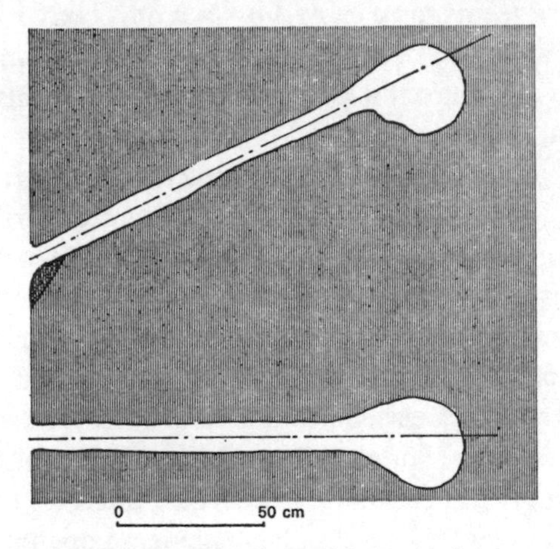

Figure 56. Sections (vertical above, horizontal below) of the nest of an Andean Flicker, *Colaptes rupicola puna*, a terrestrial woodpecker of the Andes, showing the nesting chamber at the end of a tunnel.

4 *Nests in hollow trees:* A rather similar line of evolution has led birds to nest in tree cavities; trunks attacked by fungi rot locally, to produce hollows with narrow openings. The bird has only to improve such a cavity slightly to produce an effective shelter against extremes, and many forest birds have adopted this way of nesting: most owls and parrots, some pigeons, the rollers, hoopoes, hornbills, toucans and trogons, some ducks (especially

mergansers and wood ducks), and quite a number of passerines, such as flycatchers, tyrant flycatchers and starlings. Some better-equipped birds, such as tits and treecreepers and tropical barbets, themselves excavate the holes which form their nests from the rotten parts of trunks. Such nests seldom contain any inner lining. Woodpecker eggs merely lie on a layer of chips, whereas treecreepers, tits and starlings arrange an insulating mat at the bottom of the cavity by piling up vegetable matter.

Treecreepers restrict the openings to their nests by plastering them up with mud, a practice carried much further by most hornbills, which imprison the female on the nest. Both male and female begin to plaster up the entrance to the selected hole, and then the female settles inside and from within helps her mate to complete the closure. She can poke the tip of her bill through a narrow slit, to take food brought to her by the male. He feeds her throughout laying and incubation (which last for an average of twenty-eight days), and continues unaided to feed her and the young for a further three weeks. After this the female breaks down the partition, frees herself and helps the male to forage for the brood, while the young sometimes repair the barrier behind her. Her departure seems to be caused by lack of room within the nest, and by the increasing appetites of the young which require both parents to feed them. This curious nesting habit is probably a defence against nest predators, especially arboreal snakes.

The indisputable advantage of hole nesting is reflected in the success rate of the broods. Statistics on the birds of the temperate North America (Nice 1957) show that only half of 22,000 eggs counted in open nests hatched, as against two-thirds of 94,000 eggs in hole nests. The lower rate of loss is explicable in terms of obviously less intense predation and much more effective protection against unfavourable physical factors of the environment. Apart from anything else, the maintenance of a very constant temperature is advantageous to the young before their thermoregulation is perfected.

All birds which lay in holes have white eggs. Such an absence of cryptic coloration would obviously be a disadvantage in an open nest, where visual selection has been effective, whereas economy has been the factor determining the colour of eggs in holes. White may even be directly advantageous in such a situation, by helping the brooding parent to see the eggs in the gloom of the nest.

Since sites are limited, there is intense competition between and within the species which use holes. The nest site is more important than the area of the territory, so that the position of the nest is occupied first and the defended area defined later, instead of the other way round as in birds nesting in the open. Defence of the nest involves frequent fights, especially by the Pied

Flycatcher *Ficedula hypoleuca* which wards off not merely members of its own species but also other birds such as tits. It has been estimated that Great Tits fall victim more often to flycatchers than to predatory raptors. In turn the Wryneck *Jynx torquilla* ousts the flycatchers – entering a nest and throwing the whole contents out through the entrance hole – while the Starling is another formidable competitor. Thus the number of holes suitable for sheltering nests is an important limiting factor, and populations of flycatchers and tits increased sevenfold and more when suitable nestboxes were provided (Haartman 1957).

5 *Nests built on the ground:* Birds arrange a cup in which to lay the eggs. The arrangement of twigs and other vegetable matter constitutes the beginning of a nest, which has been successively perfected. From a simple platform it has evolved into a more and more perfect cup, carpeted with finely-woven materials providing good thermal protection. Cranes, divers, gulls and albatrosses gather grass, seaweed or reeds to form rough nests, usually unprotected, in the form of flattened or conical platforms, topped by hollows in which the eggs lie. Other birds such as harriers and rails hide their nests in vegetation. In order to protect the eggs better against cold, the inside of the nest is usually made of finer material – especially the down and feathers which some birds such as ducks and geese pluck from their own bodies. Such cups are much more highly perfected by certain ground-nesting passerines, whose nidicolous young demand more efficient protection. Ground-nesting larks and finches build very complicated nests, surrounded externally with coarse material and lined within by twigs, moss, horsehair and wool.

These nests are always camouflaged with great care. Saharan larks (*Ammomanes*, *Galerida*) stick their nests against stones in such a way that the cups are shaded during the hottest part of the day, while Andean finches hide theirs in tufts of grass which make them completely invisible. Nevertheless ground nests are so much easier for predators to reach that nestling mortality in them is high, and the parent birds take special care of them: camouflaging the nest to the best of their ability, laying eggs which blend with the background, and approaching the nest with infinite precautions. It is among such birds that 'wounded bird' distraction displays are most highly developed.

Some aquatic birds have ceased to nest on solid ground, and build nesting platforms which float on the quaking soil of marshes, or even on the water to which they are driven by their ecological preferences and search for security. Some make floating nests by piling up reeds and other water plants – the platform being simply moored to growing stems, on which it can slide so as to follow variations in water level and avoid being swamped. Some coots build huge constructions in shallow water, such as the platforms of the Giant Coot *Fulica gigantea* of high Andean plateaux, which are several metres across and can support the weight of a man. These are used from year to year, and constantly maintained so as to make good the losses caused by decomposition at the bottom of the structure.

6 *Nests among branches:* Many birds have put their locomotor abilities to use in choosing nesting sites which are inaccessible to predators. Able to build platforms or more elaborate constructions, they are well fitted to take advantage of ledges and especially of trees, where their nests are safe from terrestrial predators. A recent example of this evolutionary step was provided by the Tooth-billed Pigeon *Didunculus strigirostris*, a largely terrestrial Samoan pigeon, which abandoned its habit of nesting on the ground for the trees after the human introduction of predators (especially pigs). Furthermore differences in nesting sites within species have been noted which agree with the presence or absence of carnivores. Arboreal nesting is very widely adopted by birds whose young are nidicolous, and therefore subject to nest predation for a long time. This is especially true among the passerines, which certainly evolved as an arboreal group. Those passerine species which nest on the ground have probably returned secondarily to the terrestrial state.

Some birds, such as certain raptors and ibis and many pigeons, nest on narrow ledges of rocky cliffs, their nests often shaded by overhanging slabs. However, most nest in trees, where they build more or less complex structures at various heights above the ground. This type of site demands that the nest should provide a strong platform firmly attached to the branches, so that the eggs and young do not fall to the ground. Nevertheless, some birds nest in trees without making any nest. A few nightjars balance their single eggs on branches and brood them by stretching themselves out along the boughs, their cryptic plumage making them difficult to distinguish from the bark. Similarly the White Tern *Gygis alba* sticks its egg to a branch with a little mucus, so that it remains fast although the tree is shaken by the wind. However, these are exceptional cases, and all other birds which nest in trees build real constructions.

The nest is usually built of vegetable materials collected locally. Raptors and herons make rough platforms from interlaced branches, on which they put material which is scarcely finer. Such nests are often of enormous size, especially the eyries of raptors which they use from year to year, piling on new material each season in order to remake the platform. The nest of a Bald Eagle was 4m high by 3m across, and weighed more than two tons (Herrick). In contrast, passerines and some others such as hummingbirds build much better-finished nests. Still somewhat primitive among the crows, the nest is perfected until among small passerines such as finches and warblers it becomes a remarkable assemblage of interlaced twigs, enclosing a deep cup lined with very fine material such as leaf debris, vegetable threads, moss, horsehair, feathers and wool, sometimes consolidated with spider web. The outside of the nest is decorated with bits of lichens and other vegetation so as to blend it into the background, and the nest is often wedged into a fork or among epiphytes. The Tailor-Bird *Orthotomus atrigularis* of tropical Asia gets its name from its method of building its nest in a large leaf, whose edges it has brought together by sewing them with a strong thread. It then piles material into the cone thus formed and arranges it into a cup, which is admirably protected and camouflaged inside a leaf which remains green.

Provision of a deep cup lined with a thick layer of heat-insulating material is certainly related to the pursuit of an especially favourable thermal environment. It is noteworthy that the nests of small birds are the best protected against cooling so that the young develop most rapidly, benefiting from the heat of their parents' bodies. Furthermore, birds in cold climates build nests with thicker walls than representatives of their kind in warm countries. Some passerines even go much further and build entirely closed nests, an arrangement which guarantees effective protection against environmental extremes, cooling and predators. This type of nest probably represents an improvement on the original cup, since it is said that some birds begin by building an open nest and then cover it with a roof. Magpies build a crude cup as a base over which they arrange interwoven thorny branches, and the resulting loose dome serves as protection from plunderers which would otherwise be tempted by a nest exposed to the sky at the top of a tall tree. This trend is further developed in those birds which build their incubation chamber within a huge mass of vegetation, opening through a side entrance, like the Hammerhead *Scopus umbretta*, the only wading bird which does not nest on a simple platform. The Creamy-breasted Canastero *Asthenes dorbignyi*, an ovenbird of the Peruvian high Andes, wedges into a tree an enormous mass of interlaced branches up to 50cm in diameter, out of all proportion to this sparrow-sized bird, in the middle of which is an incubation chamber reached through a lateral passage and carpeted with fine material. Other birds nest in pouch-shaped nests, entirely closed but for narrow openings, which are hung by fine strong threads from the tips of branches, and so protected against most predators. Sunbirds build this type of nest from twigs, leaves, spider webs, feathers and vegetable fluff. The Penduline Tit *Remiz pendulinus* forms its pouch from a dense felt made up of vegetable fluff taken from willows, poplars or reeds, supported by a web of long fibres and spider webs. Weavers do actually weave their nests from long vegetable fibres which they collect from palms and banana trees. Seizing the tip of a fibre, the bird lets itself fall with all its weight, so as to pull the strand away from the rest of the leaf and finally from the stem. Then it crosses the fibres on one another, plaiting them so as to form a sort of fine basketwork. In some species which nest in low vegetation the nest forms a rough ball with a side entrance, while in those which nest in trees high above the ground the nest is conical and opens downwards by an entrance spout which may be more than 25cm long, giving access to an incubation chamber in the top of the nest. American caciques build nests up to 80cm long on the same plan, and unlike most nests these are very conspicuous at a distance because of their exposed situation. The safety of the brood is ensured by their inaccessibility and the arrangement of the nest, which opens only through a long narrow spout directed downwards.

It is noteworthy that covered and closed nests are especially widely distributed in the tropics. This may be partly an adaptation against predators, since snakes and other arboreal reptiles are particularly common in hot climates, but has to be considered primarily as a better protection against physical environmental factors. Closed nests are remarkably well protected against rain, and against sunshine whose rays are dangerous to the young. The interior of the nest is always cooler than the outside, where temperature differences may reach 8° C. Since the climate is thermally more favourable in hot than in cold or temperate countries, the birds there do not have to brood continuously. They freely leave their eggs for long periods, since the temperature within the nest remains relatively constant and free from extremes thanks to the effective thermal insulation provided by its walls.

7 *Plastered nests:* Other birds have become plasterers, using earth and mud as the raw materials for the walls of their nests. Sometimes these materials are only an addition strengthening a structure of vegetable matter, as in the nests of thrushes, but they may form the principal component. The best-known example of this is provided by the swallows, which gather earth and mix it with saliva to form a very binding mortar with which they form cup-shaped nests plastered to walls. American Cliff Swallows *Petrochelidon pyrrhonota* even make almost entirely closed mud nests, like flowerpots glued to walls. The Red Oven bird *Furnarius rufus*, characteristic of the Argentinian pampas, builds a huge nest in the shape of a spherical furnace, from which it gets its name. A side entrance opens into an outer chamber, shut off at the bottom by an incomplete partition which encloses the brood chamber. Although the bird weighs only 75g the nest may be 5kg in weight, which gives some idea of the work involved in building it. A still greater disproportion is shown by the rock nuthatches such as the central Asian *Sitta neumayer*,

which builds enormous constructions cemented into crevices in the rocks: a bird of about 40g may accumulate 4kg of soil.

The same habit of mixing saliva with earth to form a binding mortar is found among swifts, whose salivary glands grow large and secrete copiously in the nesting season. This tendency is most marked in the swiftlets (*Collocalia*) of tropical Asia, the greater part of whose nests is formed from dried saliva, which sets like thick mucus. Some species mix in a variable proportion of feathers, twigs and algal fragments, whereas others use nothing but solidified saliva – a unique example of a bird 'secreting' its own nest. Swiftlets' nests are little cups about ten centimetres across, fixed to rocky walls, especially in sea-caves.

8 *Communal nests:* The available food supply, and especially the suitable nesting sites, for some species may be concentrated within a restricted area, so that the birds are bound to nest in close proximity to one another. This is true of swallows, whose mud nests are sometimes even built touching one another, and of several starlings. Colonies of the African Wattled Starling *Creatophora cinerea* comprising 400 nests have been built in single acacias (Liversidge). Similarly weavers establish their colonies in trees where dozens of nests may be found side by side. This tendency has led to the building of communal nests belonging to several pairs, each of which has its own chamber within the mass. Though this way of nesting allows economy of nesting materials, and hence of energy, it can be shown only by gregarious species, since (although each couple has its own nest) its territory is exceedingly reduced, and communal nesting implies the constant close proximity of other birds and very limited independence for each pair. The Palmchat *Dulus dominicus* of Hispaniola builds such nests under the leaves of palm trees, about thirty pairs combining to build each nest. However, the largest of such communal nests are those of the Sociable Weaver *Philetarius socius* of South Africa. Made of dried grass arranged like a thatched roof, one of these forms a kind of dome beneath which open as many as a hundred brood chambers. The nest may be 5m across, and when several are side by side they entirely fill the tree chosen by the colony. All the birds combine to build the roof, after which each pair makes its own chamber. Such a method of nesting gives participating birds effective protection against environmental extremes and predators.

Nesting associations

Whereas some birds nest in colonies made up of individuals of the same species, others associate with members of different species during the nesting season. For example, the eyries of large predators, made up of roughly interlaced branches, sometimes provide nesting sites for small passerines. Such small birds have nothing to fear from predators of the size of eagles. On the contrary, the large raptors deter both the smaller ones which would prey on the passerines, and nest-robbers such as crows. The enormous nest of the Hammerhead, similarly made up of a heap of branches up to 4m in diameter, is regularly used as a nesting site by many small birds, which build between the nest materials of the wading bird. This habit has given rise to the legend that the Hammerhead employs servants, forcing the small birds to build its nest for it.

Some nests are hollowed out of the tough material which forms arboreal termitaria, a habit which is found in all tropical areas but is especially common in Australia and America. The distribution of the Orange-fronted Parakeet *Aratinga canicularis* across Mexico and central America corresponds to that of the termite *Eutermis nigriceps*, in whose piles this bird habitually makes its nest holes (Hardy). At least fifty species of parrots, king-fishers, woodpeckers, barbets and various passerines regularly nest in termitaria (Hindwood 1959). In order to get a cavity of the right size, the bird attacks the substance of the insect structure, and the insects carefully stop up the opening of their damaged galleries and continue to live in the intact part, without there being any contact between the bird and the insects.

Particulars of nesting

The variety of nests demonstrates the remarkable adaptations of birds to their environments. The most important of the nests' many functions is to receive the eggs which the bird, as an oviparous animal, has to deposit; and to protect it effectively against predators by being inconspicuous or inaccessible, and against heat loss through its insulating properties. The importance of these various functions varies from species to species, and the variety of possible solutions explains the multiplicity of nest types, reflecting the adaptability of birds.

Despite the great differences in modes of nesting, they show many characters in common. Nesting sites are always selected with a view to the safety of the brood. There can be no doubt that tree nesting has developed under the pressure of selection by this need, and it is the same with the siting and form of the nests. These are usually most carefully camouflaged, birds never building conspicuous structures but placing them so that they merge into the background. The birds themselves try never to give away the positions of their nests by their behaviour. While they noisily make their presence known in other parts of their territories, they are remarkably discreet near their nests, which they approach furtively by hidden and indirect routes. At the approach of a potential enemy a brooding bird tends to 'freeze', so as to merge into the background with the help of its cryptic coloration. However, once its nest is discovered it immediately changes its behaviour, seeking to attract the attention of the intruder by its frenzied movements and calls, so as to lead it away from the brood.

Though the nest almost always serves solely for reproduction, a few birds do use their nests in display or as resting places. This is especially true of wrens, whose polygamous males build relatively large numbers of nests during each breeding season. A male House Wren *Troglodytes aedon* or European Wren *T. troglodytes* begins by building up to six or seven nests, each in the form of a rough sphere. By its song and posturing it tries to attract a female, which takes possession of one of the nests and completes it. Some of the nests remaining unoccupied, the male spends the night or leads his young there. The Ploceidae show similar behaviour, male weavers sometimes feverishly building several nests as roosts. Outside the breeding season, tyrant flycatchers, honeyeaters and woodpeckers very often shelter for the night in their nests, though such use is of minor importance.

Nest building is controlled by hormonal influences which are still poorly understood. The sexual hormones play a predominant part, as has been demonstrated by injecting oestrogens into Turtle Doves (Lehrman) and androgens into weavers (Collias). Once brought into breeding condition a bird may be induced to nest by external stimuli, especially sites and materials suitable for building its nest. Sometimes these influences are so strong that they result in spontaneous activity, as though the bird could not resist the attractions of suitable circumstances for nest building.

The male often chooses the position of the nest, behaviour which is in some way integrated with the definition and delimitation of his territory. Sometimes he even begins to build as part of his sexual display. However, building is often carried out co-operatively. Sometimes, as in the Rook, both collect the material but only the female actually builds. In other birds the division of labour is still more marked: only the male pigeon gathers and transports the materials, while only the female arranges them. In species such as hummingbirds, grouse and pheasants, in which the male takes no interest in the brood, the female necessarily undertakes all duties. More rarely the roles are reversed, as in frigate birds in which the female gathers material while the male arranges it, and in some shrikes and weavers (including the Baya Weaver *Ploceus philippinus*) in which the whole work of building the nest falls upon the male.

Nest building involves a series of complicated behaviour patterns, which vary very much according to the type of nest. The bill and feet are used to gather the nesting material and then to arrange it, interlacing the strands so as to produce finely wrought cups in which the fibres are woven into a fabric or basketwork. The lining of the nest, built within a framework of coarser material, is formed by the bird pressing its breast against the edges and turning itself about so as to hollow and round out the cup and make it the right size. Weavers can make very complicated knots and splices. Holding the fibres between their toes, which move independently, they pass through another which finally produces a tight knot, or else is passed under the others, which are stretched in different directions to form a neat network.

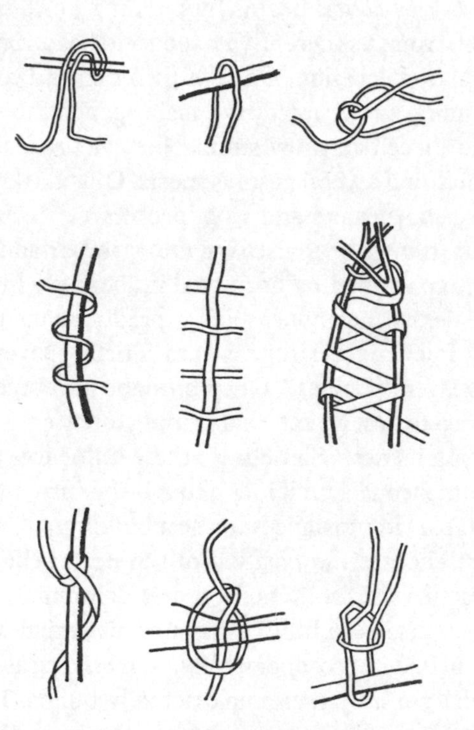

Figure 57. Various knots and bows used by weavers (Ploceidae) in building their nests.

Building nests as complicated as those of the passerines involves considerable work, and the number of components used may be huge. 2,457 feathers have been counted from the nest of a Long-tailed Tit *Aegithalos caudatus*, and 3,387 vegetable items (leaves, grasses and fibres) from that of a Lichtenstein's Oriole *Icterus gularis* (Wing). Many journeys are needed to

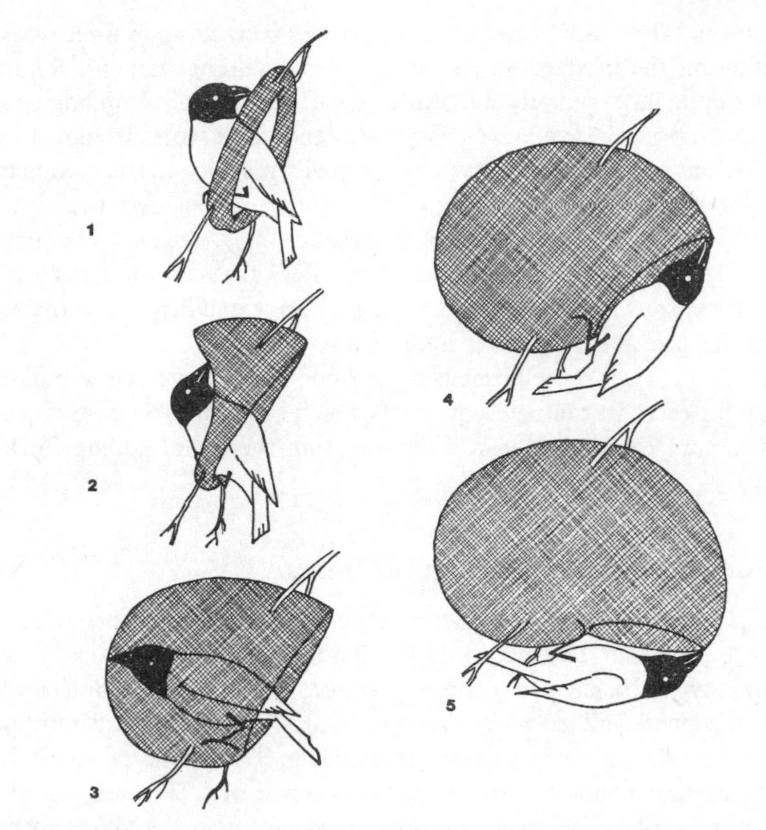

Figure 58. Successive stages in nest building by a Black-headed Weaver *Ploceus cucullatus*. 1. ring. 2. roof. 3. brood chamber. 4. porch 5. entrance.

gather so much material: 1,300 visits to the nest by a Chaffinch (Marler) and 1,200 trips to collect mud by a Swallow. Birds which bore in wood or burrow in the earth to form nesting shafts deliver an incredible number of blows with their bills in completing the task. The available amount of nesting materials may also be a limiting factor. This is strikingly true of weavers, which need enormous masses of grasses and therefore wait for the season of maximal plant productivity. Birds which plaster their nests cannot build during prolonged droughts, when they cannot gather mud. The time spent in nest building depends of course on the complexity of the structure, as well as on individual differences and the supply of material. Many small passerines take two to six days, the Penduline Tit twelve to fourteen days, the Long-tailed Tit a month, and the Swallow a week. Swiftlets take up to forty-one days to complete their nests of dried saliva. The Black-capped Chickadee *Parus atricapillus* takes eight to twenty-five days to bore its nest-hole, and the Black Woodpecker *Dryocopus martius* up to three weeks, depending on the hardness of the wood. Nest building may last for two months among large raptors, and four in the Hammerhead. Building tends to slow down from the beginning towards the end of the work. In some birds such as herons and raptors, the nest is continually rearranged and completed during incubation and the raising of the young. When nest building is accompanied by displays, it takes correspondingly longer. The male, returning to the nest with materials, often offers them to the female in a ritual manner, this behaviour forming part of the sexual display, so that the materials have an added symbolic significance.

When a pair raises several broods during one breeding season, it builds a new nest for each. In contrast some birds which build durable nests use the same structures year after year, after repairing them and adding further material.

Instinct and learning in nest building

The parts played by instinct and learning in such complicated behaviour is difficult to establish. There is no doubt that nest building is an instinctive and stereotyped behaviour pattern, elicited by various stimuli under hormonal control. Influenced by external factors, especially temperature, it is released by the sight of a suitable nesting site. The phases of actual nest building are then regulated in a wholly stereotyped way. Weavers, raised in captivity and kept for four generations from any nesting material, are not as a result any the less capable of weaving the nests characteristic of their species, when suitable vegetable fibres are put at their disposal. Thus nest building is inherited in these birds, built in to their genetic endowment.

This does not however mean that learning has no part to play. Young birds are less adept than older ones, and experience gained alongside those which have already nested helps them to improve their building techniques. Young Village Weavers *Ploceus cucullatus*, hand-reared without contact with adults, indicate a preference for green building materials rather than for similar ones coloured yellow, blue, red or black, and prefer flexible blades to stiffer ones. At a year old they can build normal nests, although more slowly and with less care than normal birds, and they gradually acquire personal experience and become as dextrous as wild birds.

Chapter 13

Eggs and Young

ONCE the nest is built the female starts to lay. All birds are of course oviparous, since this class has never, like the reptiles and fishes, evolved viviparous forms.

The structure of the egg

The egg consists essentially of a fertilized ovum, wrapped in very diverse layers which protect and nourish it. Once entirely enclosed in its hard shell, the egg has to rely on its own contents for all the ingredients necessary for the growth and development of the embryo – except for oxygen, since the shell is permeable to gases. The symmetry of the egg as a perfect solid of revolution results from the peristaltic movements of the genital tract, and its own resulting rotation within the shell-gland. The egg before laying is a complex structure, of the same plan in all birds. Embryogenesis begins before laying, but is then halted and only recommences on incubation.

The relative volumes of the components differ from species to species. The shell is sometimes very thin (as in waders, pigeons and some woodpeckers), sometimes very thick (as in francolins, where it accounts for 28 per cent of the weight of the egg). The relative volume of the yolk varies from 15 per cent in cormorants to 50 per cent in some ducks. Generally speaking, the eggs of nidicolous birds have less yolk (20 per cent) than those of nidifugous ones (35 per cent).

The egg contains many and very varied chemical substances, whose qualitative and quantitative composition, adapted to the species and the circumstances of its development, is specifically characteristic. A hen's egg consists of 65·6 per cent water, 12·1 per cent proteins, 10·5 per cent lipids, 0·9 per cent glucides and 10·9 per cent mineral salts (Romanoff & Romanoff 1949).

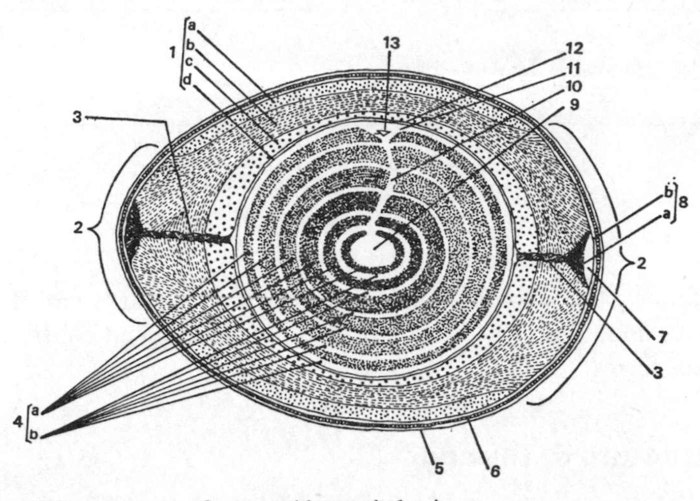

Figure 59. The structure of an egg (domestic hen).
1. albumen: 1a and 1c liquid layers. 1b dense layer. 1d chalazeal zone.
2. ligamentum albuminis
3. chalazea
4. yolk: 4a yellow layer, 4b white layer
5. cuticle
6. shell
7. air-cell
8. shell membranes: 8a egg membrane, 8b true shell membrane
9. latebra
10. neck of the latebra
11. nucleus of Pander
12. vitelline membrane
13. germinal disc.

Shapes and sizes of eggs

The most universally known eggs, especially that of the hen, have so regular a shape that they have been given the name 'ovoid'. The shapes and mean dimensions of eggs, though they differ considerably between diverse species, are very constant within each one. In pigeons, swifts, humming-birds, nightjars and swallows the difference between the two ends is reduced, and the shape of the egg approaches an ellipsoid. In contrast the eggs of raptors are decidedly larger at one end, while others such as those of the kingfishers, gulls, bee-eaters and parrots tend to be subspherical. The shape is to some extent adapted to the type of nest. The eggs of birds which nest in holes are in no danger of rolling out, and their rounded shapes are therefore no disadvantage. In contrast, the most clearly conical or pyriform eggs offer the greatest resistance to slipping or rolling, and a tendency towards this shape is especially marked in guillemots, which nest without any protective structure on narrow overhanging ledges. Their eggs are remarkably conical, pointed at one end and much enlarged at the other, and when disturbed and set in motion they do not roll in a straight line but in a small circle which keeps them on the ledge. The conical shape of other eggs allows them to be more closely packed in the nest, to form a compact global mass which is easier to incubate. Plovers arrange their eggs in a circle with the narrow ends in the middle, so that they occupy the least space, and the brooding bird can thus warm them more evenly. The shape of the eggs is also related to that of the pelvic girdle. Birds with a deep girdle (such as gulls and raptors) lay spherical eggs, while those with narrow flattened girdles (such as grebes) have elongated eggs.

Eggs vary considerably in size, both absolutely and relatively to the size of the bird. The largest eggs known are those of the elephant-birds *Aepyornis*, subfossil Madagascan ratites, which measure about 35 × 24cm and contain almost eight litres. The smallest is that of the Vervain Hummingbird *Mellisuga minima* which measures less than 10mm long. In general the relative volume of the egg is inversely proportional to the weight of the bird. Ostrich's eggs are 1·7 per cent of the bird's weight, those of sea-eagles 2·4 per cent and those of wrens 13 per cent. This relationship is confirmed more rigorously when birds belonging to the same systematic group are compared. Among the Procellariformes, the egg represents 6 per cent of the bird's weight in albatrosses (7,500g), 15 per cent in the Fulmar (700g), and 22 per cent in storm petrels (45g). Proportionally the largest egg is that of the kiwi, a hen-sized bird whose egg, at more than 400g, weighs about a quarter as much as the bird itself. Generally speaking, nidifugous birds have larger eggs than nidicolous ones of similar adult size. Thus the nidifugous crane lays eggs of 4 per cent its own weight, while this proportion is 2·8 per cent in the nidicolous eagle, though the adult birds are of much the same weight. Young nidicolous birds, hatched almost as embryos, need less nutrients for the development up to this stage than young nidifugous birds, which hatch as chicks. Birds which lay several eggs to the clutch generally have smaller eggs than those which lay only one. Even within a species, eggs in large clutches are smaller than those in small ones.

The egg represents an encumbrance and an excess weight during the later stages of its formation in the oviduct, so that this formation must be rapid if the female is not to be handicapped for too long.

Textures and colours of eggs

Most eggs are unpolished and more or less rough to the touch whereas others are shiny and appear polished, like those of woodpeckers and tinamous which resemble porcelain.

Eggs are also very variously coloured. White is perhaps the primitive colour, which has been retained by hole-nesting species and by some more primitive birds such as petrels, albatrosses, pelicans and cormorants. In contrast the eggs of many other species are very diversely coloured, by pigments which are deposited at the shell-gland into the superficial layer of the shell, or rarely into deeper layers. These pigments, which originate by breakdown of red blood corpuscles in the walls of the genital tract, belong to only two chemical groups: porphyrins, derived from haemoglobin and responsible for red, brown and black colours (ooporphyrin); and cyanins, derived from bile pigments and responsible for green and blue colours (oocyanin, which is identical with bilirubin). Chalk, humic acids from the soil, and the products of renal excretion and of blood, can also colour eggs.

Some eggs are uniformly coloured, for example, the greenish colours of cormorants' eggs, green-blue of some herons, blue of the Dunnock, brick-red of Cetti's Warbler, or black of the Chilean Tinamou *Nothoprocta perdicaria*. Others bear reddish, brown or black spots – one or more colours being represented on a given egg – contrasting with a background which varies from whitish to blue or green. The shapes of the spots and their distribution on the surface of the shell are very variable.

Every species lays eggs of a characteristic coloration, with the occasional aberrant egg which may be gigantic or dwarfed, albinistic, melanistic, cyanistic or erythristic. Apart from such abnormalities, there is variation between populations of the same species, individuals, and even eggs in the same clutch. The eggs are extremely polymorphic in some species such as certain tinamous – or the Guillemot *Uria aalge*, whose eggs vary individually from deep greenish blue to white through every tint of blue, rufous and ochraceous, with wide variation in the distribution of dark markings. It seems as though in this species the polymorphism serves as a means of recognising the individual eggs, which rest without any nests on narrow ledges where many birds are crowded together. A bird will accept a strange egg only if it is coloured like those it lays itself (Tschantz). It is more difficult to explain the dimorphism in the clutches of Moussier's Redstart *Diplootocus moussieri* whose eggs are sometimes blue and sometimes white; and of the American Eastern Bluebird *Sialia sialis* among whose normally blue eggs white ones occur at a frequency of about 10 per cent. Such dimorphism may involve nothing more than interference with the production of pigment. Finally, the polymorphism found in the eggs of cuckoos constitutes an adaptation to reproductive parasitism.

Egg colours may sometimes be interpreted as adaptations to the environment, cryptically blending with the background. Thus the colour and distribution of markings on the eggs of waders help them to merge into the soil. The Indian Yellow-wattled Lapwing *Lobipluvia malabarica* has polymorphic eggs adapted to the colour of the substrate, so that on the Malabar coast where the soil is brick-red the eggs are also of that colour (Baker). Colours of eggs are of scarcely any taxonomic importance, since closely related birds often lay very differently coloured eggs, and are much more closely associated with the ways of life and modes of nesting of the birds concerned.

Numbers of eggs and of successive clutches

Dates of laying, which are integrated with the whole sexual cycle, vary greatly according to the species and the area under consideration, so as to synchronize with the ecological optimum.

Depending on the species, a clutch may consist of one or several eggs. There is only one among Procellariiformes, some penguins and large raptors, most auks, and other species belonging to very diverse groups. There are two among divers, pigeons, hummingbirds and many penguins, three among gulls and terns, four among waders. Many passerines lay four to six eggs, while the figures are higher among geese and ducks which lay up to twelve, tits whose clutches may easily run up to sixteen, and game-birds which lay as many as twenty-two eggs. In general large birds lay fewer eggs than smaller ones belonging to the same systematic group. Birds which nest in holes have more eggs than those with open nests. The number of eggs in the clutches of some birds is remarkably constant: thus hummingbirds always lay two eggs. In contrast the number is decidedly variable in many others. Variation in clutch size is greatly affected by the major ecological factors.

In general the size of the clutch is adjusted to the number of young a pair can raise (Lack 1954). The same species may lay clutches of different sizes in different geographical areas. Thus the Common Buzzard lays three eggs in central Europe, but only two in France. Birds of cold climates generally lay larger clutches than those of warm ones, a generalization known as Hesse's rule: the Robin lays an average of 3·5 eggs per clutch in the Canaries, 4·9 in Spain, 5·8in the Netherlands, and 6·3 in Finland (Lack). Many tropical birds lay only two eggs, although close relatives lay from four to six in temperate zones. These differences are explicable chiefly in terms of the longer summer days at higher latitudes, which allow a pair to hunt for longer each day and thus to feed more of their nidicolous young.

Clutch size is certainly adjusted genetically to the losses which a species sustains in a given environment, since it represents the positive factor in population dynamics. All other factors act negatively, through environmental pressure. Among tits the sedentary species (*Parus atricapillus*, *P. palustris* and *P. cristatus*) lay clutches averaging seven to eight eggs, while the migratory ones (*P. major*, *P. caerulus* and *Aegithalos caudatus*) lay eight to ten or more, often with two clutches a year (Steinfatt). Predation too has an obvious effect, which must act through heredity, producing a higher reproductive rate in compensation where predation is especially intense.

Besides these differences between species or populations, variations can be observed which are directly linked to the immediate conditions of the environment, among which the food supply is the most important. In the Snowy Owl *Nyctea scandiaca* clutch size doubles in years when lemmings are abundant, while in the Nutcracker the mean clutch increases from three in years when conifers produce few cones to four in years of abundance. Clutch size decreases in years when meteorological conditions are unfavourable. It is always greater than the number of young the parents can normally raise, providing a margin of which the species can take advantage under favourable circumstances.

The eggs of a clutch are laid at intervals which differ from species to species: twenty-four hours in most passerines, many ducks, woodpeckers and small waders; forty-eight hours in herons, storks, pigeons, some raptors and owls, hummingbirds and kingfishers; from four to five days in large raptors; and from five to seven days in certain gannets and hornbills. Laying stops when the required number of eggs is attained, thanks to a stimulus released by the clutch itself – no doubt through sight and brooding contact. Some species continue to lay if their eggs are removed as they are laid. In this way birds have been persuaded to lay in quick succession: thirteen eggs in fifteen days by a Jackdaw (with a normal clutch of four to five); sixty-two eggs in sixty-two days by a Wryneck (normal clutch seven to ten); and seventy-one eggs in seventy-three days by a Golden-shafted Flicker *Colaptes auratus* (normal clutch six to eight). The number of eggs laid by certain birds has been increased under domestication. Thus a hen lays an average of 240 eggs a year, and sometimes as much as 300 or more (Sturkie). However, this is not possible with birds such as pigeons, gulls, and waders, in which all the oocytes ovulate simultaneously (although their growth is desynchronized), whereas in the first group the oocytes are liberated progressively as laying proceeds.

The volume or weight of a complete clutch is often large in comparison with the bird. The twelve eggs of the Firecrest represent 120 per cent of the bird's weight, those of the Spotted Crake *Porzana porzana* 125 per cent. However, this proportion is distinctly less in large birds, falling to 4·8–7·2 per cent for the two to three eggs of sea eagles.

Many birds raise only one brood a year, while others have two – the second clutch immediately following the flight of the first brood, and often containing fewer eggs than its predecessor. Some birds, such as the Mourning Dove *Zenaidura macroura*, lay up to five clutches in a breeding season. As a general rule, the same species lays fewer clutches in high than in low latitudes, since the favourable season is shorter. If the nest and the eggs which have been laid are destroyed, laying stops immediately, since in order to lay the female requires the stimulus provided by the sight of the nest. Most birds then build a new nest and lay a replacement clutch, often with a slight delay varying from four days in the Red-backed Shrike *Lanius collurio* to four weeks in the Golden Eagle. If incubation was well advanced, laying may not begin again, because of difficulties in the reproductive cycle; while birds of low fecundity and rigidly controlled breeding cycles, such as the albatrosses, do not produce replacement clutches.

Incubation

In order for the embryo to develop, the egg needs to be maintained at a high temperature, which is the reason for birds incubating their eggs, warming them by contact with their own bodies. A brooding bird keeps to the nest, and applies its ventral surface to the eggs in order to conduct its body heat to them. Some birds begin to brood from the laying of the first egg, and thus protect it from adverse environmental factors. This is especially true of grebes, divers, pelicans, gannets, herons, raptors, parrots, owls and some passerines. Nevertheless, true incubation does not necessarily begin then, for the heat given out in brooding may still be insufficient to allow normal embryogenesis. If sufficient, it results in staggered hatchings, with the same intervals as between the layings, so that the brood consists of young of different ages, and thus of very variable vigour and degrees of development. Some birds abandon the eggs which have not yet hatched when the first young have appeared. Thus parent grebes leave the nest with the two first young, abandoning the remaining four or five eggs, their instinct to feed the young here overcoming the instinct to incubate the eggs. In other birds such as owls and raptors, the competition to be fed by the parents is unfair, the furthest developed young (having lived a fortnight longer than their brothers and therefore being stronger) receiving all the food, so that the others are progressively eliminated. This is seen especially among eagles, where in many broods only one of the two young survives to complete its growth, the weaker often being eaten by the stronger. This natural regulation must be favourable to the species, since the eggs which are laid later are despite everything capable of developing and can replace the earlier ones in case of accident. Thus the chances of producing an average brood is spread over a greater number of eggs.

Other birds in contrast begin to incubate only when laying is complete, as in almost all passerines, ducks, geese and game birds. This means that the young all hatch at the same time (a brood of a dozen ducklings usually hatches within two hours) and thus at the same age. Among nidicolous passerines this situation has the advantage of giving all the young equal opportunities in their competition to be fed; while among nidifugous birds the chicks are all equally fit to follow their parents, so that the homogeneous brood behaves as a unit.

Brooding is controlled by internal hormonal factors (prolactin inhibiting ovulation and provoking incubation by the female) and by external factors such as the sight of the clutch, with here again a chain of interactions between the two groups (Lehrman 1961). Birds often abandon their eggs when disturbed, building a new nest and beginning to lay again. In general they show powerful attachment to their eggs. Penguins quarrel over eggs and later over chicks which they wish to incubate, to such an extent that the eggs may be broken during the dispute, while some unmated penguins even brood stones. Experiments have shown that substitutions are possible (Tinbergen) and that very diverse objects, such as light bulbs, golf balls and shells are brooded when placed in the nest. Oystercatchers prefer a copy, the larger the better, to their own eggs. Whereas some birds such as guillemots recognize their own eggs individually, others seem not to pay the smallest attention to the colour of the eggs. Experiments have shown that the Mourning Dove *Zenaidura macroura* will accept eggs dyed in various colours (McClure). This raises a problem in relation to cuckoos' eggs, which birds seem to recognize very accurately. It is noteworthy that passerines are concerned especially with the combination of the nest and eggs, and that a single egg outside the nest does not hold their attention and is abandoned.

A bird engaged in brooding applies its body to the eggs so as to achieve the best possible contact. Since feathers are poor conductors – their function being precisely that of retaining heat – *brood patches* develop in the ventral apterygia of a brooding bird. These are naked areas not covered by feathers, and richly vascularized. The downy feathers of the apterygia fall before the first egg is laid, the dermal capillaries extend to produce a local oedema, and the dermal muscles atrophy. The condition of such a patch is a kind of inflammation, resulting in a raised temperature favourable to the eggs. This differentiation, which regresses after incubation, is under hormonal control: administration of oestradiol to birds outside the breeding season produces brood patches. Experiments on birds from which the pituitary has been removed show that several hormones, such as oestradiol and prolactin, are actually necessary for the patches to differentiate fully (Bailey 1952). In the Sparrow and cowbirds oestradiol causes the feathers to fall, while prolactin is solely responsible for the changes in vascularization. External stimuli arising from contact with the eggs are necessary for the complete differentiation of the brood patches (Selander and various authors). However, a close correlation between the development of brood patches and brooding behaviour should not be assumed. Patches may be developed by the males of species, such as the tyrant flycatchers, in which only the female broods; while inversely among many songbirds the male broods without developing brood patches, and simply applies its ventral apterygia to the eggs.

The arrangement of the brood patches varies between species. There is only one median patch in many passerines, raptors, grebes and pigeons,

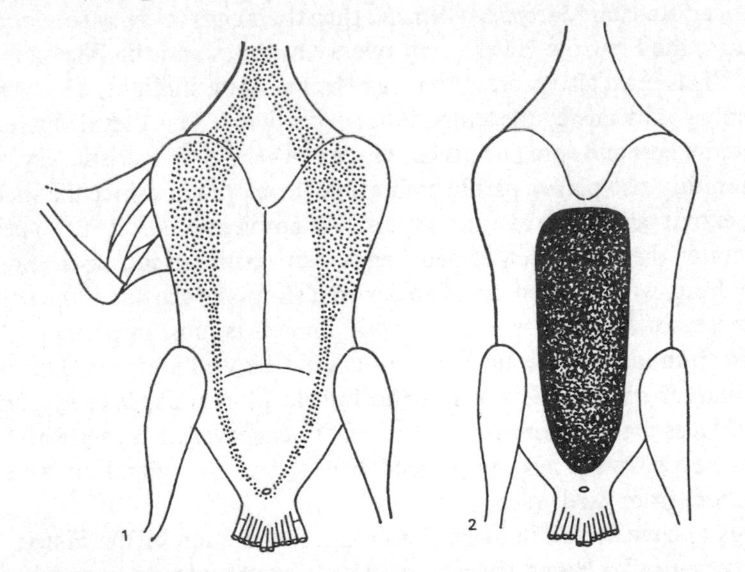

Figure 60. The underparts of a White-crowned Sparrow *Zonotrichia leucophrys* (an American finch). 1. arrangement of the pteryla. 2. position of the brood-patch (in black). Note the correspondence between the patch and the central apterygum.

while there are two symmetrical patches in auks, and three in waders, gulls and game birds (Tucker). In contrast ducks, geese, pelicans, cormorants, gannets and penguins have no patches. Ducks and geese pluck their own bellies, in order to leave part naked and also to surround their eggs with an insulating layer of down, while gannets brood by standing on their eggs, spreading over them the highly vascularized webs of their broad feet. The male Emperor Penguin balances the single egg upon his feet, enclosing it in a pouch formed by a fold of ventral skin.

The heat given off by the body of the brooding bird causes a considerable increase in the temperature of the eggs, which reaches an average of 34° C (Huggins). The surface of the egg in contact with the body is naturally warmer than the lower part – measurements on pheasants have shown temperatures of 39·5° C at the brood patch, 35·1° C at the top of the egg and 25° C at the bottom (Westerkov) – while the eggs at the outside of the clutch are cooler than those at the centre. Therefore the bird periodically alters the positions of the eggs and turns them, so as to distribute the heat throughout the whole mass. This also serves to prevent the embryonic membranes from adhering to the shell. Hens' eggs in artificial incubators which were turned only twice a day showed 58 per cent success in hatching, while those which were never turned showed only 15 per cent success, although the temperature is uniform throughout an incubator. Some passerines such as the American Redstart *Setophaga ruticilla* turn their eggs on average every eight minutes, the Sparrow Hawk every twenty minutes and the Pheasant every hour. Birds must also protect their eggs from strong sunlight, which can kill the embryos by raising the temperature excessively. This they do by remaining on the nest and spreading their wings in a very characteristic way.

Humidity also plays a part in embryonic development, since the shell is to some extent permeable to water vapour. An ambient humidity of 80 per cent accelerates the incubation of hens' eggs, but results in increased mortality at hatching, while a humidity of 40 per cent slows the growth of the embryos but increases the success of the brood. Thus it is most important that the relative humidity of the environment should be well adjusted. During hot dry weather some birds, such as the Purple Martin *Progne subis*, arrange green leaves in their nests to achieve this. Others cover their eggs with foliage when they leave the nest, to protect them from cooling and drying and to hide them from predators.

Eggs behave as poikilothermal animals. Incubation of the House Wren *Troglodytes aedon* lasts thirteen days at 35° C, but increases to eighteen days at 32·2° C because its development and metabolism slow down. Such differences explain why, in the same type of bird, incubation as well as post-natal growth lasts longer in cold than in warm climates. Thus among hummingbirds incubation lasts for eleven to fourteen days in the Ruby-throated Hummingbird *Archilochus colubris* of eastern North America, but for twenty-two to twenty-three days in the Andean Hillstar *Oreotrochilus estella* of the high Peruvian plateaus where the climate is much more rigorous. However, eggs can stand a considerable drop in temperature for a longer or shorter time, especially at the beginning of incubation. Like adults, they can resist cooling better than overheating. They commonly drop to the ambient temperature for an hour after the parents have left the nest. The eggs of the Manx Shearwater *Puffinus puffinus* can be abandoned for seven days and yet hatch successfully, though naturally they need an extended period of incubation. This resistance to cooling meets an ecological need by allowing the bird to leave its nest.

The rhythm of incubation varies widely from species to species. Penguins, woodpeckers, pigeons, trogons, hornbills, hoopooes and some passerines incubate continuously. In contrast other birds, and especially those in which only one sex undertakes incubation, occupy the nest for only 60 to 80 per cent of the day. Their attendance depends on the ambient temperature, and increases when this is low so as to avoid undue cooling of the eggs. Thus the brooding bird exercises very precise heat control. In some birds attendance at the nest is irregular, while in others it follows a well-defined rhythm. Thus in the Redstart *Phoenicurus phoenicurus* the female stays on the nest for fifteen minutes and leaves it for eight to twenty-five minutes, thus breaking up its brooding into twenty-three to thirty periods per day (Ruiter), while the female Song Sparrow *Melospiza melodia* remains on the nest for an average of 28·5 minutes and leaves for an average of 8·8 minutes, dividing its daily incubation into twenty-three. A hummingbird may leave its nest to feed 140 times a day, and a Robin a score of times. In contrast petrels can incubate night and day for two to five days, and albatrosses from six to thirty days, before they are relieved by their mates. These differences are related to very different resistances to fasting in different species. The male Emperor Penguin incubates the egg left him by the female for sixty-two to sixty-six days, fasting throughout this time and losing from 12 to 15kg of his initial mean weight of 36kg (Prévost).

Study of the division of labour of incubation between mates shows that every possible arrangement is realized (Skutch 1957). It seems that incubation by both sexes is the primitive situation, from which specialization towards one sex or the other has evolved. In many birds both mates undertake incubation. The incubation of the Red-legged Partridge *Alectoris rufa* is unique, since the female lays two complete clutches in quick succession, one of which is looked after by the male and the other by herself (Goodwin). Relief of the brooding bird may take place every twenty-four hours, as in the Sooty Tern, or at shorter or longer intervals which may extend to several days in penguins, shearwaters and albatrosses. In some birds, such as pigeons and certain trogons, the female broods by night and the male by day, or as in warblers the male may alternate with the female during the day. In other birds, especially the Ostrich, the male broods by night and the female by day. Water birds divide the task equally, but among passerines the female plays the main part, and may even be fed on the nest by her mate. In other birds only one of the mates, usually the female, incubates. This is true of course where the pair is ephemeral, as in hummingbirds and game birds, but also in species with lasting pairs such as most passerines. The female may starve while brooding as in pheasants, or leave the nest from time to time in order to feed, and is often also fed by the male. In other species it is the male who alone undertakes the incubation, as in the Emperor Penguin, kiwis, and birds with inverted sexual behaviour such as painted snipe, button quails, phalaropes and jacanas. The female Pheasant-tailed Jacana *Hydrophasianus chirurgus* has about ten clutches of four eggs each, each one in the charge of a male which later looks after the young (Hoffman). The females of a rhea harem lay twenty to fifty eggs in the same place, which the male alone tends. Similarly in certain tinamous, such as the Argentinian Brushland Tinamou *Nothoprocta cinereascens*, several females lay about ten eggs in a common nest, and the male incubates and later takes care of the young (Lancaster). Finally, the tropical South American anis (*Crotophaga*) nest communally. The birds of a flock build a common nest where the females lay up to twenty-four eggs, which several individuals incubate simultaneously or in relays, while the young are raised by all the members of the flock.

The duration of incubation varies a great deal, from eleven to twelve days in the Cowbird *Molothrus ater* to eighty-one days in the Royal Albatross *Diomedea epomophora*. Most passerines incubate for twelve to fourteen days, raptors for twenty-nine to forty-five days, and owls for twenty-six to thirty-six days. The table below sets out the incubation periods of a number of birds. Passerines show accelerated embryogenesis because of their especially fast metabolism. There is also a relation between the duration of incubation and the safety of the clutch. Birds which nest in holes incubate for one or two days longer than similar birds which use open nests. Among tropical tanagers the duration of incubation is eleven to thirteen days in those which nest close to the ground, fourteen to twenty days in those which nest higher, and seventeen to twenty-four days in those with closed nests (Skutch).

	Duration of incubation (in days)
Ostrich *Struthio camelus*	42
Emu *Dromiceius novaehollandiae*	56–60
Brown Kiwi *Apteryx australis*	75–80
Emperor Penguin *Aptenodytes forsteri*	62–66
Great Crested Grebe *Podiceps cristatus*	26–29
Cormorant *Phalacrocorax carbo*	23–25
Gannet *Sula bassana*	43–45
Manx Shearwater *Puffinus puffinus*	51–53
Yellow-nosed Albatross *Diomedea chlororhynchos*	78
Heron *Ardea cinerea*	25–26
Mallard *Anas platyrhynchos*	22–28
Grey Lag Goose *Anser anser*	27–29
Quail *Coturnix coturnix*	17–20
Partridge *Perdix perdix*	21–25
Pheasant *Phasianus colchicus*	23–25
Domestic fowl	21
Redshank *Tringa totanus*	23–25
Herring Gull *Larus argentatus*	25–27
Black-headed Gull *Larus ridibundus*	23–24
Lammergeyer *Gypaetus barbatus*	52
Common Buzzard *Buteo buteo*	28–31
Sparrow Hawk *Accipiter nisus*	35–38
Barn Owl *Tyto alba*	30–32
Eagle Owl *Bubo bubo*	33–36
Feral Pigeon *Columba livia*	17–19
Great Spotted Woodpecker *Dendrocopus major*	12–13
Silvery-cheeked Hornbill *Bycanestes brevis*	40
Hummingbirds	11–23
Swift *Apus apus*	17–22
Eastern Phoebe *Sayornis phoebe*	16
Skylark *Alauda arvensis*	11–12
Swallow *Hirundo rustica*	12–14
Pied White Wagtail *Motacilla alba*	12–14
Wren *Troglodytes troglodytes*	13–20
Robin *Erithacus rubecula*	12–14
Ovenbird *Seiurus aurocapillus*	12
Garden Warbler *Sylvia borin*	12–13
Great Tit *Parus major*	11–15
Starling *Sturnus vulgaris*	12–13
Rook *Corvus frugilegus*	16–18
Magpie *Pica pica*	17–18
House Sparrow *Passer domesticus*	12–13

Hatching

The young bird takes several hours to free itself, sometimes a day or even three or four among albatrosses and shearwaters. While within the egg the young bird develops two temporary structures which it uses in breaking the shell. The first is a hard projection, the *egg-tooth*, at the tip of the upper mandible, which pierces the shell by rubbing against its inner surface. The second is a special superficial muscle (*M. complexus*) which develops at the back of the head towards the neck, and whose contractions force the egg-tooth against the shell. The mass of this muscle, at least among grebes and ducks, is proportional to the hardness of the shell. It reaches its maximum just before hatching, and then regresses (Fisher 1958, 1966).

Eggs tend to hatch towards morning, probably as a result of a rhythm in the freeing movements of the young, which are most active at this time. Thus the newly hatched chick can take its first meal at once, without having to wait through the coldest hours for daybreak.

The parents of nidicolous young remove the shells to a distance as soon as they have hatched, so that these conspicuous fragments will not attract the attention of predators. Sometimes, as among flamingos, the young eat the shells. In contrast nidifugous birds do not take the trouble, since their nests are deserted soon after the last chick has hatched.

The incubation of megapodes

Megapodes, relatives of game-birds which are confined to the Australo-Papuan region, have a mode of nesting which is unique among birds. They neither build nests nor incubate their eggs, but trust them either to soft soil warmed by the sun, or to decaying organic matter – which releases enough heat to maintain a temperature suitable for their development.

The Maleo *Megacephalon maleo* of Celebes lives in the jungle of the interior, from which it comes down to the beaches of black volcanic sand. There it digs a slanting hole, in which the female buries an egg at a depth of about a metre before returning to the scrub, from which she emerges again to lay the next egg. The black sand absorbs enough heat to maintain the eggs at a favourable temperature, while these birds sometimes also use the warm soil near volcanic springs. The young free themselves and immediately make for the scrub. Other megapodes use the heat produced by decomposing vegetation. The Brush Turkey *Alectura lathami* of the dense forest of Australia each year heaps up a bulky mass, sometimes over 1m high and 3m across, of the decomposing vegetation scattered on the ground. In this the female lays her eggs, which are warmed by fermentation while the male keeps continuous guard. The Mallee Fowl *Leipoa ocellata* of dry areas in Australia shows still more complex behaviour. During the four months before laying the birds dig a depression in the soil, originally about 3m across and 60cm deep, and fill it with dried leaves to form a mound about 40cm above the ground. Here the female lays her eggs, and the mound is covered with a layer of sand. These mounds are used again year after year, so that their height increases progressively. Junglefowl (*Megapodius*), which are widely distributed from the Philippines to Samoa and Australia, also build mounds – which are sometimes enormous, reaching 12m across and 3m in height – of a mixture of earth and foliage or even seaweed. The eggs are laid in tunnels which drive obliquely into the substance of the mound. These mounds too are re-used annually, after reconstruction.

Megapodes obtain the heat for the incubation of their eggs from several sources: sometimes from the sun or volcanic activity, in others principally from the fermentation of vegetable matter with sun heat in addition. Those species which nest in humid areas have the advantage of a favourable and relatively constant climate, so that the temperature of the mounds is well balanced and does not need to be controlled by the birds. They have merely to maintain fermentation at a sufficient rate, so that enough heat is released throughout incubation. This is not true for the Mallee Fowl, whose habitat is far from regular in climate. The bird must continually regulate the temperature of its mound, to prevent its eggs from cooling or rising to a lethal temperature. The male, which alone controls the temperature of this natural incubator, has to tend the mound assiduously for the two months of incubation. When the heat of the sun, added to that of fermentation, warms the heap unduly, he opens it up during the chilly mornings by removing the layer of sand, and then closes it again as the sun grows warmer, having allowed the excess heat to dissipate. When the sun becomes too hot, he scrapes up thicker layers of sand to give better protection. When fermentation slows down at the end of incubation because the vegetable matter has largely decayed, he opens the heap during the hottest hours to let the sun's rays warm the eggs, and then covers it with sand again to prevent them from cooling during the night. All this represents a considerable labour, and it has been estimated that the male bird digs and shifts the materials of the mound for up to thirteen hours a day during the summer, when warming by the sun's rays is more important than that due to fermentation. The Mallee Fowl can accurately gauge the temperature of the mound, since its behaviour can be experimentally modified and even reversed by altering the natural temperature, either by warming the mass electrically or by removing vegetable matter and so cooling it. Its thermal sensitivity is so fine that it can maintain the eggs within one degree of 33° C. This sense is probably located in the mouth, since the bird tests material by holding it in its bill (Frith 1956). At hatching the young open a passage through the material of the mound, and from the moment they emerge can live without help from their parents, which ignore them.

Raising the young

After the eggs hatch the parents of all birds except a few unusual species such as the megapodes, take continuous care of their young to ensure their growth, development, and 'education'. There is a fundamental distinction in the developmental stage reached at hatching. Some birds hatch almost as miniature adults. Covered in down, with well-developed limbs and open eyes, they are very soon able to walk and follow their parents (and even to swim) very briskly. These chicks are called *nidifugous*, and belong to terrestrial and aquatic groups which are good walkers or swimmers, such as ratites, tinamous, game-birds, sand grouse, cranes, rails, bustards, waders, grebes, ducks and geese. In contrast are the birds which hatch virtually as embryos. Naked, with closed eyes and poorly developed hind limbs, these nestlings cannot leave the nest, since their sole reflex action is to lift their heads and open their bills so as to be fed by their parents. They are called *nidicolous*. With few exceptions, they belong to birds which are arboreal or which nest in holes or inaccessible sites, such as petrels, albatrosses, pelicans, gannets, cormorants, herons and egrets, diurnal and nocturnal raptors, pigeons, trogons, cuckoos, woodpeckers and other piciform birds, parrots, rollers, kingfishers, swifts, hummingbirds and all passerines.

Between these two categories there are a few intermediates. The chicks of flamingos, coots and moorhens remain in the nest for several days before they are capable of following their parents. Young penguins, gulls and terns can leave the nest almost immediately, yet for a long time they remain dependent on their parents for food, which they cannot find or ingest for themselves. They have the locomotor behaviour, but not the foraging behaviour, of nidifugous chicks. The distinction corresponds also to different ecological conditions, especially the mode of nesting.

In birds in which the union of the sexes is brief, the female alone cares for the young (in nidicolous hummingbirds as in nidifugous pheasants) while in the few species such as button quails (*Turnix*) and painted snipe in which the sexual characters are reversed, it is the male alone which does so. In other birds where both mates occupy themselves with the young (as much among nidicolous passerines as nidifugous waders), the part played by each sex is very variable. Sometimes the male only plays a minor part in feeding the young, as in the Black-whiskered Vireo *Vireo altiloquus* whose females bring 75 per cent of the food, or as in the Chiffchaff *Phylloscopus collybita* he only defends the territory. Sometimes on the other hand the male alone feeds the young, the female being engaged more particularly in brooding the nestlings (and sometimes in making ready for a second brood as in the European Robin and the House Wren *Troglodytes aedon*). The kind of care and the types of behaviour involved in raising the young vary widely between nidicolous and nidifugous birds.

1 The parents of *nidifugous* young brood them when newly hatched in order to dry and warm them. The chicks do not need to feed for one to three days, their embryonic yolk reserves not yet being wholly resorbed – domestic chicks still have 5g left at hatching. For the next few days their parents still need to warm them, especially at night, since their thermoregulation is not yet fully established. Then it adjusts itself and they have less and less need of external heat, except at night and during bad weather when their small size still makes them vulnerable.

Chicks (except those of kiwis which are like miniature adults) do not closely resemble their parents, since they are covered in down rather than feathers. Although they can walk, and those of aquatic birds can swim and dive, this means that they are incapable of flight. The adults watch over them, lead them to supplies of food, and teach them to find these for themselves and to react with the appropriate behaviour to the situations which they meet in their environment. Sometimes as in geese and swans both parents look after the young, or as in cranes each may take care of one or more chicks. Sometimes, as in certain ducks, the male takes part only at the beginning of the rearing and then loses interest. The females of polygamous species such as bustards, pheasants and grouse raise the young single-handed whereas the male does so among ostriches, casuaries, rheas, tinamous and species with reversed sexual behaviour.

The great variety of sounds at the disposal of birds allow the parents to communicate with their young, especially by alarm calls and calls advertising the finding of food, while the young in turn have contact calls and calls of distress and satisfaction, which allow the parents to keep watch on the state of their flock and to intervene in case of need. Nidifugous birds also have visual means of communication, especially devices which induce the young to follow the adults automatically. For example the tail of the moorhen held upright, black edged with pure white, stimulates the young to follow, as is clearly shown by experiments with artificial models showing the same pattern.

In case of danger the parents' alarm calls bring about a reaction by the chicks, which depending on the situation may disperse, huddle together or 'freeze' where they sit, their cryptic coloration effectively hiding them against the soil or among vegetation. Parent birds usually defend their young vigorously.

The chicks can soon feed themselves. Some birds such as grebes and bustards feed their chicks beak to beak during their first days of life. Young gulls are fed with predigested food regurgitated by their parents, and show a remarkable reflex response called into play by visual stimuli as studied in the Herring Gull (Tinbergen 1953). The bill of this species is bright yellow, marked with a red spot on the lower mandible, which acts as a specific releaser, causing the young to tap at it with its own bill – a stimulus which in turn provokes the adult to regurgitate. Models of the bill and head of a gull were made, varying in shape, colour and the contrast between the coloured patches. These showed that the frequency of responses by the chick was considerably reduced when the red spot was omitted, and to a lesser degree when it was replaced by a spot of another colour. It is largely the contrast with the colour of the bill which matters, and not the colours of the bill and head. The general shape of the head has no effect at all on the intensity or frequency of the responses, whereas the shape of the bill is very important and the greatest response is obtained with models incorporating bills which are longer than normal. It is also important that the bill should be presented fairly low, and as vertical as possible. Very strangely, by exaggerating natural characters – especially the contrast of the terminal spot – a kind of 'super-gull' can be obtained, to which the young will react more strongly than to their parents. This supernormal behaviour recalls that of Oystercatchers, which prefer large artificial eggs to their own. These releasers of feeding behaviour seem to have a specificity related to the diversity of bill coloration among gulls. Thus in experiments on the red-billed Franklin's Gull *Larus pipixcan*, a model bill of this colour produced seven times as many responses as a similar but white model, and five times as many as a green model (Collias & Collias). In the American Laughing Gull *Larus atricilla*, the stimulus to feeding is provided by the whole head and bill, of particular proportions and colours (Hailman).

2 Young *nidicolous* birds are raised in a fundamentally different way. They depend completely on their parents, and can only gape widely and take the food brought by the adults. Those of passerines and owls are both blind and deaf, those of pelicans and petrels only blind, while those of raptors and herons can both see and hear.

Young nestlings are incapable of controlling their own temperature, and their thermoregulation only develops gradually. At first they behave as true poikilotherms, while homoiothermy is established in from two to twelve days after hatching depending on the rate of development of the species. For this reason the parents continue to brood assiduously for several days – up to eight in pigeons, and longer in penguins because of the rigorous climates under which they live. In some birds this poikilothermy of the young is truly adaptive – especially in the swifts. The poikilothermy of nidicolous young also explains their slower growth in cold climates, other things (especially the length of day) being equal. Thus Andean Hillstars *Oreotrochilus estella* take thirty to forty days to fledge, whereas Mexican White-eared Hummingbirds *Hylocharis leucotis* take twenty-three to twenty-eight and western North American Rufous Hummingbirds *Selasphorous rufus* about twenty days.

The parents are entirely responsible for feeding nidicolous young, a task which sometimes falls to both sexes, and sometimes to the female alone. In the latter case the male either plays no part, or hands his prey over to the female to distribute it to the nestlings – notably among raptors and such passerines as the Song Sparrow *Melospiza melodia*. The food brought in this way by many birds, especially passerines, is loaded into the gaping bills of the nestlings. It consists largely of animal matter, such as insects (including caterpillars and other larvae) and small worms, even for species which are vegetarian when adult. This more carnivorous diet is in accordance with the demands of growth, since birds need animal proteins to form their own tissues. Thus the House Sparrow feeds its nestlings a diet consisting of 68·1 per cent of animal matter and 31·2 per cent vegetable matter, while the values for the adult diet are 3·4 per cent and 96·6 per cent respectively (Kalmbach).

Here again a series of stimuli releases the feeding reflex. The approach of an adult and its landing at the nest excite the young, which begin to wriggle, while auditory stimuli are added to these visual ones which are effective only when the eyes have opened. Sometimes the parents announce their arrival by special calls, such as the shrill notes of swallows, while the sound of the wings during a landing and the shaking of the nest when the bird settles on it have a semantic significance to the young. The darkness which falls in a closed nest when a bird is passing through the entry hole, as in woodpeckers, similarly releases the reflex in the young, causing them to gape widely, waving their heads and giving shrill calls. Among visual releasers shown by the nestlings the lining of the buccal cavity, brightly coloured yellow or orange, is among the most important. Different species have developed complicated and highly contrasting patterns with marks of various colours – red, orange, black or white – which attain their maximum complexity in the Ploceidae. Nestlings in closed nests, which are very dark within, have rounded reflective protruberances inside their mouths which show the parent bringing food where the nestlings are, thus providing a releaser which can be seen in the poor illumination of the nest. In addition the edges of the mandibles often carry pads coloured bright yellow or sometimes white. These too are strongly reflective in birds such as woodpeckers and rollers which nest in holes. A nestling which has been fed by its parent as a result shows less lively reflexes than its hungry siblings, and therefore has less chance of being fed next time, which automatically ensures a more or less fair distribution of the food.

Other species feed their young differently. Female raptors tear up the prey at the nest and distribute the shreds to the young. Some birds share out foods which are already partly digested, or at least impregnated with digestive juices. At a late stage storks and ibis leave food on the edge of the nest, from which the grown young take it. In crows, kingfishers and hummingbirds, and in the Fringillidae for the first few days, the adult thrusts its bill deeply into that of the young in order to regurgitate the contents of its alimentary canal. In contrast young pelicans and cormorants plunge their own bills and heads entirely into the throats of the adults to find partly digested fish. Regurgitation, however it is accomplished, allows feeding to be more widely spaced, since more food can be carried in the alimentary canal than in the bill.

Among pigeons, some species of *Columba*, *Melopelia* and *Zenaidura* feed their squabs on 'pigeon milk', a cheesy substance rich in proteins and fats which is secreted by their crops under the influence of pituitary hormones, and which the young seek within the throats of their parents. It is their sole food for the first five days, after which it is mixed with half-digested grain, drawn back spasmodically into the crop.

The frequency of feeding is always regular, but varies greatly from species to species. It is relatively low among species which bring a large amount of food at each visit. Thus among raptors an eagle will on the average bring prey only twice a day to its infants, while procellariiforms feed their chicks only one large meal every other night. In contrast passerines feed their nestlings very frequently, since they can carry only a small amount in their bills, while the young need a large quantity of food. The Pied Flycatcher feeds its young thirty-three times an hour – an average of 6,200 times to raise each nestling. A pair of Eastern Phoebes *Sayornis phoebe* made 8,942 visits to the nest in raising its young (Kendeigh). The Great Tit feeds its young up to sixty times an hour, while a Blue Tit has been counted visiting its nest twenty-seven times an hour. The first feeds are brought before sunrise and the last before sunset, while in general their maximum frequency is reached late in the morning (Keil). The frequency increases as the young and their need for food grow, although they eat progressively less in proportion to their weight. Feeding becomes suddenly less frequent immediately before they leave the nest, as though the parents wished to underfeed their young to persuade them to leave. This behaviour is very marked in procellariiforms, which leave their young without food for a fortnight.

The parents' labour also includes cleaning the nest, although some birds dispense with this task. This is true of trogons, of kingfishers whose tunnel nests are choked with the excrement of the young and with debris of rotten fish, and of hoopoes. In contrast the cup nests of small passerines are kept remarkably clean (with some exceptions like the European Goldfinch). While they feed, the young eject their faeces enclosed in remarkably tough mucous sacs. These are quickly seized by the parents, which swallow them or carry them away. This difference in behaviour seems to involve a question of security. Nest sanitation also prevents invasion by parasitic or commensal insects, which would be harmful to the young.

In case of danger the nestlings flatten themselves against the bottom of the nest and count on their immobility to remain unnoticed. This behaviour is triggered by many stimuli, both auditory such as the alarm calls of the adults, and optical such as the silhouette of a bird of prey. If more seriously disturbed they try to escape from the nest.

Thus care and raising of the young requires much work from the adults. The Pied Flycatcher devotes more than nineteen hours a day to these tasks in the Arctic, because of the length of the day, while a tit travels about 1,000km while raising its young. It is not surprising that parent birds lose up to 20 per cent of their weight during the breeding season.

Sometimes 'helpers' (birds other than the mates) come to the aid of the brooding pair. Thus the young of an earlier brood may take part in raising their younger siblings, while in other cases immature or non-breeding birds are involved. These 'helpers' are principally concerned in bringing food and in nest sanitation, while secondarily they protect the nestlings by giving warning of danger, and sometimes help in nest building and incubation. These individuals usually belong to the same species as the birds which they help, but some cases of interspecific associations have been recorded. Thus Blue Honeycreepers *Cyanerpes cyaneus* have been seen to help in the raising of young tanagers; American wood-warblers of various species to assist one another; and Rufous-sided Towhees *Pipilo erythrophthalmus* to feed young Northern Mockingbirds *Mimus polyglottus*. This behaviour is associated with sexual drives which cannot be satisfied normally (Skutch 1961). It contrasts strongly with the hostility which many birds show to strange young. In colonies of gregarious birds, young are often killed by adults on to whose territories they have intruded – not a rare event among gulls.

The young usually grow rapidly, at a speed inversely proportional to the duration of incubation – as though the rate of cellular multiplication which determined the latter continued throughout growth – and to the size of the bird. Thus young finches reach their adult weight in about ten days, the Kestrel in twenty days, while some large raptors have not yet reached two-thirds of this weight at the age of three months. In pelicans and many procellariiforms the young exceed the adults by up to 30 per cent in weight, but then grow thin during a period of fasting before they take flight. Nidicolous young grow faster than nidifugous ones. The rate of growth, slow at first, accelerates and then slows down again, and during the last few days the maximum weight is attained.

In most birds the young reach the adult size and weight when they are ready to fly, while a few actually exceed it: thus the adult White Pelican weighs about 10kg and the young nearly 14kg. This phenomenon is very general in the Procellariiforms – the young Wilson's Petrel *Oceanites oceanicus* weighs twice as much as the adult – and occurs also among penguins, parrots, owls, kingfishers and swallows. It is due no doubt to the accumulation of reserves with which to face shortages of food, which the young are able to do while they are flightless. These reserves are used towards the end of the period of growth, in making good the considerable expenditure of energy involved in the growth of feathers and the specialization of flight muscles. Some birds, especially procellariiforms, leave the fattened young alone for a period of fasting. Different parts of the body and its members of course grow allometrically, according to the general and specific requirements of the bird's life. Towards the end of development the feathers grow steadily, and the young achieve their definite size. Usually their plumage is very different from that of the adults, which is important in social relations, allowing the young to pass unnoticed without becoming the objects of competition and hostility from the males. This juvenile plumage only gives way to the species' definitive plumage at sexual maturity, which follows after a longer or shorter interval according to the species. Sometimes a long series of intermediate plumages intervenes, especially in some sea birds such as frigate birds and procellariiforms and in raptors.

At this stage the young often leave the nest before they can really fly, remaining perched nearby or hiding in the bushes and gradually acquiring their powers of flight, while they are still fed for a time by their parents. This care outside the nest lasts only a few days in most small passerines, but in others such as the Blackbird for several weeks, or even months in some raptors. The parents and young of some birds form a small flock which may remain together until the next breeding season, as in the tits, while in geese and swans the family group survives through the post-nuptial migration. Nidicolous young at this stage have the advantage in some ways over newly-hatched nidifugous chicks, since the parents continue to educate them. They progressively acquire experience in hunting and detailed knowledge of their habitat. Thus they become independent and often separate from their parents, especially where the latter are occupied with a new brood. They either remain in the region where they were hatched or they scatter – demonstrating post-juvenile dispersal on a grand scale, such as is well known in herons and starlings. The migratory species prepare to leave for their winter quarters, and for all the life of an independent bird has begun.

Chapter 14

Nest Parasitism

THOUGH most birds devote themselves with jealous care to building their nests and raising their young, a few entrust these tasks to others. This behaviour is what is known as *nest* or *brood parasitism*, in which a bird delegates to another, usually of a different species, the building of a nest, incubation of the eggs and raising of the young. This involves not only the loss of some behavioural patterns involved in various phases of reproduction; but also the acquisition of new ethological characters, which reflect the adaptation of the parasite to its hosts, and a high degree of reproductive specialization. Parasitism has evolved independently in several avian lines: the Anseriformes (Anatidae), Cuculiformes (Cuculidae), Piciformes (Indicatoridae), and Passeriformes (Icteridae and Ploceidae). Although in most it has attained a very advanced stage of evolution, some primitive stages do persist which show how the habit has arisen in a normally reproducing group.

Laying in already completed nests may be considered as the first stage, as when House Sparrows take possession of a Swallow's nest, Starlings drive other birds from the holes in which they are established, or birds of prey settle in the nests of other raptors or of crows. Joint nesting is no doubt another early stage of parasitism, which is exceptional among most birds (though a Buzzard for example sometimes lays in another's eyrie, which explains the clutches of six eggs occasionally recorded). However, it has become regular among the anis (Crotophaga), a group of tropical American cuckoos. All the birds of a flock, up to a score in number, build a single nest in which the females lay. Only a proportion of the eggs hatch, probably because they form such a large mass that it is difficult to incubate them efficiently. The members of the flock combine to look after the young. Thus this mode of nesting involves the loss of territorial behaviour and of proprietorial instincts concerning the young, which are raised in true crèches. Parasitic tendencies are further developed in other members of the Cuculidae, a family which shows all stages from facultative parasitism to a highly evolved form – as well as species with normal reproductive biology such as *Guira*, *Piaya* and *Centropus*. American *Coccyzus* cuckoos, and especially the

Yellow-billed Cuckoo *C. americanus*, usually build their own nests, though always in a rather rudimentary way; but certain females lay their eggs in other nests, usually of their own but sometimes of other species, such as the American Robin and the Catbird *Dumetella carolinensis*. This behaviour shows the essential characteristic of nest parasitism: the ability to lay in the nest of another species, to which the eggs and young are confided; a habit which becomes the rule in the most highly evolved parasites.

The physiological mechanism of parasitism is still very poorly understood. From information on the Brown-headed Cowbird *Molothrus ater*, it seems that the absence of normal components of reproductive behaviour is not due to hormonal deficiencies, but rather to insensitivity of the nervous system to the action of hormones which release such behaviour in other birds; for the pituitary of a cowbird secretes as much prolactin as that of a normally-reproducing Red-winged Blackbird (Selander & Yang 1966). This subject requires exhaustive researches, which will no doubt allow the origin of parasitism to be explained in terms of physiological disturbances, preventing the normal development of the various phases of reproduction.

Parasitic ducks

It occasionally happens that female ducks of various species lay their eggs in the nests of other members of their own species, which then raise the two broods as one. However, parasitism has become strict and obligatory only in one species of the Anatidae – the Black-headed Duck *Heteronetta atricapilla*, a bird found from Chile to the Argentine and southern Brazil, which is still very little known. Its eggs have been found in the nests of other species of ducks, especially the Rosy-bill *Netta peposaka*, as well as in those of Coscoroba Swans, screamers, limpkins, gulls, coots, ibises, rails and even raptorial caracaras (*Milvago*).

Cuckoos

Within the Cuculidae parasitism is shown by most species of the subfamily Cuculinae, in which it is obligatory and has reached very advanced stages of evolution. The parasitic species nearly all belong to the Old World, their American counterparts (apart from the genus *Tapera*) having an almost normal mode of reproduction. The European Cuckoo *Cuculus canorus*, known longest and most thoroughly, may serve as a standard for comparisons of the parasitic behaviour within this group.

The Cuckoo forms no lasting pair, since the male is polygamous, and he does not defend a territory. The female does show some territorial behaviour, although neighbouring territories often overlap widely and several females can even live together. We shall see how this fact is explained by the adaptation of each female to a different host species. The territory has thus lost most of its importance and biological meaning. After mating the female devotes herself to seeking out passerine nests in course of construction, and the sight of one stimulates her to ovulate at intervals of about forty-eight hours. A female normally lays twelve to eighteen or sometimes as many as twenty-six eggs, which is four or five times the clutch of the parasitized species. A single egg is deposited in each nest (always where the clutch is incomplete or unincubated), whereas cuckoos of the genera *Urodynamis* and *Clamator* lay several eggs in the same nest. The fundamental condition for the success of the operation is that the laying should take place in the absence of the hosts, since the female cuckoo must pass completely unnoticed if her egg is to be adopted. This is no doubt why the egg is so often laid between two and six in the afternoon, a time when the host birds, which are not yet brooding, freely leave their nests. If they do come upon the Cuckoo, they chase, attack and put it to flight, exactly as though it were a bird of prey.

Laying is achieved in two different ways. The female may lay directly into the nest, depositing her egg very quickly and leaving immediately; or she may lay on the ground and carry the egg in her bill to deposit it in the host's nest. The egg is sometimes hidden, as though in reserve, especially if no nest is available when the female is ready to lay. When laying she removes one or more of the host's eggs, and afterwards continues to keep a watch over the various nests, which now contain her own. She removes these from nests which are abandoned, depositing them again and so giving them another chance of being accepted. This stage is crucial for the future of a cuckoo's egg. If the hosts notice the cuckoo in the act of laying, or detect the presence of her egg, they usually abandon their nest and build another. If on the other hand the cuckoo's stratagem is not noticed, the egg is adopted and incubated with the host's clutch.

The Cuckoo develops very fast, never taking more than twelve and a half days to hatch, while the host's eggs take thirteen to fourteen days. This reduction in the time of incubation is perhaps due to the very distinct commencement of embryonic development before the egg is laid, which has been observed in the European Cuckoo and the African Black-and-white Cuckoo *Clamator jacobinus*. Although very short incubation times have also been noted among cuckoos with normal reproduction, the acceleration is clearly very advantageous to a parasitic bird (Perrins 1967). Thus the young Cuckoo is the first nestling to hatch. Although naked and blind, and rather less active than the young passerines which have become its foster brothers, it shows very acute dermal irritability, especially in the region of the back, from about ten hours after hatching for at least three days. This irritability is due to a purely reflex action, by which the young Cuckoo gets rid of anything which comes in contact with its back – especially the host's eggs and any of its young which may have already hatched. It gets them between its shoulders, and by heaving with clumsy but effective movements throws them out of the nest. Thus often it comes to supplant them all and to end as the sole occupant. This behaviour is also found in other cuckoos – though it is not general, and young *Chrysococcyx*, *Eudynamis* and *Clamator* do not tend to expel the eggs or young of their hosts. The latter none the less have only a little while to live, as do the foster brothers of the European Cuckoo if they have not been expelled before its dorsal irritability, and hence its expulsive behaviour, are lost. The young cuckoo, growing and developing faster than the host young, monopolizes the food and condemns the others to starvation.

Unexpectedly, the foster parents, so suspicious of the adult cuckoo, show great solicitude towards the young parasite. Ignoring their own starving or expelled young, they assiduously feed the young Cuckoo – a considerable labour because of the huge appetite of this rapidly growing bird, which soon exceeds them in size. It leaves the nest after twenty to twenty-three days, but they continue to feed it for a further two or three weeks. The true parents have by then long abandoned it, among all except the African glossy cuckoos (*Chrysococcyx*), whose young are fed by their own as well as by their foster parents – a situation unique among parasitic birds.

Thus cuckoos show highly specialized reproductive behaviour. Because of the many conditions which have to be satisfied, a particular egg's chances of success are slight, though this is partly compensated by the birds' greater fecundity.

Just as the adult Cuckoo must pass unnoticed for the young to succeed, the egg too must not be noticed within the host's normal clutch. This often involves adaptation towards the form and coloration of the host's eggs, so that cuckoos' eggs are relatively small for birds of their size, while there is marked egg mimicry in the European Cuckoo and certain other members of the Cuculidae, whose rate of success depends on its perfection. However, it is known that the same species of cuckoo leaves its eggs in the nests of various hosts, which do necessarily lay eggs of the same colour. This is especially true of the European Cuckoo, which parasitizes about 125 species of birds – especially pipits (*Anthus pratensis* and *A. trivialis*), wagtails (*Motacilla flava* and *M. alba*), warblers (several species of *Acrocephalus*, *Sylvia borin*, *S.*

communis, *S. atricapilla* and several species of *Phylloscopus*), the Robin (*Erithacus rubecula*), Dunnock (*Prunella modularis*) and Wren (*Troglodytes troglodytes*), not to mention many other species which are less commonly parasitized. These varied passerines lay eggs which differ very much in colour, parallel to which the eggs of the European Cuckoo show remarkable and extensive polymorphism. Each female lays eggs of only one type, resembling those of the host species to which she specifically entrusts them, and there is every reason to believe that she parasitizes hosts belonging to the species by which she herself was brought up. Thus the colour characteristics of the eggs could be hereditary, and within a population ecological races of *Sylvia*-Cuckoos, *Acrocephalus*-Cuckoos and *Erithacus*-Cuckoos could arise. However, adaptation has not reached the same degree of perfection towards the various host species. The Cuckoo's egg-mimicry is far from perfect – especially of the Wren, though this species is frequently parasitized in parts of Europe. The fact that female Cuckoos may be genetically adapted to a particular species of host explains how several of their territories may come to overlap, since there can be no reproductive competition between individuals parasitizing different hosts. Thus the species is polymorphic in this respect, and each individual may be adapted towards a particular host, the well defined oological 'races' being distributed throughout the range of the cuckoo species.

The eggs of primitively non-parasitic cuckoos are white, from which colour have evolved the eggs of parasites adapted to those of their hosts. These adaptations have reached different degrees of perfection, varying between host species, and also between populations within a species of cuckoo, with marked differences between geographical regions. The adaptation of the European Cuckoo, which occurs from India to western Europe, is much clearer in Asia than in Europe, where the species has certainly arrived more recently and has not yet had time to adapt precisely. The Koel *Eudynamis scolopacea* parasitizes only one species in India, the House Crow *Corvus splendens*, and accordingly lays a single type of egg. In contrast the Australian populations of the same species parasitize very diverse hosts – birds of paradise, lories, shrikes, drongos and honeyeaters – and the females lay very varied eggs adapted to those of these species.

Honeyguides

The honeyguides – a family distributed throughout the Old World tropics and especially Africa – are all apparently strict and obligatory parasites (Friedmann 1955). They seem to have 'forgotten' normal reproductive behaviour to a greater degree than other parasites, since the males do not exclude one another during the mating season and seem to have entirely lost the territorial instinct, while their breeding displays are reduced to their simplest expression. The females lay in the nests of a fairly long series of hosts, mostly hole-nesters. Many of these are barbets (Capitonidae) and woodpeckers (Picidae) – that is, birds belonging to the order Piciformes like the Honeyguides – while starlings, swallows, bee-eaters and hoopoes are also parasitized. Some like the Slender-billed Honeyguide *Protodiscus insignis* also lay in the open cup-nests of white-eyes (*Zosterops*) and warblers of the genus *Apalis*, though only occasionally. This specialization towards hole nests avoids competition with the African parasitic cuckoos which lay their eggs in the open nests of other species, only a few starlings (the Pied Starling *Spreo bicolor* and the Cape Glossy Starling *Lamprocolius nitens*) being known to be parasitized by both groups.

Figure 61. Newly-hatched honeyguide. Note the hook at the tip of each mandible, and the heelpad (shown also by young woodpeckers).

Honeyguides' eggs are pure white without any markings, like those of the piciforms and coraciiforms which are their principal hosts. These eggs are fundamentally different from those of other hosts such as starlings, hoopoes and swallows, in whose nests they seem to score much the same percentage of successes. The female lays a single egg in each nest, often piercing one of the host's eggs and thus eliminating a competitor. Honeyguide eggs incubate rapidly, taking about twelve to sixteen days, but the nestlings develop slowly and remain in the nest for up to forty days. These times are of the same order as those of their principal hosts the woodpeckers, barbets and hoopoes, a synchronization which certainly favours the parasite by allowing the hosts' parental behaviour to develop normally. However, young honeyguides are no less successfully raised by passerines, whose own young develop more rapidly, and whose parental behaviour, especially feeding, is maintained for longer than usual by stimuli from the young parasite.

At hatching both mandibles of a young honeyguide's bill end in strong needle-pointed hooks. These temporary structures, which last until the fifteenth day, are histologically similar to the egg-teeth which other young birds use to pierce the shell. Although naked and blind on hatching, the nestling can soon use its bill – shaped like strong pincers with crossed points – as a weapon. As has been observed of the Lesser Honeyguide *Indicator minor*, the young shows great ferocity from its second day and mercilessly attacks its foster brothers, nipping, wounding and finally killing them. However, it does not seem able to pierce the eggs as was formerly believed. This ferocity lasts only a few days, but this is enough to eliminate the host's nestlings, and leave the young honeyguide alone in the nest.

Cowbirds (Icteridae)

A complete subfamily of the Icteridae, the cowbirds, has become parasitic. In American it biologically replaces the cuckoos of the Old World.

The Bay-winged Cowbird *Molothrus badius*, found from Brazil to northern Argentina, sometimes builds its own nest. None the less it shows the first signs of weakened reproductive behaviour, in preferring to use abandoned nests which it merely restores, or in expelling other birds from their completed nests and installing itself in them, while several females sometimes lay in the same nest. This rudimentary parasitism has been perfected and become obligatory in the Brown-headed Cowbird *Molothrus ater*, which is common throughout North America. In contrast to the Cuckoo this cowbird is an unspecialized parasite, showing no adaptation to a particular host. No less than 206 species (333 species plus subspecies) are known to be parasitized by it, although the frequency of parasitism varies considerably between them and some are merely occasional hosts. Friedmann (1963) has noted seventeen species which he considers as its regular hosts, among them the Yellow Warbler *Dendroica petechia*, the Song Sparrow *Melospiza melodia*, the Red-eyed Vireo *Vireo olivaceus*, the Chipping Sparrow *Spizella passerina*, the Eastern Phoebe *Sayornis phoebe*, the Rufous-sided Towhee *Pipilo erythrophthalmus*, the Ovenbird *Seiurus aurocapillus*, the Yellowthroat *Geothlypis trichas*, the American Redstart *Setophaga ruticilla*, and the Indigo Bunting *Passerina cyanea*, with some other tyrant-flycatchers, wood-warblers, vireos and finches. Thus the last three families provide the cowbirds' regular hosts, while other species belonging to such groups as the thrushes, mockingbirds and tanagers are occasionally parasitized. The frequency with which a particular species is parasitized often varies from region to region according to ecological conditions. Thus in the eastern

United States the Red-Winged Blackbird *Agelaius phoeniceus* is rarely parasitized since it nests in marshes and among reeds, whereas it is a regular host in the west where it nests in bushes, a habitat which the cowbird also frequents.

Cowbirds, especially females, have much better-defined territories than Cuckoos do, but several may still overlap even to the point that the eggs of two females are occasionally found in the same nest. However, as a rule a single egg, or more rarely a pair, is laid in each. Like the Cuckoo, the female watches over the eggs being built in her area, lays into fresh clutches, and removes one or two of the host's eggs. She lays one egg a day, and it seems that her normal laying seldom exceeds the mean value among passerines of about five eggs. However, other observations suggest that the number may be higher, reaching twenty-five eggs from a female in one season. The eggs are rather variable in size, shape and coloration, thus falling into several classes some of which mimic the eggs of the host into whose nest they are laid. However, they never show the marked colour adaptations of the Cuckoo's eggs. This is obviously connected with the hosts' indifference to the cowbird, since in contrast to the Cuckoo's hosts in Europe most of them do not consider it as an enemy nor attack it, but usually accept its eggs as though they were their own. They rarely abandon the nest, nor cover the bottom of the cup so as to isolate the parasite's egg (together with their own eggs already laid) in order to produce a replacement clutch.

The cowbirds interest in its eggs soon ends, and except in quite exceptional cases the raising of the young is left entirely to the hosts. The cowbird's incubation period is the shortest among passerines, about eleven or even ten days. The young bird therefore hatches at least a day before those of the host, although these began to develop earlier. Unlike the young Cuckoo it is unaggressive, showing not the slightest hostility towards its foster brothers. However, it grows more quickly than they do, attaining a larger size by the end of its growth. Its faster development allows it to monopolize the food supply to the detriment of the others, though again in contrast to the cuckoo it does not usually eliminate them. According to observations made on the Song Sparrow, the average success of unparasitized broods is 3·4 young, as against 2·4 for parasitized broods, so that there is a simple substitution of the young cowbird for one young of the host. However, this is only true of fairly large hosts, since small birds seem to suffer more.

Although some females show a clear tendency to lay in the nests of particular host species – Friedmann (1963) reported that one female laid eighteen eggs, all in the nests of Yellow Warblers – this does not seem to be the general rule. Thus nineteen eggs of a single female, easily recognizable by their distinctive coloration, were distributed in the nests of Song Sparrows (eleven eggs), Yellow Warblers (six) and Alder Flycatchers (two).

Another species, the Shiny Cowbird *Molothrus bonariensis*, which replaces the preceding species in South America, is also an obligatory parasite having lost all normal reproductive behaviour, although it does begin to build a rudimentary nest during its mating displays. Unlike the North American species this one scarcely parasitizes vireos and wood-warblers, but mainly finches, tyrant flycatchers, icterids, tanagers, mockingbirds and wrens. At least 148 species are known as its hosts. Many of its eggs are lost because the females frequently lay them on the ground or in abandoned nests, and they sometimes lay in one nest many more than the number of young the host could raise: as many as thirty-seven have been found in one ovenbird's nest. This habit, which seems to be related to the complete disorganization of reproductive behaviour, must affect the success of the broods and the dynamics of the populations. The young itself seems rather abnormal in its behaviour, since it does not show the response of most nestlings to the distress calls of its foster parents. Instead of flattening itself and freezing, it does not seem to 'understand' the danger signal, and hence often falls victim to nest robbers such as the Chimango Caracara *Milvago chimango*.

In contrast to the Shiny Cowbird, the Giant Cowbird *Somocolax oryzivorous*, found from Mexico to southern Brazil, parasitizes only those icterids known as caciques (*Xanthornus* and related forms), and its range corresponds exactly to theirs. It has to overcome the defensive reactions of its potential hosts, which guard their nests the more effectively since they nest in colonies and their elongated purse-shaped nests are easy to defend. Finally the Screaming Cowbird *Molothrus rufoaxillaris*, considered to be the most primitive in the group, parasitizes exclusively and very curiously its close non-parasitic relative the Bay-winged Cowbird *M. badius*. There is no detectable difference between the eggs and young of the two species, which are thus raised by only one of them.

Whydahs (Ploceidae)

A latent tendency towards parasitism is shown by many weavers, especially among the Estrildinae, which sometimes use ready-made nests. This foreshadows the considerably more highly evolved parasitism which has appeared in two distinct lines, the Viduinae or viduine whydahs and the Cuckoo Weaver *Anomalospiza imberbis*. The former, which have been the subjects of many studies (Neunzig 1929, Friedmann 1960, Nicolai 1964 &

1967), have certainly reached the most perfect stage of reproductive parasitism, especially in the specificity of their adaptations to their hosts. Whydahs seem from many points of view to be closer to the Euplectinae than to any other branch of the Ploceidae, their display behaviour, innate components of song and eclipse plumages linking them clearly with this group. In contrast other more obviously adaptive characters – especially those of the nestlings and fledgelings – recall the Estrildinae, the group which they parasitize and mimic in the most remarkable way. Thus the parasites and their hosts show parallel evolution, with the resemblances extending to the minutest details.

The first similarity is in their vocalizations. The song of each viduine species consists of one component which seems to be a conservative foundation resembling euplectine song, and another component mimicking the song of its estrildine host and thus showing kinship by adoption. Sound spectrographic analysis has shown this so clearly that Nicolai (1967) holds that one can use the songs of viduines to determine their hosts, where this is not yet known. Thus the Senegal Combassou *Hypochaera chalybeata* mimics the various songs and calls of its host the Senegal Firefinch *Lagonosticta senegala*, its vocalizations varying geographically according to the races of its host. Host and parasite understand one another perfectly, and can thus begin a kind of dialogue.

Female viduines lay in the nests of estrildines without having to face the show of hostility which most other parasites encounter. They usually lay only one, or more rarely two, eggs in each nest, and remove as many. One can scarcely suggest egg mimicry, since the eggs of almost all these hosts and parasites are pure white, but there is a remarkable resemblance in the mouth markings shown by the nestlings. In viduines as in estrildines, the mouth lining presents a strongly contrasting pattern of dark spots forming a geometrically symmetrical design, while round the edges are rounded tubercules which appear luminous, but are in fact merely reflectors, giving the effect of minute light sources in the semi-darkness within the closed nests of these birds. Histological examination of these tubercules shows that they consist of connective tissue supporting a layer rich in melanin pigment, topped with fibrous connective tissue of high refractive index. The pigmented layer acts like the metallic layer behind a mirror, and causes the whole structure to reflect and concentrate the faint incident light. These organelles, whose specific arrangement has been used by systematists in studying the classification and phylogeny of these birds, combine to form sign-stimuli which are very important in the behaviour involved in feeding young. Recent experiments (Nicolai) have shown that estrildines have developed a high degree of selectivity in this response, so that the young of another species, showing a different buccal pattern and behaviour, are not normally fed. In order for the young parasites to be cared for and fed by their foster parents, it is thus essential for them to show the same buccal pattern as the young of the host species.

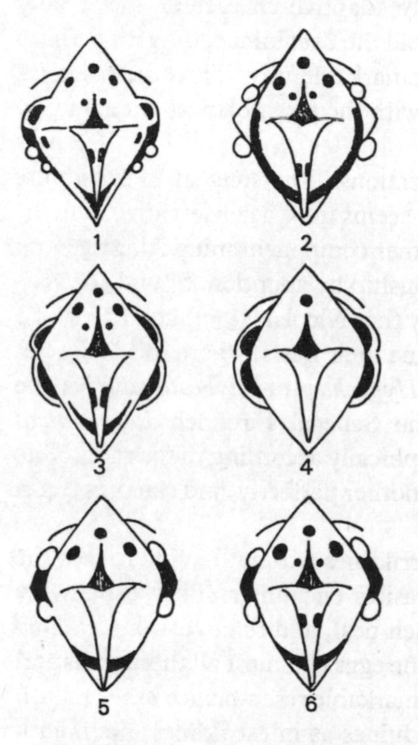

Figure 62. Nestling throat patterns in which whydahs resemble their host species.
1. *Estrilda astrild* and *Vidua macroura*.
2. *Granatina ianthinogaster* and *V. fischeri*.
3. *Estrilda erythronotos* and *V. hypocherina*.
4. *Pytilia melba* and *Steganura paradisea*.
5. *Lagonosticta senegala*.
6. *Hypochera chalybeata*.

Adaptation of the parasite to its host is also shown in the juvenile plumage, those of viduines being very like those of the species which each parasitizes. Young viduines and estrildines as a whole have single-coloured and rather dull plumages in tones of brown and grey, but detailed comparison shows that the adaptation is very accurate. The almost identical plumages of the young are all the more striking, since the adult plumages of the two groups are totally different. These resemblances to the hosts' own young are indispensable because the young viduines are brought up side by side with them. Unlike cuckoos, honey-guides and even cowbirds, the young parasite shows not the smallest hostility towards its foster brothers, but simply takes one of their places. During evolution exact relations have developed between species of viduines and estrildines, each of the former becoming progressively adapted to the host on which it is dependent and which it most frequently parasitizes, so that one can draw up a table of agreements between hosts and parasites:

Parasite	*Principal host*
Senegal Combassou *Hypochera chalybeata*	Senegal Firefinch *Lagonosticta senegala*
Queen Whydah *Tetraenura regia*	Violet-eared Grenadier *Granatina granatina*
Fischer's Whydah *T. fischeri*	Purple Grenadier *G. ianthinogaster*
Paradise Whydah *Steganura paradisea*	Green-winged Pytilia *Pytilia melba*
Broad-tailed Paradise Whydah *S.o. orientalis*	Green-winged Pytilia *Pytilia melba*
Broad-tailed Paradise Whydah *S.o. obtusa*	Orange-winged Pytilia *P. afra*

During evolution the morphological and ethological specificity of the hosts has forced the viduines to specialize, so that they form unique evolutionary doublets with the estrildines.

Most viduines are able to parasitize species other than their principal hosts. Thus the Queen Whydah *Tetraenura regia* particularly parasitizes the Violet-eared Grenadier *Granatina granatina*, yet its eggs have been found in the nests of fifteen other weavers and even of a warbler, the Black-chested Prinia *Prinia flavicans*. Similar statements have been made about other species.

The Ploceidae include one other parasite besides the viduines: the Cuckoo Weaver *Anomalospiza imberbis*, which is a case of an isolated parasite within a group which otherwise has normal reproductive behaviour. Distributed right across the African savannahs, it lays almost exclusively in the nests of grass warblers (*Cisticola*). A single egg is laid in each nest, from which hatches a nestling which occupies the nest alone, probably eliminating the young warblers by competition for food. Its plumage is rather ordinary and not very different from that of the host's young, although it shows no signs of true adaptation to them.

The evolution of nest parasitism

Nest parasitism has arisen independently in several groups of birds. It implies the gradual disappearance of normal reproductive behaviour, during evolution which has left traces in the few 'imperfect' parasites whose parasitism is not obligatory. Parasitism can only be favourable to a species if it is already almost 'perfect' – that is if it has passed beyond the preliminary stages, as it will have done very rapidly during evolution.

Parasitism presents a series of negative characters, resulting from the abolition of certain types of behaviour. Parasitism also involves the acquisition of new adaptive characters, which are very remarkable and precise. The essence of the problem for a parasite is to succeed in laying in the nest of a host and to persuade it to adopt the parasite's egg. Next the egg itself has to remain undetected, a need which has resulted in the evolution of egg mimicry reaching its highest level among the cuckoos. The nestling parasites equally show adaptations, co-ordinated with the reproductive details of their foster parents. It is also necessary for the diets of the young parasites and hosts to be the same. Quantitatively, on the other hand, the parasite's dietary demands for development explain why it may or may not evict the host's young through aggressive behaviour. The obvious aggressive behaviour in cuckoos is linked to ecological necessities concerned with food, as the result of the difference in size between the cuckoo and the host. Weighing 3g at hatching a young Cuckoo reaches 90g at fledging, whereas young passerines are much lighter: a complete brood of the Redstart *Phoenicurus phoenicurus* weighs 8g at hatching and 93g at fledging, that of a Pied Wagtail *Motacilla alba* 9g and 110g respectively.

This competition, during which the host's own young are at least partly eliminated, leads us to consider the impact of the parasites on the host populations. It is noteworthy first of all that only common species with accessible nests are parasitized. The populations of some parasites are themselves large, so that in many cases a large proportion of nests is parasitized. The consequences vary greatly, depending on whether or not the young parasite eliminates the young of the hosts. Where it evicts the whole brood, as the European Cuckoo does, the effect can be at least locally serious. Records of a population of the Reed Warbler *Acrocephalus scirpaceus* showed the number of nests in the study area, and the proportion parasitized, changing inversely from year to year: from fourteen nests with 29 per cent parasitized, to fifteen with 40 per cent, twelve with 50 per cent, eleven with 73 per cent, nine with 67 per cent, and finally eight with 88 per cent (Schiermann). Thus the increasing abundance of the cuckoo involved reduction in the numbers of the warblers, and such a parasite, by eliminating a high proportion of its hosts, can admittedly contribute significantly to the control of their populations. However, this reduction in the host population leads to more and more intense competition between the cuckoos, as the rising proportion of parasitized nests shows. Thus the cuckoo population is limited by the increasing difficulty of ensuring the success of the eggs. It in turn diminishes greatly, whereupon those of the potential hosts begin to increase, and the cycle begins again. Thus a parasite which eliminates its host's broods, and is dependent on a few species, will cause the populations of the parasitized species to fluctuate widely.

The effect of a parasite also depends on environmental conditions, and the impact on populations of certain species may therefore be significant. In this way the Cowbird is partly to blame for the reduced population of Kirtland's

Warbler *Dendroica kirtlandi* in Michigan, the success of whose broods is reduced by 36 per cent through parasitism. Such action can reduce a population the more effectively, since because of its non-specificity the density of the parasite is not significantly dependent on that of the host, so that there is no auto-regulation (Mayfield). The host specificity of viduines must involve a more definite reduction in the chances of success of the parasitized broods, and the most important cause of mortality among young Senegal Firefinches *Lagonosticta senegala* is parasitism by the Senegal Combassou *Hypochera chalybeata*. In parasitized nests the percentage of eggs which give rise to a fledging young falls from 33 per cent to 21 per cent, and the mortality rate rises from 25 per cent to 53 per cent (Morel).

Chapter 15

Bird Populations and Their Regulation

BIRD densities, whether calculated for a single species or for the whole community in a particular habitat, are for the most part remarkably stable. Yet the annual rate of growth of such populations is relatively high, as is shown by a classic experiment conducted on Protection Island, an islet of about 160ha off the coast of the state of Washington, USA. In 1937, two cocks and six hens of the common Pheasant *Phasianus colchicus*, until then absent from the island, were introduced. The population was counted again every year, and after five years had reached 1,850 individuals. The stock was reduced each winter, by mortality resulting from bad weather, but the general trend of the curve representing the growth of the population was sigmoidal, showing exponential growth in the early stages, as might have been expected from first principles (Einarsen 1942). The natural or artificial introduction of birds to large areas where they were unknown similarly shows their potential for increase. Cattle Egrets *Bubulcus ibis* invaded the whole American continent, starting from a small population which reached Venezuela. Starlings, of which a hundred or so were released in New York in 1890–1, established themselves across the whole of North America, where their populations are now estimated at about fifty millions.

The *theoretical* potential for increase is still higher, and it has been calculated that a single pair of American Robins, though they raised only two broods of four young each every year, would have about 19,000,000 descendants after ten years, if every pair was equally fecund and if every individual lived to the end of the period. This does not of course happen. By the action of opposing forces, *reproductive potential* and *environmental resistance*, a population which has been introduced into a new environment reaches a state of equilibrium. The curve of growth stabilizes at a particular level, with fluctuations which are usually minimal as long as the physical and biotic conditions remain the same. This level is what is known as the *carrying capacity* of the habitat. If ecological conditions change the population stabilizes at a new level, higher or lower according to whether the environment has improved or deteriorated with respect to the needs of the species. This limiting capacity represents the population which a particular habitat can sustain, and is determined by the food supply and by other factors (especially the ethological needs of the species concerned) which we shall now try to specify.

Sampling methods

The assessment of bird populations, as of other animals, involves two very different approaches.

Some methods aim at producing an exact evaluation of all the individuals occupying a particular area. Only a *census* can give results of absolute validity, and usually specifies the identity, sex, age and other particulars of the individuals involved. It is commonly undertaken at breeding sites during the breeding season, when the birds are territorial so that their populations are stable, free from the fluctuations produced by migratory or erratic movements. Direct counting can only be applied to gregarious birds, either when they are nesting or at their winter quarters. It has been undertaken especially on seabirds whose colonies have been accurately localized and inventoried by ornithologists. A count of occupied nests yields the numbers of adults and of young, given the average size of clutch and their percentage success. It is in this way that a census has been taken of North Atlantic Gannets *Sula bassana* (Fisher), and that periodic counts have been made of western European populations of the Heron *Ardea cinerea*. Such counts can similarly be made of wintering populations, provided that they remain relatively stable during the count and that the birds gather in habitats where they are conspicuous to the observers. This is true of populations of ducks and geese, which collect in flocks at a limited number of well known wintering grounds. Census returns are of course subject to many errors, through incorrect identifications, inaccurate estimations, and lack of synchronization between the counts – so that a particular flock may be counted several times or not at all, depending on its movements between sectors. The most serious errors result from the fact that the network of observers covers only a fraction of the wintering grounds. Although aerial surveys involve an appreciable margin of error, they allow huge areas to be covered and give good views of concentrations of birds to which access on the surface may be difficult. The use of aerial photographs greatly reduces the observers' counting errors. These methods have been widely employed in the United States and in Europe, where studies on a vast scale have been made for many years on wintering ducks and geese. They have also been used during the breeding season in determining nesting populations; though here the margin of error is much greater, because the birds are dispersed on their territories in less open habitats where they are more easily overlooked.

In order to determine the populations of birds which occupy closed habitats, all the individuals occupying a specified area must be carefully counted. The method consists in selecting a sample area or *quadrat* – a small area of known size representative of the area to be studied – and in counting all the birds (and signs of their activities such as nests, display grounds, regurgitated pellets, etc.) found within it. The method is particularly difficult to apply, since the areas of most bird territories are such that the quadrat must be large, while closed habitats are difficult to survey. It only remains, after many sessions in the field, to sum up all the observations, to establish the limits of territories, and to estimate precisely the numbers of the various species. This method demands considerable work in preparing the ground, and afterwards in making frequently repeated observations throughout the breeding season. However, it is the only one which yields very accurate information on the actual densities and precise composition of bird communities.

The methods already considered aim at obtaining absolute values, but they demand a great deal of time and are not always practicable, so that ornithologists are often satisfied with estimates. They no longer seek to know the total populations in a community, but only the number of birds seen, heard or trapped under particular conditions. This value, considered as an index of abundance, is taken to be proportional to the absolute density of the population, and the results obtained from such estimations are merely relative. They are none the less very useful, giving an index of population size and its fluctuations in relation to various factors, since they are comparable with each other if obtained by similar methods. The best known of such methods consists in making *line transect counts*. The observer traverses a predetermined distance in a straight line through a homogeneous environment, noting the birds he sees or hears, and in theory the density of each species is proportional to the number of contacts recorded. The method only yields an index of abundance, and not an absolute density per unit area. Various formulae have been given for obtaining absolute figures from these relative values, but none are wholly satisfactory (Yapp, Ferry-Frochot).

Another method which has been used is that of *capture-recapture*, which consists of marking a known number of individuals and releasing them in their own area. If these birds redistribute themselves randomly and evenly throughout the population to which they belong then a particular proportion of the birds taken in a second operation will already have been caught and marked. The numbers captured are related to the size of the population by a simple proportion: $\frac{N}{M} = \frac{n}{m}$, where N is the total number of individuals making up the whole population; M is the number of birds marked and released; n is the number caught in the second operation; and m is the number of marked birds among these. This method can be used in the estimation even of huge populations – such as those of ducks, when the numbers of rings used and recovered are compared by the 'Lincoln Index'.

In studying the size and structure of a bird population, it is often necessary to recognize each individual so that its movements and growth may be followed. The classic procedure consists of attaching lightly to the bird's leg a metal ring bearing a number, which allows it to be identified by means of a code. Ringing nestlings yields particularly interesting results, since one knows both the precise age of recaptured birds and their exact place of origin. It is of course advisable to mark a large number of individuals, since the chance of recovering a particular one is usually very small. Variously coloured rings may also be used, allowing individuals to be recognized and studied without being recaptured, though this method is only useful within restricted populations.

Factors of population growth

It is appropriate to consider first the positive factors, which lead to increase in populations. The growth potential results primarily from fecundity (number and size of the clutches), the age of sexual maturity, and the longevity of individuals.

Fecundity varies greatly between different types of bird – adjusting to their ways of life, to the predation they suffer, and to their death rates – and there is a balance between clutch size and the number of clutches per breeding cycle. The number of annual cycles during which a given individual breeds depends on the age of reaching sexual maturity and that at which sexual activity ends – which in birds coincides roughly with the end of life. Sexual maturity is attained at very different ages in different species. It is precocious in small birds, and especially in passerines, owls, game-birds, pigeons and ducks, most of which can breed from the year after they hatch. In contrast some gulls and waders, many raptors and the geese only breed from the age of two; large gulls, gannets, divers and storks only after three years, and eagles not for four to six years. Longevity too varies very widely among birds. Our information on the longevity of birds comes mainly from ringing, which provides valuable data on birds in the wild. The following table is based on a number of studies of marked birds.

Species	*Maximum age recorded (years)*
Yellow-eyed Penguin *Megadyptes antipodes*	18
Chinstrap Penguin *Pygoscelis antarctica*	11
Red-throated Diver *Gavia stellata*	23
Cormorant *Phalacrocorax carbo*	17
Heron *Ardea cinerea*	24·5
White Stork *Ciconia ciconia*	19
Canada Goose *Branta canadensis*	23
Mallard *Anas platyrhynchos*	20
Teal *Anas crecca*	20
Golden Eagle *Aquila chrysaetos*	20
Buzzard *Buteo buteo*	24
Peregrine *Falco peregrinus*	14
Osprey *Pandion haliaetus*	21
Black Vulture *Coragyps atratus*	11
Coot *Fulica atra*	19
Curlew *Numenius arquata*	31
Woodcock *Scolopax rusticola*	21
Herring Gull *Larus argentatus*	32
Arctic Tern *Sterna paradisea*	27
Puffin *Fratercula arctica*	21
Mourning Dove *Zenaidura macroura*	9
Barn Owl *Tyto alba*	10
Swift *Apus apus*	21
Crested Flycatcher *Myiarchus crinitus*	6
Skylark *Alauda arvensis*	6
Swallow *Hirundo rustica*	16
American Common Crow *Corvus brachyrhynchos*	14
Rook *Corvus frugilegus*	20
Blue Jay *Cyanocitta cristata*	15
Willow Tit *Parus atricapillus*	9
House Wren *Troglodytes aedon*	5
Catbird *Dumetella carolinensis*	9
Blackbird *Turdus merula*	9
American Robin *Turdus migratorius*	10
Robin *Erithacus rubecula*	11
Starling *Sturnus vulgaris*	20
Brown-headed Cowbird *Molothrus ater*	8
Purple Grackle *Quiscalus quiscula*	16
Scarlet Tanager *Piranga olivacea*	9
Red Cardinal *Richmondena cardinalis*	13
Slate-coloured Junco *Junco hyemalis*	11
Song Sparrow *Melospiza melodia*	8

These data on wild birds are confirmed by others on birds in captivity, which generally live longer. The *potential longevity*, the maximum age which can be attained under ideal conditions, is seldom reached in nature. Thus the greatest age recorded for a Herring Gull in captivity is forty-nine years, against thirty-two years in nature. The following table, after Flower (1938) gives the ages of captive birds.

Species	*Maximum age recorded (years)*
White Pelican *Pelican onocrotalus*	52
Canada Goose *Branta canadensis*	33
Andean Condor *Vultur gryphus*	52
Batealeur Eagle *Terathopius ecaudatus*	55
Eagle Owl *Bubo bubo*	68
Domestic fowl	30
Domestic pigeon	30
Sulphur-crested Cockatoo *Cacatua galerita*	56
Raven *Corvus corax*	24
Starling *Sturnus vulgaris*	17
Garden Warbler *Sylvia borin*	24
House Sparrow *Passer domesticus*	23
Red Cardinal *Richmondena cardinalis*	22

The sex ratio also affects the fecundity of the population. Though it is usually well balanced, one sex occasionally shows a marked preponderance over the other. Thus in some populations of the Bobwhite Quail introduced to New Zealand there are twice as many males as females at maturity. Such a departure from the balance shown by populations of normal structure involves difficulties in social and sexual relations, and hence a reduction in fecundity.

Factors of population decrease: predation

Predation is one of the most important mortality factors. Even in the egg, birds are threatened by a great number of predators which destroy nests despite the precautions that parent birds take to protect them. Birds which nest on the ground are especially vulnerable, since their nests are accessible to many mammals.

Arboreal birds are equally threatened by various predators, though the predation is less intense on those which nest in holes than on those which occupy open nests. Certain birds pillage other species' nests: gulls take advantage of parental carelessness to eat young gannets, terns and auks. Growth in the Herring Gull populations has caused a reduction in such birds – which have to be protected by controlling the gulls, through destruction of adults and clutches – in various European countries. Rather unexpectedly, some passerines also feed on eggs: for example the House Wren *Troglodytes aedon* robs the nests of various passerines, while Galapagos Mockingbirds (*Nesomimus*) feed partly on the eggs of Darwin's finches.

Adult birds are preyed upon by many specialized predators, including raptors such as the Peregrine and Sparrow Hawk whose very rapid flight and highly developed powers of manoeuvre enable them to take other species of birds. Cases of cannibalism have been noted, as in flocks of Giant Petrels' which immediately dispatch and eat a wounded member. Among raptors the young one which is hatched before its brothers, and is therefore the strongest, sometimes kills and eats them, especially in bad years. However, this is not the general rule, and carnivorous birds do not usually eat carcases of their own species which are protected by their specific plumage characters. Terrestrial birds are the special prey of carnivores such as foxes, jackals, mustelids and cats: the Serval of Africa in particular is specialized as a bird-hunter. Some marine mammals such as Killer Whales live partly on penguins, while even carnivorous fish may catch aquatic birds while they are swimming or diving. Pike freely take ducklings, whose remains were found in 7 per cent of the stomachs of pike autopsied in Saskatchewan (Sprungman). Similarly angler fish regularly catch grebes, divers, cormorants, ducks and auks. Occasionally crabs feed on young seabirds such as terns. Large spiders may catch small birds, trapdoor spiders especially having been recorded as taking flycatchers; but the evidence is purely anecdotal, and the incidence of such predation must in any case be very low.

Birds try to protect themselves against predators, especially through nest camouflage, choice of nesting site, and crypsis of eggs, young and adults. They are sensitive to many signals which announce the approach of a predator. The defensive reaction of flying adults consists in gathering into a compact flock, from which the predator has difficulty in selecting and isolating a prey. In case of an attack it is usually an isolated lagging individual which falls to the claws of the pursurer. Some birds, even small ones, freely attack the predator, mobbing it in groups which assemble in response to characteristic calls of precise semantic significance. This is especially true of defence against owls and other nocturnal predatory birds. Parent birds teach their young to know their enemies, though much of this behaviour seems to be innate.

The impact of predation on a particular population of birds varies considerably from case to case, and no general law can be established. When broods have been unusually successful and the level of the population is higher than average at the beginning of the unfavourable season, regulation tends to return the population to a virtually constant level by the end of winter. This rule is very important in the rational management of game stocks.

The population density of the birds serving as prey greatly affects the effect of predation. If they are overpopulated predation is much more intense, which has the effect of returning the population to a level compatible with the limiting capacity of the environment. Thus remains of Bobwhite

Quail *Colinus virginianus* are found in 5·5 per cent of the pellets of the Great Horned Owl when the populations are at only 35 per cent of the winter limiting capacity, rising to 19 per cent as the population increases to 141 per cent of the limiting capacity. This difference is doubtless explained by the greater ease with which predators find their prey when the density is higher. Predation acts more intensely against the young, which are not territorial and lack the experience of their parents. The effect of predation on a population of a given species is also affected by the other species present in the community, and their densities. When the principal prey is scarce the predators switch to other species, and predator pressure on supplementary prey increases. In northern North America the Great Horned Owl feeds mainly on the Snowshoe Hare; but in years when this mammal – which varies greatly in abundance – is rare, the owl turns to Ruffed Grouse which thus suffer greater predation pressure. Similarly the principal diet of the Snowy Owl is lemmings, but it hunts grouse when these rodents are at their minimum. Finally, the intensity of predation depends upon the prey species and its place in the ecological community. Predation has a great effect on Turtle Doves, Jays, Great Tits and House Sparrows, but only little on birds of prey, cormorants, herons and gulls. Its effects are very different on species such as pelagic birds, whose fecundity is low and which renew their populations only slowly, and those such as most small passerines, which are very fecund and renew rapidly. The contradictory opinions put forward by various authors may largely be explained by the fact that they based their conclusions on species with different demographic characteristics.

Predation actually forms part of a huge complex of factors – among which are other causes of mortality, the nature of the biotope, climatic conditions and population densities – which go to produce an equation of multiple parameters which themselves vary as functions of one another.

Birds of prey in particular are all termed 'harmful', because they take prey which is considered useful, especially the game animals which man conserves and keeps for his sole use. In fact the question is far from simple, depending especially on the species of predator concerned. Thus in Europe the Goshawk is obviously more 'harmful' than the Buzzard and smaller raptors, since its diet consists especially of large birds such as pheasants and partridges and mammals such as hares. Harmfulness also varies between individuals, since each bird specializes to some degree in hunting a particular prey. In natural communities raptors are uncommon, since they occupy the summits of food chains, and their ecological demands lead them to exclude one another from large territories. Their numbers have been still further reduced by the action of man, transforming their habitats and directly destroying them. Thus their impact has become negligible, and it is proper to protect them to the same degree as other birds.

It must be stressed that predation is actually beneficial to the population of the potential prey as a whole. Following the law of minimum effort or maximum efficiency, predators feed especially on handicapped, sick, malformed or weak individuals, and thus help to eliminate those which risk spreading epidemic illnesses and parasites. Furthermore they feed freely on young individuals, of which the surplus is in any case destined to disappear, under the pressure of other causes of mortality, during the unfavourable season. Excess productivity at low trophic levels allows consumers at higher levels to live without harm to the animal community as a whole.

The part played by the birds of prey is minimal because of their low density, and in most cases may be ignored. The beauty of raptors must in itself constitute an argument – sentimental perhaps yet none the less valid – for the protection of these animals, which form an integral part of communities which we wish to conserve in their totality.

Other factors in the reduction of populations

Predation is far from being the sole cause of reduction in bird populations. The *physical factors of the environment* may act directly, as well as more importantly through shortage of food. A sudden cold spell which lasts throughout the breeding season entails an increased mortality among young birds, while rain has disastrous effects on the young and adults. It is well known in Europe that a cold wet spring reduces the success of broods to a marked degree. In their wintering grounds, Swallows are severely harmed by tropical tornadoes: a third of the wintering swallows have disappeared after heavy rains in the eastern Congo (Verheyen). Wind too can cause serious losses among migrating birds, its effects on sea birds being particularly clear. Many are driven towards the coast and even inland, so that there are repeated records of 'invasions' of the land by specifically marine birds: Wilson's Petrels, Manx Shearwaters and Kittiwakes in Europe, and Little Auks in the United States. During the winter Little Auks *Alle alle*, which nest in the arctic regions around the North Pole, migrate southwards particularly along the Atlantic coasts of North America. Violent storms may drive them towards the coasts and even inland, as happened most spectacularly during the winter of 1932–3 (Vogt). The birds caught or picked up dead show every sign of considerable wasting, with the loss of fatty reserves and even muscular tissue. The ineffectual struggle against the wind and the inability to feed exhausts the birds, and condemns them to a more or less rapid death. So it is with Leach's Petrel *Oceanodroma leucorrhoa*, which storms periodically drive inland in western Europe when they are on migration (Boyd, Jouanin). In 1952 at least 6,700 of them were taken under these conditions in Great Britain alone, which gives an idea of the scale of the catastrophe. All showed signs of extreme exhaustion, and under these conditions losses are always high.

In winter long spells of cold and snow result in significant losses. In western Europe ducks, geese and waders are dramatically reduced by severe winters, such as those of 1956/7 and 1963/4. In March 1904, millions of Lapland Buntings *Calcarius laponicus* were killed by a cold spell in Minnesota and Iowa, when 750,000 corpses were counted on frozen lakes (Roberts) The spring populations of Herons in Great Britain, which are counted regularly, are inversely related to the severity of the preceding winter, and were found to be reduced to 40 per cent after the winter of 1946/7 which was very severe in western Europe (Alexander). Naturally, cold and rain act not only directly, but also by preventing birds from feeding just when their energetic needs arc increased.

Birds are also subject to numerous diseases, being attacked by a very large number of micro-organisms. Among these are blood parasites, especially the mosquito-borne protozoa responsible for avian malaria (*Plasmodium* and *Haemoproteus*), which is found to be carried by a quarter to one-third, or even two-thirds, of a population. In California 97 per cent of the California Quail analysed were found to be carrying these pathogens. Other protozoa cause coccidiosis, while bacteria result in digestive poisoning, tetanus, diphtheria, tuberculosis, cholera and botulism. This last is particularly common among ducks, which are killed in millions after contamination by the anaerobic bacterium *Clostridium botulinum*, proliferating in the decaying vegetation among which they feed. Viruses are responsible for many diseases, notably ornithosis – a disease transmittable to man – as well as several forms of encephalitis. Fungi can also cause serious illness: aspergillosis produce proliferation of mycelia which invade the respiratory apparatus, especially the air sacs, and causes death, particularly among penguins and ducks.

Birds are hosts to many external parasites: lice, bugs, fleas, flies (blowflies, louse-flies, mosquitoes and black flies), ticks and other acarines. Biting lice (Mallophaga) are actually confined almost entirely to birds, their numerous forms showing such host specificity that they are useful to systematists in determining the relationships of their hosts. These parasites draw blood, some of them penetrating under the skin or into the nostrils and buccal cavity, and many transmit illness by injecting pathogens. Nestlings suffer the attacks of a rich fauna of arthropod parasites which swarm in nest materials. Endoparasites are equally numerous, including trematodes (flukes), nematodes (roundworms and filaria), cestodes (tapeworms) and acanthocephala (thornyheaded worms). These parasites have complex life cycles, passing through several hosts in turn, and are transmitted to birds by their prey – whether worms, molluscs, crustaceans, fish, amphibia or even mammals. Of Mallards autopsied in Washington State 95 per cent were found to contain parasites, and a single bird sheltered 1,600 cestodes of six species. These parasites are found mostly in the digestive tract and lungs, but sometimes in the eyes and other organs.

The influences of parasitism and of pathogenic micro-organisms cannot be dissociated from the amount of food available. Lack of food thus causes severe losses among birds, which succumb to the attacks of micro-organisms and parasites. The latter are the immediate cause of death, but in fact the ultimate causes are to be found in the food supply and the physical conditions of the environment which control it.

Accidents also cause losses among birds, which may strangle themselves in vegetable fibres, or break a wing or leg – a relatively frequent accident, since up to 4·5 per cent of wild birds autopsied showed fractures, especially of the limbs and clavicle. Young birds are sometimes clogged with mud, clay, wet with rain, gathering on their feet into masses which weigh them down, immobilize them and condemn them to death – an accident to which Partridges are especially susceptible. A similar disaster involving calcareous concretions overtook Lesser Flamingos *Phoeniconaias minor* in Kenya in 1962. The enormous colonies which normally nest on Lake Natron, in the extreme north of Tanzania, emigrated as a result of floods and established themselves on Lake Magadi in Kenya. The waters of this lake are extremely saline (so that extraction of the sodium carbonate is profitable), and the salts rapidly crystallized round the feet of the young flamingos to form heavy casings. Despite the efforts of ornithologists to free the chicks from these encumbrances, the losses resulting from the 'error' in the choice of nesting place amounted to hundreds of thousands of birds.

Competition also causes losses among birds, and limits the growth of their populations. Territorial behaviour may also cause fatal fights, although territorial defence consists mainly of visual and auditory display.

Finally the food supply can directly and considerably affect populations. Famine causes considerable losses, as much among the adults as among the young – whose success depends very largely on the available food, to which it is to some extent proportional. Many physical factors of the environment act indirectly through the food supply, controlling its abundance and availability. So too with pathogenic micro-organisms and parasites, which are normally in equilibrium with a population which can satisfy its need for food, but which proliferate uncontrollably in birds debilitated by famine.

Man has added to the causes of mortality among birds, by building obstacles with which they (especially nocturnal migrants) collide. Towers, electrical pylons and radio and television aerials constitute so many obstacles, whose effects are increased by the network of electrical cables which densely cover industrialized countries. The losses are considerable: 35 per cent of White Storks ringed in Denmark in 1952–4 were later found dead under electric wires. Powerful lighthouses and airport ceilometres are lures which attract migrants to crash around them: 50,000 birds of fifty-three species have thus been killed at the ceilometer of an air base in Georgia, USA. Man as a hunter also behaves as a predator. In many cases bad management of hunting has resulted in the depletion of reproducing stocks. This is the more serious because of man's impact on the environment, as by draining wet lands or by changing methods of cultivation.

Population structure

It is interesting to know the structure of a particular population: that is, the proportion of individuals in the various age classes, from which various demographic parameters can be calculated. Most of these can be extracted from the *life table*, which yields especially:

the *survival function* (l_x), or number of individuals surviving at a given age

the *distribution function* of deaths (d_x), which is the number of individuals dying during successive time intervals

the *mortality* rate (q_x), which is the proportion of individuals dying during successive intervals

the *life expectancy*, or mean expectation of life (e_x), which is the average time which individuals of a given age have still to live.

Such information has been obtained by marking with numbered rings, allowing the individual recognition of very great numbers of birds, especially within colonies of sea birds followed from year to year.

Among work undertaken in this field is that on Herring Gulls, which have been the objects of many studies in the United States (Paynter 1947, 1966, Hickey 1952) following systematic ringing programmes on colonies whose progress has been followed for several years, from which the following life table for a theoretical population is deduced (Hickey):

Age (x) (*years*)	*Number surviving at beginning of age interval* (l_x)	*Number dying in age interval* (d_x)	*Mortality rate* ($1{,}000q_x$)	*Life expectancy* (e_x)
0–1	1,000·0	599·2	599·2	2·00
1–2	400·8	117·1	292·1	3·25
2–3	283·7	59·5	209·7	3·39
3–4	224·2	61·5	274·3	3·15
4–5	162·7	41·7	256·2	3·16
5–6	121·0	45·6	376·8	3·08
6–7	75·4	16·1	213·5	3·63
7–8	59·3	7·9	133·2	3·48
8–9	51·4	11·9	231·5	2·95
9–10	39·5	9·9	250·6	2·69
10–11	29·6	5·9	199·3	2·42
11–12	23·7	9·9	417·7	1·90
12–13	13·8	4·0	289·8	1·91
13–14	9·8	2·0	204·1	1·48
14–15	7·8	5·9	756·4	0·73
15–16	1·9	1·9	1,000·0	0·47

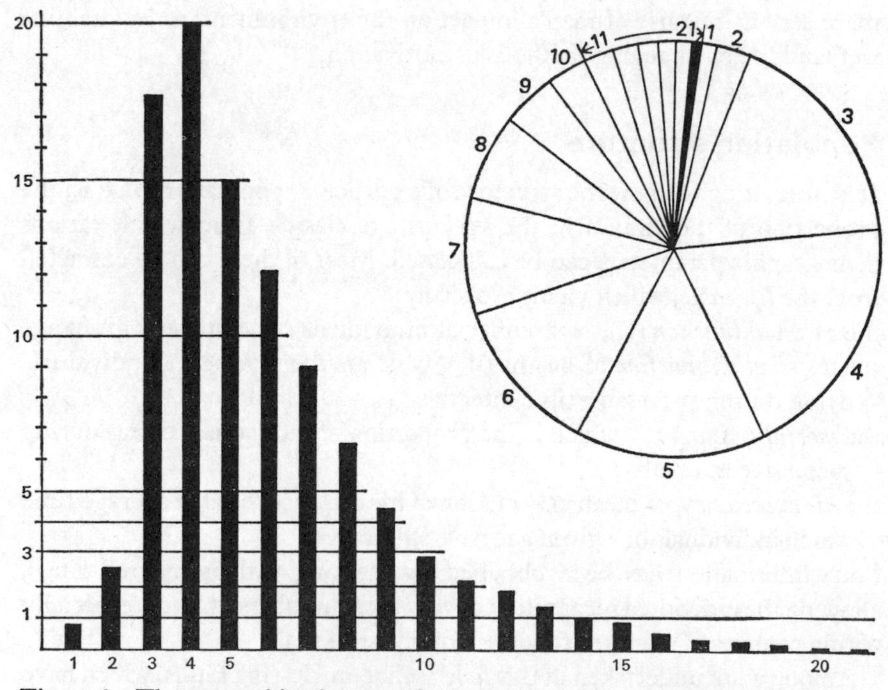

Figure 63. The composition by ages of a population of Common Terns *Sterna hirundo*, studied at Cape Cod in the USA. In the graph the ordinate is the percentage of breeders, the abscissa is the age in years. In the circular diagram (which reads clockwise) the segments are proportional to the age classes. (Data from recaptures of 6965 birds ringed in the colony.)

The greatest age recorded on Kent Island, New Brunswick (where Paynter ringed 31,694 young gulls from 1934–9, of which 1,099 were recaptured up to 1963) was twenty-eight years. A heavy mortality in the first year was followed by a very marked slowing down. Breeding began at three to four years old. The average clutch being 2·5 eggs and producing 0·92 nestlings, it can be calculated that from 1,000 eggs laid 98·4 gulls reach breeding age. Having a life expectancy of 3·3 years, they can lay 405·9 eggs, or only 41 per cent of the number required to maintain a stable population. However, the population did remain stable for the whole period of observation, which shows how important it is to be cautious with estimates of this kind.

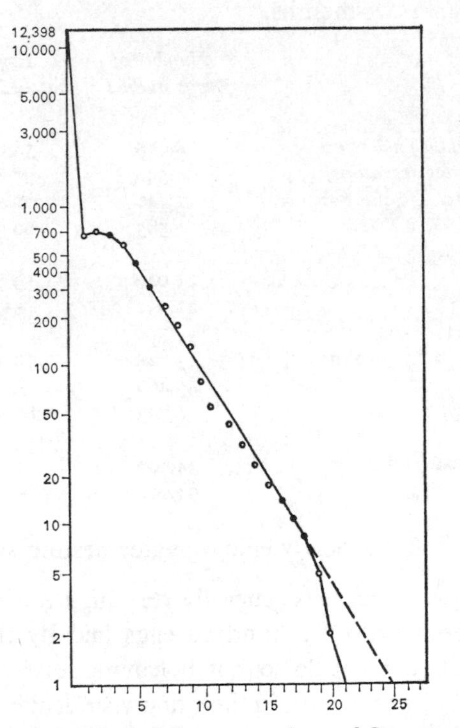

Figure 64. The age composition of the same colony of Common Terns. Ordinate, number of birds; abscissa, age in years.

Data of this sort have been collected on a reasonable number of bird species, allowing survival curves to be drawn up which show great differences between them. The high mortality of most birds during the early part of their lives must be stressed, though even that varies from species to species.

Of one hundred eggs laid by Adelie Penguins *Pygoscelis adeliae*, twenty are lost during incubation and four to five do not hatch, so that seventy-five chicks are produced of which only twenty-five to thirty survive even to the immature stage, by which time the mortality has already reached 70 to 75 per cent (Sapin-Jaloustre). Among Emperor Penguins *Aptenodytes forsteri* mortality may reach 90 per cent between laying and the moult of the immature birds – principally through bad weather and its effect on food supply, and to a lesser extent through predation by Giant Petrels. In contrast the success rate of procellariiform broods is high, since their eggs are well protected. In galliforms, with nidifugous young, the rate of hatching of the eggs is about 44 per cent and the mortality of the chicks 20 to 50 per cent during their first two months.

Species	*Number of eggs studied*	*Eggs hatched* (%)	*Young fledgings* (%)
Species with open nests			
Mourning dove *Zenaidura macroura*	8,018	54·6	46·6
American Robin *Turdus migratorius*	548	57·8	44·9
Red-winged Blackbird *Agelaius phoeniceus*	1,140	72·2	59·2
Song Sparrow *Melospiza melodia*	585	66·5	41·5
values for:			
26 studies	21,040	59·8	—
29 studies	21,951	—	45·9
Species nesting in holes			
Pied Flycatcher *Muscicapa hypoleuca*	3,724	70·7	62·2
Great Tit *Parus major*	45,466	—	64·9
House Wren *Troglodytes aedon*	6,733	82·3	79·0
values for:			
23 studies (8 species)	34,000	77·0	—
33 studies (13 species)	94,000	—	66·0

Table after Nice 1957

Birds which nest in holes clearly enjoy greater nesting success.

The mortality rate of birds is generally very high during their first year, and until they are adult. Of a hundred eggs laid by the Song Sparrow *Melospiza melodia* twenty-six do not hatch, leaving seventy-four nestlings at hatching. Of these 80 per cent die in their first year, leaving only ten birds in reproductive condition the following year, among which the annual mortality rate will then be 43 per cent (Johnston 1956). Among terns the mortality rate is as high as 95 per cent (Austin & Austin 1956), while among Blackbirds it is 72 per cent in the first year, 64 per cent in the second, and about 40 per cent in succeeding years (Halvin). Among Buzzards it is 51 per cent the first year, 32 per cent the second, and then falls to 19 per cent (Mebs). In ducks and game-birds the females suffer heavier losses than the males, but it is not certain that this is a general rule.

After the birds have become adult, the mortality rate stabilizes at a constant level for the remainder of their lives, so that the survival curve is *diagonal*. The rate varies very much between species, as this table shows (from various authors in Lack 1954 and Farner 1955, and more recent data).

Species	*Annual mortality rate* (%)
Yellow-eyed Penguin *Megadyptes antipodes*	10
Manx Shearwater *Puffinus puffinus*	5
Pintado Petrel *Daption capense*	5–6
Snow Petrel *Pagodroma nivea*	4–7
Sooty Shearwater *Puffinus griseus*	6·6
Mallard *Anas platyrhynchos*	40–64
Redshank *Tringa totanus*	25
Ringed Plover *Charadrius hiaticula*	25–30
Common Tern *Sterna hirundo*	30
Heron *Ardea cinerea*	31
Mute Swan *Cygnus olor*	38·5
Pheasant *Phasianus colchicus*	72
Bobwhite Quail *Colinus virginianus*	87
Herring Gull *Larus argentatus*	30
Buzzard *Buteo buteo*	19
Mourning Dove *Zenaidura macroura*	69
Swift *Apus apus*	18
Swallow *Hirundo rustica*	63
White-bearded Manakin *Manacus manacus*	11
Redstart *Phoenicurus phoenicurus*	56
Robin *Erithacus rubecula*	57–66
Blackbird *Turdus merula*	42
American Robin *Turdus migratorius*	48
Starling *Sturnus vulgaris*	50–63
Great Tit *Parus major*	46
Blue Tit *Parus caeruleus*	41–73
Song Sparrow *Melospiza melodia*	44
Red-billed Firefinch *Lagonosticta senegala*	70–75

Thus the mortality is generally between 40 and 60 per cent, or even up to 75 per cent, among passerines, game-birds and ducks. Elsewhere it may be much lower: around 20 per cent in swifts, 10 per cent in penguins and 3 to 7 per cent in procellariiforms. Even some passerines fall into this category, such as the White-bearded Manakin *Manacus manacus* whose annual mortality is only 11 per cent.

A population's structure and stability depend essentially on three values – fecundity, mortality rate, and potential longevity – which enter into an equation with three parameters interlinked so as to ensure the survival of the species and its component populations. High mortality is compensated by higher fecundity or prolonged longevity, or both.

Mortality of the first category is very high throughout life. The success rate of its broods is variable but may be low, while both life expectancy and potential longevity are low. However, fecundity is very high. Among the tits, whose annual mortality rates reach 72 per cent and are compensated by high fecundity, some species raise the young from two clutches of ten eggs every year. Sexual maturity is reached very rapidly. This is the condition in certain tropical passerines such as the Red-billed Firefinch *Lagonosticta senegala* studied in Senegal (Morel 1964). This ploceid lays up to four successive clutches per year, each containing an average of 3·4 eggs, and the birds breed from their second year. Thus the potential fecundity rate is 13·6 eggs per pair per year. However, mortality is very high at each stage. Only 45 per cent of the eggs hatch, and 28 per cent give rise to young which reach the flying stage, while the annual mortality rate of adults is not less than 70 per cent. The mean age of a population in this category is low, older age groups being represented by small numbers of individuals, as is life expectancy. The survival rates of a few such species have been established, as follows (in percentages):

Age	Robin *Erithacus rubecula*	Blackbird *Turdus merula*	Swallow *Hirundo rustica*
0	100	100	100
1	28	60	32
2	15	36	13
3	3·9	18	6·1
4	1·5	11	4·0
5	0·8	7	1·0
6		4·3	1·0
7		2·3	1·0
8		1·5	1·0
9+		0·4	1·0
Life expectancy at hatching (years)	1·0	1·9	1·1

The replacement rate of such birds is very high, which in ecological terms implies a very high productivity. This population structure is found in many passerines, in game-birds and ducks.

In the secondary category on the other hand, the success rate of the broods may be very high while the mortality rate of adults remains low throughout their lives, whereas their fecundity is very low. The structure of such populations is thus entirely different, the mean age being higher because of much greater life expectancy and potential longevity. Such birds obviously show very slow replacement of their populations, implying very poor productivity, so that the part they play in the transfer of energy must be less important, despite the sometimes notable deductions they make from the environment. This is true of raptors and petrels.

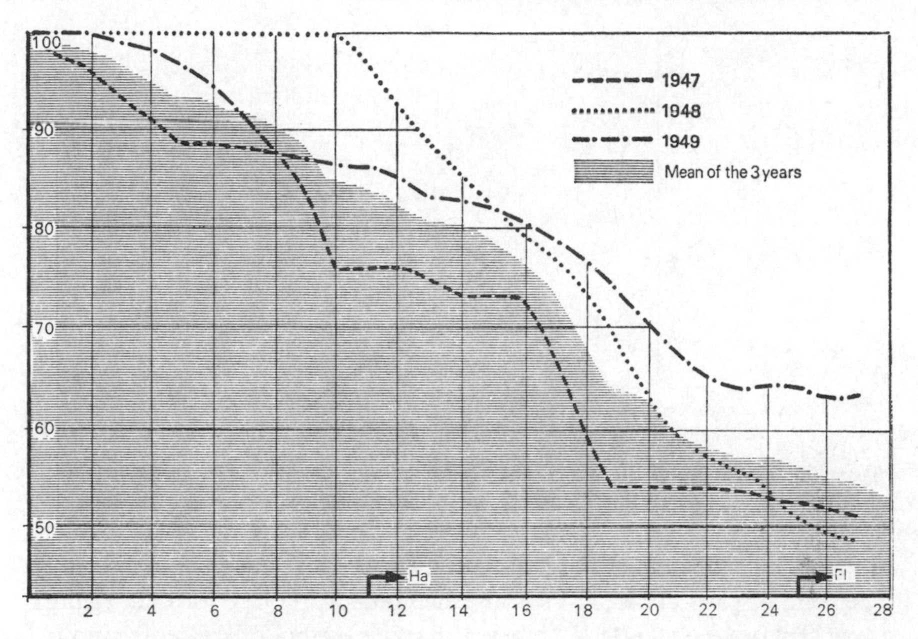

Figure 65. Survival curve of the eggs and young of the Common Grackle *Quiscalus quiscula* (Icteridae). Ordinate, percentage survival; abscissa, days in the nest. Ha=hatching, Fl=fledging.

Prolonged life and a very low mortality rate among adults may sometimes compensate for low fecundity combined with very high mortality among the young. This is notably true of several penguins, which suffer very high mortality before reaching the immature stage. They live to an advanced age and the mortality of adults is very low – the sole causes of death being rare diseases and parasites, accidents, and predation while at sea by certain seals and the Killer Whale – mortality at the colonies being negligible. This may also be the case with tropical forest passerines such as the White-bearded Manakin *Manacus manacus*. Mortality at the nest is high, and it has been estimated that each female only succeeds in raising one young a year. However, the adult mortality is very low, of the order of 11 per cent, which compensates for the losses at an earlier age (Snow 1962).

The population equilibrium is thus maintained in very different ways, mortality at the successive stages being variously compensated by fecundity, so that productivity varies widely between species.

Population structure certainly depends on the species, each of which is adjusted to its ecological niche; but there are also differences within the range of a single species according to the environment, produced by variation

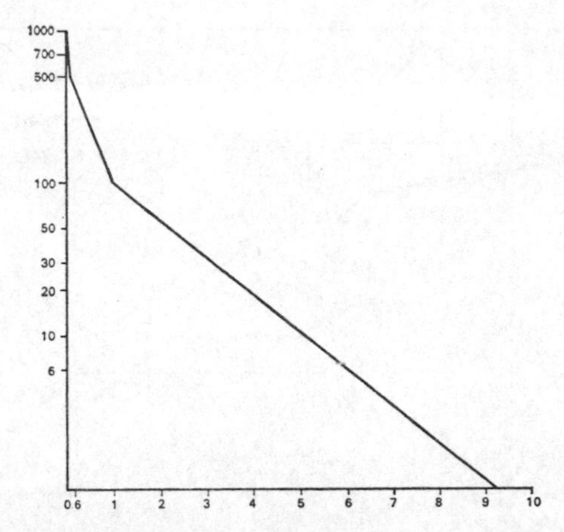

Figure 66 The curve of life-expectancy for the Song Sparrow *Melospiza melodia*, a North American passerine. Ordinate, number of birds; abscissa, age in years.

in essential parameters of the population structure. Thus for example only 7 per cent of pairs of Great Tits raise two broods in Great Britain, against 35 per cent in the Netherlands, where the species is thus of greater fecundity. These differences probably correspond to adjustment to the environment, especially to different predation-pressures. Blue Tits *Parus caeruleus* show a mean clutch size of 11·6 and an annual mortality rate of 73 per cent in Great Britain – whereas in Spain the clutch is only 6·0 but mortality falls to 41 per cent – while in the Canaries the clutch is 4·3 and the mortality 36 per cent.

Dynamic equilibrium and the regulation of populations

The structure of a population at a given moment may give an entirely false impression of being static. The level at which a population of birds is established results from the interplay of two very powerful forces acting in opposite directions: the reproductive potential of the species, and the environmental resistance. As a result populations fluctuate considerably, though in most species the variation is held within fairly strict limits.

First of all there are the annual fluctuations resulting from the fact that – in regions where there is a well marked seasonal cycle – the breeding season occupies only a part of the year. At the end of this there is a population surplus, correlated with a profound change in the numerical ratios between the age classes, a high proportion of the individuals being less than a year old. From this high peak the population decreases after the end of the breeding season because of the high juvenile mortality, and then during the hard season, and reaches its minimum just before the beginning of the next

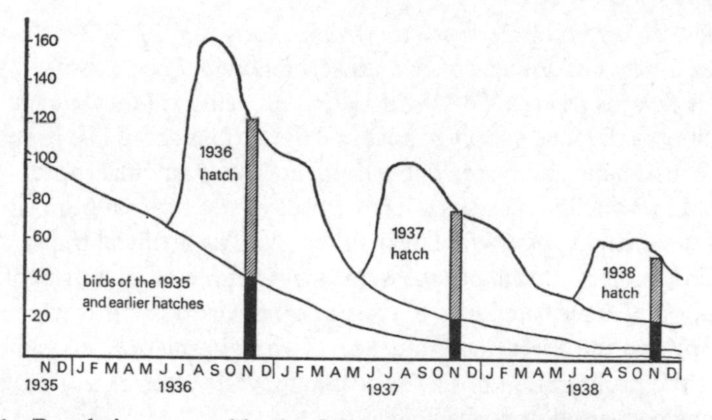

Figure 67. Population renewal in the California Quail *Lophortyx californicus*. The vertical bars represent the age-structure of the population in November: adults in black, immature birds shaded (eggs and chicks not represented).

breeding season. Such annual fluctuations are well shown by the classic study on California Quail (Emlen 1940).

Environmental resistance deflects the population curve on to a plateau, whose level precisely represents the carrying capacity of the environment. For sedentary birds this is the capacity at the most unfavourable phase of the annual cycle (whereas migrant populations can occupy a habitat at the most favourable phase, so as to exploit important but temporary resources). The carrying capacity corresponds to the maximum population which the habitat can sustain throughout the year.

At each annual cycle the population maxima and minima reach different levels, according to the brooding successes and to the losses suffered during the period when the limiting factors act most intensely. In Europe brooding success varies widely according to the climatic conditions of the spring, and a hard winter results in excessive mortality among sedentary populations. In this way the fluctuations in population may attain considerable amplitudes. Thus Dutch stocks of Great Tits may vary from one to four (Kluijver), while census returns from an area in Saxony show that the numbers of Starling pairs vary from forty to 100, and of Pied Flycatchers from thirty-six to sixty-nine (Berndt). It is the very fecund species which show these characteristic saw-tooth fluctuations, varying from year to year within wide limits. In contrast the variation is damped in species of low fecundity, such as storks and herons, whose populations no more than double and halve themselves. However, stocks of such birds take several years to recover their normal level, after reduction due to abnormally high mortality.

These diverse facts, taken together, pose the difficult problem of the *regulation* of bird populations, to which many recent works have been devoted. The most important are those of David Lack (1954, 1966) and Wynne-Edwards (1962). To David Lack, the principal factor which limits populations is the food supply available. Populations should be proportional to the consumable biomass, depending on the particular needs of each species. The wide fluctuations of some birds of the arctic tundra rigorously parallel those in the quantity of available prey. The artificial transformation of habitats for agricultural purposes, puts much greater quantities of food at the disposal of some categories of consumer. Swarms of birds which devastate crops are the direct consequence of the abundance of grain, which explains the proliferation of the Red-billed Dioch *Quelea quelea* in African savannahs converted into rice fields.

A pair does not always raise the maximum number of young which it could feed, taking the available food supply into account. This is proved especially by the raising of nest parasites, which demand more food and more effort than the hosts' normal brood as has been shown for cowbirds (Nice) and whydahs (Morel). However, very many examples are available to show how general is the relationship between food supply and breeding success.

Certain factors which limit populations act with an intensity proportional to the density of the population, so that population increase automatically brings into play the factors which regulate it. Population pressure too increases in proportion with the populations of potential prey. However, it would be wrong to attribute population regulation solely to the food supply. The surplus should be assessed within the framework of the ecosystem as a whole, since it may be used by other consumers which are of the same ecological community yet belong to very different groups. Thus only part of a grain crop is used by birds, other parts being taken by other graminivorous animals such as rodents.

In most cases bird populations do not 'fill up', only rarely attaining such densities that the least increase would exceed the carrying capacity and produce overpopulation. As Wynne-Edwards and several others have stressed, strictly ethological factors seem to maintain them at levels clearly below this threshold, so as to maintain a safety margin avoiding needless losses from overpopulation. Among the most important of these factors is territorial behaviour. Thus the numerical increase of a population produces, through a strictly ethological mechanism, an automatic density-dependent slowing of its growth (Ribaut 1964).

Factors of yet other kinds intervene to limit bird populations, even when the food resources of the habitat could support greater numbers. This is especially true of nesting sites, whose number strictly limits the number of nesting pairs. Populations of birds which produce guano on the coasts of Peru, increased when new nesting grounds were artificially prepared (by enclosing certain peninsulas with walls which made them inaccessible to predators).

It is thus impossible to invoke a single factor to explain the regulation of bird populations. Beyond immediate causes such as the number of nesting sites, food supply is certainly the ultimate factor whether it acts directly or not. According to a general ecological rule, the number of consumers is proportional to the quantity of food available, a relation which thus defines the *maximum* level of population which the habitat can sustain, but which is seldom attained. Ethological factors – especially important in animals as highly developed psychologically as birds – cause dispersal of the individuals, maintaining the population at a lower level which may be termed *optimal.* Using an admittedly anthropocentric word, one could express this in terms of the 'wellbeing' of an individual or population of a given species, though it is difficult to specify this concept entirely objectively.

Thus linkage which brings psychological factors to bear on a purely material relation, connecting the food with the consumer, allows populations to preserve a certain permanent margin, against reaching the limits of overpopulation and of overexploitation of the environment. There always is, or at least appears to be, a surplus, which should be interpreted in relation to the whole animal community and ecosystem. Furthermore, the population dynamics of a given species must be studied within the framework of the whole ecological community, where the ecological interactions are innumerable.

Population cycles

The populations of almost all birds fluctuate only moderately. Most tropical and temperate ecosystems are so complex that the number of factors which intervene (whether directly or not) in the regulation of populations reduces these fluctuations to a minimum. However, this is not true of arctic terrestrial communities, which are considerably simplified in their extreme reduction in number of components. The intervals separating successive cycles are often regular, but not their amplitudes nor the ways in which the increases and decreases take place. These cycles, characteristic of certain arctic birds, are also shown by mammals of the same zones, with periods of from four to ten years.

The regular cycles of predatory birds are easy to explain. Snowy Owls *Nyctea scandiaca*, found in the arctic tundras of the Old and New Worlds, feed principally on lemmings. It is known that when food is abundant owls lay more clutches which are more successful, so that the populations of Snowy Owls increase during good lemming seasons – once again showing the strict relation between species density and the available food supply. Stocks of the rodent then rapidly fall towards a minimum, leaving too many owls dependent on too small a population. They suffer severe losses and some of them try to find better feeding towards the south. During their massive emigrations they erupt into temperate zones in western Europe and the United States, where they are normally unknown. These emigrants soon die, leading the owl populations back to balance with the density of prey in their countries of origin. This phenomenon is usually repeated every four years, the rhythm of the lemmings' cycle of abundance. The example clearly shows how a bird's rhythm of abundance is established as a function of its food supply, and how overpopulation leads to disordered emigration without hope of return, very different from true migrations, during which the surplus population is eliminated. Other carnivorous arctic birds show a similar phenomenon. The Great Grey Shrike *Lanius excubitor* and the Rough-legged Buzzard *Buteo lagopus*, which feed at least partly on lemmings and northern voles, follow more or less accurately the four-year cycles of abundance of their prey. In contrast, the populations of the Goshawk *Accipiter gentilis* nesting in northern North America follow a ten-year cycle, since they feed mainly on Snowshoe Hares, which show a ten-year cycle of abundance.

Regular cycles are similarly shown by herbivorous arctic birds. Thus in Canada the Ruffed Grouse *Bonasa umbellus* has a cycle of abundance with a rhythm of nine to eleven years, and of such amplitude that the density of population at a particular place may vary in the ratio ten to one. Far from being simultaneous throughout the species' range, these fluctuations are out of synchronization, and depend on the particular local population. Other

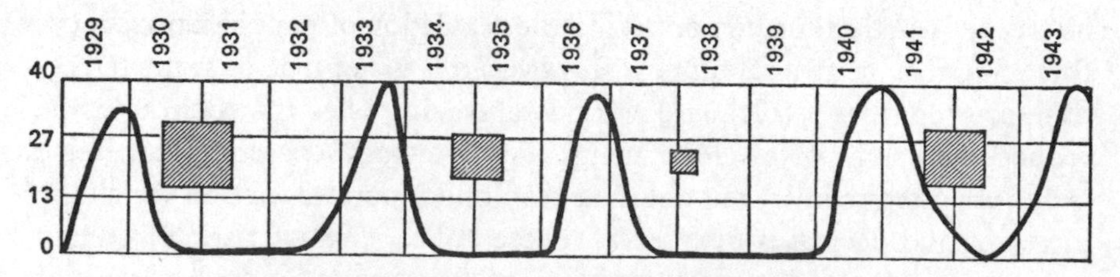

Figure 68. Fluctuations in the Canadian populations of lemmings, related to appearances of the Snowy Owl in New England. Ordinate, the number of lemmings per hectare. The sizes of the squares are proportional to the irruptions of owls. Whenever the lemmings become rarer, owls move southwards.

grouse show similar rhythms – especially the Willow Grouse, which has a four-year cycle in Norway and a ten-year one in Canada. The Red Grouse is distinguished by a rather unusual cycle of six to seven years. The annual fluctuations of herbivorous arctic birds as a whole are parallel to and synchronous with those of the rodents and lagomorphs which occupy the same areas and habitats, though the control of these cycles is still almost unknown.

Oscillations arising from a density-dependent relation between producers and consumers have also been invoked (Lack 1954). When herbivores are very abundant they take a greater quantity of vegetable matter than is made good by plant growth, causing damage with reduction in the plant cover. Reduced food supply leads to a reduction in the stock of consumers, allowing the plants to regenerate, which initiates a new proliferation of the animals and the beginning of a new cycle of abundance. The poverty of arctic environments results in a simplification of the producer–consumer relations. Since the latter cannot find alternative foods as they can in richer habitats, the relation is more immediate and lacks the regulatory mechanism which could stabilize the populations at a constant level. Attempts have also been made to explain these large fluctuations in terms of predator–prey relations. Predators which feed on mammals with regular cycles – lemmings or arctic hares – substitute birds for these principal prey when the latter become scarcer, resulting in increased predation pressure leading to a reduction in the bird populations. Because of the simplification of arctic ecosystems this relation too is established more directly, causing fluctuations without any regulation to damp them down. None of these hypotheses has yet been verified, and it is not impossible that an intrinsic rhythm controls fecundity in a regular cycle. However, the determination of these regular fluctuations is probably complex, resulting from more than the play of a single factor. Cycles shown by many game birds – even in populations of California Quail *Lophortyx californicus* introduced to New Zealand – show that the phenomenon is not necessarily linked to the environment.

Irregular fluctuations in population densities, on the other hand, are much more easily explicable, since they are directly linked to the amount of foodstuffs available. This is notably true of Crossbills *Loxia curvirostra* – finches characteristic of the conifer forests of the boreal zone – which invade eastern Europe in massive irregular irruptions. These usually follow a massive build up of the Scandinavian populations, resulting from a good crop of the cones (especially of the Common Spruce *Picea abies*) on which these dietary specialists feed. When the crop is suddenly poor the populations shift massively, in an attempt to find elsewhere the food supply which

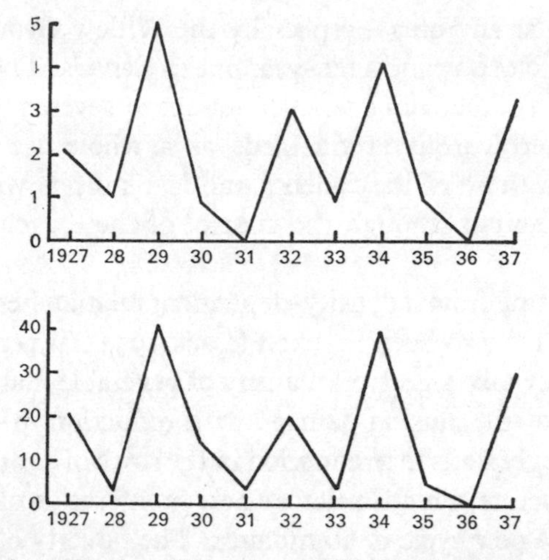

Figure 69. Above, fructification of the Spruce *Picea abies* in Finland (0–5, fructification nil to considerable).
Below, density of Crossbills *Loxia curvirostra* in the same area (ordinate: numbers of pairs identified on a traverse of 120km). The curves parallel one another remarkably closely.

has failed in their own area. Most do not return, and the movements can be considered as a means of eliminating a surplus of population which has passed the carrying capacity of the habitat. So too with the Siberian Nutcracker *Nucifraga caryocatactes macrorhynchus*, which feeds mainly on the seeds of the Siberian Stone Pine *Pinus cembra sibirica* whose crops are exceptionally variable (from 111·5 to 133·5 kg/ha). Its populations fluctuate in parallel with the *Pinus* crop, and emigrate massively when a dearth follows years of plenty. It is notable that the Nutcracker takes a distinctly polyphagous diet in the territories it invades, though the berries and insects which it now takes may be considered as replacement foods rather than its free choice. Pallas' Sandgrouse *Syrrhaptes paradoxus*, of the desert steppes of central Asia, also varies very greatly in density. It feeds mainly on seeds, especially those of one of the Chenopodiaceae *Agrophyllum gobicum*, which varies very much in abundance from year to year. When the crop is poor, or when snow is too deep or floods too prolonged during the winter, the sandgrouse which have multiplied during the years of abundance leave in massive emigrations for Russia and even western Europe. The stocks thus displaced disappear without hope of return, and (despite sporadic attempts to nest) without the smallest chance of establishing themselves in the areas they temporarily colonize.

A complication may be seen among certain birds which simultaneously show regular migrations, irregular emigrations, and irruptions with a definite rhythm. This is true of Waxwings *Bombycilla garrulus*, passerines which constitute a specialized circumboreal family. Searching studies have been devoted to the European populations, which nest in the Scandinavian forests and feed especially on the fruits of the Rowan *Sorbus aucuparia*. Every year Waxwings make regular migrations leading to winter quarters in central Europe; but when their food becomes rare in Scandinavia after years of abundance, these regular movements become massive and disordered, taking on the quality of one-way emigrations. In addition the breeding range of Waxwings widens and their populations increase with a regular periodicity of ten years, apparently independently of the food supply. In the following winter there is a massive invasion of western Europe, earlier than those just considered and also larger in scope, since Waxwings reach the Mediterranean basin. These invasions sometimes take place in years of abundance, so that they are not determined by food supplies. The displaced populations do not return to their nesting areas, and as in all birds making movements of this type their behaviour is disordered and not related to conditions, so that they cross areas where food is abundant without halting there.

Thus these various movements are very complex, and include very different types. The cycles of abundance and erratic movements of herbivorous birds remain basically unexplained. The various examples involve very different types of movement which are linked to large fluctuations in population of which they are the final results, and which are without any doubt controlled by a great variety of mechanisms.

The irregular irruptions and cycles of abundance of carnivorous birds, and the unrhythmical movements of herbivorous birds, are easier to explain. They are certainly controlled by diet, for they are strictly dependent on the food supply and its fluctuations. They show themselves either in extremely simplified ecological communities such as arctic habitats, or among stenophagous birds with highly specialized diets. Simplified ecosystems are subject to considerable fluctuations in their productivity, and this simplification prevents any substitution in the choice of foods. Birds of specialized diet depend on a single source of supply, making them dependent on fluctuations in the latter, and here too there can be no compensation as there is by omnivorous birds.

At the onset of famine, surplus populations attempt to survive through massive emigration; but leaving their habitual environment the birds show strongly disturbed behaviour, and have no chance of surviving. The population returns to a distinctly lower level compatible with the carrying capacity of the habitat. Here we have a true auto-regulation, with a massive and brutal reduction in stocks which benefits the species by avoiding overpopulation.

Densities and numbers of birds

Censuses and counts, undertaken throughout the world, have yielded quite precise information on the densities of bird populations, sometimes of a single species and sometimes of the whole avian community within a particular biome. Naturally the number of individuals of various species per unit area varies very widely between habitats in relation to their food production, and to their capacity to supply the ecological needs of an avifauna, whose diversity will depend on the complexity of the environment. The range of this variation is illustrated by the following table (from various authors):

Biome	*Locality*	*Number of birds: adults/40ha*
Alpine zone	N. Finland	9
Desert steppe	Colorado	10
Rocky tundra	Canada	44
Grassy tundra	Canada	84
Prairie	Michigan	112
Prairie	Washington	246
Grassy scrub or veld	South Africa	65
Scrub	Zambia	310
Tropical savannah	Tanzania	4,000
Dwarf forest	Lapland	36
Fir forest	Switzerland	576
Coniferous forest	N. Finland	120
Taiga	Russia	150
Young deciduous forest	Virginia	522
Mature deciduous forest	Virginia	724
Mixed forest	Netherlands	896
Forest	Germany	5,600
Fallow	Great Britain	200
Cultivation	Great Britain	200
Orchards	Maryland	468
Gardens	Great Britain	3,000
Bird sanctuary	Great Britain	5,800
Mediterranean bush	Southern France	184

Varied habitats like open forest and gardens have a higher total density, because of a greater consumable biomass at different trophic levels. Accordingly their faunas are much more diverse, occupying very varied ecological niches. Nevertheless, increase in the number of species is not necessarily accompanied by increase in the total number of individuals. An environment with little diversification of ecological niches, but where food is abundant at certain trophic levels, can sustain more individual birds than one which is more varied and ecologically more complex, but poorer in total food supply. This fact is clearly shown by the comparison of bird densities in tropical forest and in savannah. The number of consumers depends essentially on the amount of food available, while its variety has the greatest influence on their faunistic diversity.

Enormous densities are attained by certain seabirds at their nesting sites. Sooty Terns sometimes congregate three pairs to the square metre, as do Guanay Cormorants on the Peruvian coast. In a colony of 750,000 Guillemots nesting on Three Arch Rocks, Oregon, each bird had only a square foot.

The total density of birds in certain communities often varies widely with the time of year. Birds which frequent habitats subject to great seasonal variation, such as the arctic and seasonal tropical zones, are partly migratory. Their rhythms of movement and settlement agree with fluctuations in the quantity of food available in the various places they occupy during their cycle.

The many counts recently undertaken allow the stocks of certain species to be estimated. The place of each species in the ecological pyramid strictly limits its numbers, which are further limited by the available area. This limitation may be geographical where land surface is restricted, or ecological if the species is dependent on highly specialized environmental conditions which are rarely met. The small numbers of birds endemic to certain minute islands in the Antilles and in Polynesia are easily explained. In North America Kirtland's Warbler *Dendroica kirtlandi* nests only in woods of young pines within an area about 130km by 100km in Michigan, and its total population must be less than a thousand individuals. In contrast other species have enormous stocks because of their wide distribution, their specialization for a particularly favourable ecological niche, or their great flexibility in adapting to very varied natural or artificial environments. A notable example was the Passenger Pigeon *Ectopistes migratorius*, a bird of deciduous forests in eastern North America which has become extinct, but which was once no doubt the most abundant land bird: a flock said to be of 2,230,272,000 individuals was observed by Wilson around 1810. The size of these populations was related to the abundance of beechmast, acorns and other forest fruits.

Some species have benefited from cultivation, such as the House Sparrow in temperate zones and the Red-billed Dioch *Quelea quelea* in tropical Africa. Others have been considerably reduced, especially among insular populations. The Japanese Crested Ibis *Nipponia nippon* is now represented by only about a dozen individuals, the Mauritius Kestrel *Falco punctatus* by about ten. The most dramatic example is that of the American Whooping Crane

Grus americana, whose world population is estimated at forty-two individuals. Waterbirds such as ducks, geese and small waders, though still very numerous, have none the less been reduced to a disturbing degree.

Recent censuses allow the population occupying given regions to be estimated. The total for Great Britain has been estimated as 120 million birds of all species, of which the commonest are Chaffinches (10 million), Blackbirds (10 millions), Robins (7 millions) and Starlings (7 millions). The total land-bird population of the United States amounts to between 5 and 6,000 millions, of which the most abundant species (at least in the East) is the Red-eyed Vireo, although the Song Sparrow, Chipping Sparrow, American Robin, Mockingbird, Red-winged Blackbird, Meadowlark and Cowbird are scarcely less numerous (Peterson).

Data on population structure and dynamics are of great scientific interest, and are equally fundamental to bird conservation and the rational management of stocks of game birds. Among other things which may be calculated from a knowledge of the numerical sizes of populations is the *biomass*, or total weight of the individuals occupying a given area in a particular habitat, whether calculated for a single species or for the whole avifauna. This value is from some points of view more important ecologically than the number of individuals, since it enters directly into calculations on the energy balance of an environment, and the transfer of energy to the various trophic levels.

Chapter 16

Ancestral Forms and the Evolution of Birds

IF during vertebrate evolution no group had colonized aerial space, a great number of ecological niches would have remained unoccupied. Birds appeared as a branch of the reptiles – highly evolved, 'exclusive' and remarkably well defined, so as to constitute a perfectly distinct class of vertebrates separated from all the rest by a large gap.

Birds have often been compared with mammals on account of several characters common to both. However, the true relationship is relatively slight, between two groups which have evolved in parallel from the same class of vertebrates – the reptiles – but from quite distinct ancestral forms.

Certain structural affinities incontestably associate birds with reptiles, and even with modern reptiles, of which they represent one of the ultimate evolutionary stages.

None the less the birds form a unique group. The demands of flight have modelled avian morphology, and forced birds to acquire a rigidly defined group of characters. No doubt many 'prototypes' appeared and disappeared without leaving descendants, though within a relatively short time there appeared an animal which was already very like a bird, and no doubt it quickly developed towards the modern avian type during the crisis of mutability which all rapidly evolving groups undergo during their palaeontological history. The rapidity of this evolution is no doubt one of the explanations for the exceptional rarity of fossil remains of birds. Another lies in the fragility of bird bones, which also fossilized only rarely and very poorly because of conditions in their habitats. Bird remains are found only in lake deposits, accumulations in caves, and tar pits. Apart from a few exceptional beds like the Rancho La Brea near Los Angeles, from which more than 100,000 fossils representing 125 species have been taken, and Quaternary caves too recent to be of great interest in the history of birds, fossiliferous strata are discouragingly barren of avian fossils. This is especially unfortunate, since a detailed reconstruction of the history of birds would be of exceptional interest for understanding certain aspects of evolution. In all some 787 fossil forms are known, most of them dating from the upper Tertiary (Wetmore).

Birds seem to have appeared during the Mesozoic, more precisely during the Jurassic, as forms which have no modern representatives. After this first 'rough draft' – or at least the only one which has been preserved – other avian types appeared from the Cretaceous onwards, when as far as we can tell from the known remains they were still rare. The group 'exploded' during the Tertiary, when all existing major forms appeared. This evolution is very briefly summarized below.

Archaeopteryx

This most primitive of all birds is known from four world-famous specimens, one of which is no more than the print of a feather. The first skeleton, preserved in the British Museum (Natural History), was discovered at Solenhofen in Bavaria in 1861, in lithographic limestone. The second, acquired by the Berlin Museum, was discovered in 1877 at Blumenberg near Eischstadt, about 20km from the first, while a new specimen was recently discovered in the same Solenhofen quarry. These crow-sized birds date from the Jurassic, when they seem to have drowned in the sea and been buried in mud. Although palaeontologists have sometimes considered them as two different genera, recent researches have shown that they all belong to a single species *Archaeopteryx lithographica* (=*Archaeornis siemensi*).

Archaeopteryx shows a remarkable mixture of reptilian and avian characters. As a whole its skeleton resembles those of certain small bipedal dinosaurs, and consists of unpneumatized bones. Its skull, at once bird-like and reptilian, carries teeth implanted in sockets of both jaws, while its brain (of which a natural cast has been preserved by calcareous concretion) is clearly reptilian. All three digits of the hand have claws. The sacrum extends over not more than six vertebrae, of which five are fused together. Furthermore, unlike all other birds this animal had a long tail made up of twenty distinct vertebrae. Other characters are clearly avian. The pubis, elongated and directed forwards, is of the form characteristic of birds, as are the clavicles fused in the mid-line to form a furcula. Above all, this animal was covered with feathers, arranged as in modern birds and showing the same structure of barbs and barbules. The wing – supported by a strong though keelless sternum, and with the bones of the hand not yet fused like those of modern birds but already reduced in number – carried sixteen incurving remiges, divided into six primaries and ten secondaries, while the tail was covered by rectrices. The leg showed a mixture of characters, some reptilian (well-developed fibula, unfused metatarsals) and some avian (four toes, of which one was opposable).

Archaeopteryx must without any doubt be considered as a true bird whose characters of skeleton and brain having somehow lagged behind the whole, retaining reptilian characters. It is the first animal to have evolved so as to occupy the ecological niche of existing birds. It was arboreal, and no doubt flew badly – its pectoral muscles were probably poorly developed (absence of a sternal keel) and their movements poorly coordinated (small size of the cerebellum); its spine was flexible, and its pelvic girdle vulnerable to landing shocks. It was perhaps merely a good glider, but had none the less conquered aerial space and succeeded in the 'avian metamorphosis'.

Cretaceous toothed birds

There is a considerable gap in geological time before the next fossiliferous strata showing the remains of birds. These are the two types found in 1870–2 in the Upper Cretaceous of Kansas, *Ichthyornis* and *Hesperornis*, each of which includes several forms. Their structure is that of true birds, particularly in their pelvic and pectoral girdles and brains. However, the presence of well-formed teeth in *Hesperornis* justifies separation in a group known as Odontornithes. *Ichthyornis* were gull-like in size and general appearance. Their wings, built like those of modern birds, permitted flapping flight. It was long believed that they also had been proved to be toothed; but the mandible found near a skeleton of *Ichthyornis* (with which it was not anatomically connected) belongs not to this bird but to a mosasaurian reptile, so that their possession of teeth is hypothetical. *Hesperornis*, some of which were nearly two metres long, were like divers, with elongated bodies, vestigial fore-limbs reduced to the humerus and inserted on a keelless sternum, and legs far back on the body. Their mandibles bore conical teeth implanted in a groove, except towards the tip of the bill (the premaxillary), which was no doubt cornified like those of modern birds. Their brains retained some reptilian features. Despite specifically avian characters these birds, together with many others known from very incomplete fragments and resembling herons and cormorants, still showed clear reptilian traces. *Ichthyornis* were strong flyers, showing that conquest of the air had already been achieved on the shores of the Cretaceous seas which then extended over North America. In contrast *Hesperornis* were aquatic birds, swimming by the aid of powerful hind limbs and long mobile tails which helped in propulsion. Both types must have fed on fish, many remains of which have been found fossilized in the same beds.

Bird remains dating from the Cretaceous have also been found in Australia and Europe, whence are known an intermediate between pelicans and cormorants (*Elopteryx*) and forms resembling cranes and ducks (*Gallornis* from France). These fossils show that during this era birds were already widely distributed and highly diversified. Thus the beginning of major differentiation among birds probably goes back to the Cretaceous.

Tertiary birds

There is another long gap, to the Palaeocene, before the next bird fossils, a few rails and cormorant-like birds from New Jersey. An extraordinary phenomenon, a real evolutionary explosion, occurred in the Eocene, when many reptiles disappeared. At least twenty-seven families of modern birds appeared, with the characters which still distinguish them. If the Eocene was the era of major evolution for birds, it was equally so for mammals, which also arose from reptilian ancestors. However, the course of evolution was notably different in the two groups. Eocene mammals were only distant ancestors of modern types which they did not resemble closely – many intermediates were to appear and disappear before modern mammals emerged from that involved pedigree – whereas the contemporary birds were already partly modern (not of course the species, nor often the genera, but the families and orders). Well before the Quaternary there were dead ends, several perhaps poorly adjusted lines destined to failure and disappearance; but the modern avifauna with flying, swimming and running types, appears from the Eocene onwards. Divers, kites of the genus *Milvus*, eagles, fisheagles, flamingos, and waders of the genera *Tringa* and *Numenius* were sixty million years ago already rather like modern forms, and there were also owls, herons, vultures, game-birds, tits (*Palaegithalus*), trogons, cuckoos and even shrike-like passerines (*Laurillardia*). Some others which have now disappeared were intermediate between well marked modern types, such as *Romainvillia* in France and *Telmabates* in Argentinia between ducks and flamingos.

This diversification continued, though slightly more slowly, throughout the Oligocene, 35 million years ago. During this time albatrosses, shearwaters, cormorants (*Phalacrocorax*), turkeys, pigeons, parrots, and many new passerines appeared: sparrows (*Passer*) are known from this era. During the Miocene and Pliocene the rate of differentiation slowed down more markedly, but new types still appear: storks, falcons, bustards, swallows, larks, crows and very many other passerines. By the end of the Pliocene the majority of modern genera was distinguishable. Several giant forms, now extinct, also appeared, such as a kind of pelican *Osteodontornis orri* described from California, which attained 5m in wing span.

The whole modern fauna was present during the Pleistocene, when birds attained their maximum diversity (Wetmore 1951). However, in California of 180 species known in the Pleistocene about thirty have disappeared, the survivors being in every way like existing species (Howard 1955). Climatic changes were no doubt responsible for these accelerated disappearances, to which man may have contributed directly by hunting, or indirectly by felling and bush fires (Martin & Wright 1967). Evolution still proceeded in detail, and most modern *species* differentiated during the Pleistocene from older species belonging to the same genera (Selander 1965). In addition the geographical distributions of many groups were profoundly altered by movements of populations. Some of these were the result of the advancing and retreating Quaternary glaciers, but these geological events did not result in any major differentiation. The major avian lines had long since become fixed.

Reconquest of the terrestrial environment

Though birds' essential conquest was that of aerial space, some seem to have then renounced it and returned to the ground. Having once flown they lost the use of their wings, which became non-functional. This has happened during recent times to species of rails, auks and even passerines; and also in the geological past to entire major groups.

This is especially true of the ratites: the ostriches, rheas, emus, cassowaries and kiwis. Some have maintained that these birds form a homogenous natural group, which separated from other birds before the power of flight was acquired. However, this theory does not seem to agree with the findings of comparative anatomy – the palaeontological evidence proves little, since the fossil relatives of these birds already showed the characteristics of the modern species. The ratites preserve many characters proper to flying birds – notably the structure of the vertebral column and wing, the reduction of the tail, the pneumaticity of bones and the development of the cerebrum – survivals which would be incomprehensible if their ancestors had not been good flyers. They must thus be considered as regressive forms, highly specialized for terrestrial life. The ratites may have derived from a primitive group of flying birds, of which the tinamous are the modern representatives. These show resemblances to the ratites in the structure of skull, sternum and pelvic girdle, and in some biological traits. The relationships are no doubt distant, and without necessarily invoking true polyphyly it is perfectly justifiable to classify each of the ratite types in a distinct order. Some of them have disappeared in recent times – like the elephant birds *Aepyornis* endemic to Madagascar which reached 3m in height; and the moas (*Dinornis*, etc.) of New Zealand, where they formed a real community of many species from which we have remains even of the soft parts.

Before this, similar evolution from the carinates had already produced flightless groups which have long since disappeared. *Diatryma* of the Eocene of North America and Europe, more than two metres in height, was a true carinate, though degenerate in its greatly reduced fore limbs which made flight impossible. It was closely related to the herons, as were the European *Gastornis* which appeared in the Eocene, and South American forms which appeared at the same time (from the Eocene to the Pliocene) as a highly specialized fauna. Among these was *Phororhacos* of the Patagonian Miocene, taller than a man with a head 65cm long carrying a powerful bill. The development of these giant terrestrial birds in the Eocene can be explained by the absence of predators. The archosaurian reptiles had disappeared, while mammals were still of small size and included no forms capable of preying on these giants. There was thus an 'empty place', which the birds attempted to fill by the development of many terrestrial forms; but the spread of mammals was not long in stopping this trend and forcing birds to confine themselves once more to aerial space. Some flightless birds disappeared long ago while others survived longer – some even into the Recent period thanks to very special conditions in insular areas, sheltered from more recent major evolutionary trends, such as Madagascar, New Zealand, Australia and New Guinea. The special biological equilibria of minute islands allow some to survive, such as the wingless rails of Polynesia and the Flightless Cormorant of the Galapagos, just as highly specialized ecological niches do for the Ostrich and Rhea.

Penguins and the marine environment

The penguins present a special case, for they are highly specialized for their aquatic life, and differ very widely from other modern birds. Their wings, incapable of flight, are transformed into paddles, and some authorities have considered them as specialized directly for swimming, evolving independently of other birds as an 'aquatic branch' since the beginning of the Tertiary. However, fossil remains – especially from the Eocene of New Zealand and the Miocene of Patagonia, some of which represent penguins which must have been 1·5m tall and weighed about 100kg – show that they are derived from flying forms. The skeletons of Tertiary penguins were actually more like those of carinates than of modern penguins. For example, the tarso-metatarsal elements were more narrowly fused in the fossil than the modern forms. The palaeontological evidence is wholly in favour of a derivation from a strongly flying form, probably from the same stock as the procellariiforms with which penguins share several skeletal characters. This derivation is very old, since the fossil forms already showed the essential characters of the group. While from some points of view their evolution might be considered as regressive, since it involved loss of the power of flight, it included a high degree of specialization which has allowed these birds to occupy an ecological niche which was formerly vacant, and given some of them an unequalled mastery of the aquatic environment (Simpson 1946).

The evolution of birds during geological time

Birds present a case of oriented evolution which may be unique among vertebrates, their origin lying in the alteration of structure and change of environment involved in the conquest of the air.

During the Mesozoic the conquest of aerial space by vertebrates remained to be achieved, and a range of ecological niches was still vacant. Certain reptiles, especially the pterodactyls of the Jurassic and Cretaceous, had made the attempt, but this was doomed to failure on first principles since the structural organization of reptiles is poorly adapted to the demands of flight.

There is no doubt that the origin of birds is to be sought among the archosaurian reptiles. During the Mesozoic, from the beginning of the Triassic, there was a group of small lizard-sized carnivorous reptiles, the Pseudosuchia, whose remains have been found in South Africa (*Euparkeria*) and in Europe (*Ornithosuchus*, *Aetosaurus* and *Saltoposuchus*). While these were probably ancestral to birds, which they resemble especially in the structure of the skull, they were still very remote from them. They showed a beginning of bipedalism, with fore-limbs distinctly less developed than the hind ones, which latter were very strong and adapted for running during which the long tail served as a balancing organ. Their bodies were covered in elongated scales provided with extensions, which may have presented the general appearance of feathers. They were thus very different from *Archaeopteryx*, a true flying bird covered in feathers and arboreal. No transitional form is known between these two evolutionary stages, so that an intermediate hypothetical being has had to be erected and named *Proavis*, which authors have conceived in various ways. It is really more appropriate to imagine a series of proaves, forming the line leading to modern birds.

One author (Nopsca) visualized *Proavis* as a bipedal animal which ran on its hind legs, just like some modern reptiles, while beating with its fore-limbs. The latter would have developed supporting surfaces, the feathers, and taken the weight off the hind legs, the animal thereafter no longer running but flying. This theory is certainly incorrect, because of a real antagonism between the demands of bipedal running and of flight. It is difficult to see how the animal could have passed from one mode of propulsion to the other. To others (notably Heilmann) *Proavis* was arboreal. Whether bipedal or quadripedal, it would have leaped among the branches thanks to strong hind limbs helped by beating of its fore limbs, soon covered in scales and later in feathers. The development of feathers from scales is wholly unexplained; though it may be accepted that these complex and conspicuous structures were developed at first, not so much as organs of flight, as to provide an insulating layer during the appearance of homoiothermy. It may have been only secondarily that some became sustaining and propulsive organs.

By way of hypothesis one can accept that the archosaurian reptile evolved into *Archaeopteryx* in several stages (Brown). A running bipedal animal would have become arboreal. Reptilian scales would (one still does not know how) have become feathers at the same time as homoiothermy developed. The fore limbs would have become more and more important during leaps, which could be steered thanks to the development of feathers, and which would have progressively developed into planing flight. Finally modifications of the skeleton and musculature would have allowed active flight, so that the stage represented by *Archaeopteryx* would have been achieved.

The latter was certainly a very peculiar bird – so much so that a separate division, the Archaeornithes, has been set up for it against the Neornithes for all other birds including more recent fossils. None the less it showed most of the essential characteristics of birds. Its wing was elliptical in shape and surprisingly modern, resembling those of some passerines though lacking an alula. The absence of a sternal keel and the form of the shoulder girdle made *Archaeopteryx* a poor flyer, and perhaps merely a good glider; but it had none the less conquered the air. It is not certain that it was on the direct line to modern birds, since *Proavis* may have given off one branch leading to it and another to the latter.

By the Cretaceous, birds had divided into two lines, one leading to aquatic forms of which *Hesperornis* were the fossil and the divers (and perhaps other birds such as the Pelecaniformes) the surviving representatives. The other line produced terrestrial or arboreal forms, some of which later became secondarily aquatic (Savile). Thus an important change took place during the Mesozoic, with the appearance of a new type of organization. In the Cretaceous true birds (some at least of which still had teeth) show that the 'prototype' was complete and satisfied all the necessary conditions for a successful aerial vertebrate. Most birds were certainly developed during the Cretaceous, and forms intermediate between modern groups are to be sought at this level, since by the Tertiary the existing forms were already fixed.

Birds began to prosper during the Tertiary, and more especially the Eocene. Their evolution expresses a real explosion in many directions, so as to occupy all ecological niches. The Eocene avifauna was already largely the modern one, including genera which have survived. The situation in birds is not by any means that such forms persisted, but that advanced ones evolved early while the other vertebrates paused in their development for several million years.

Comparison with the evolution of mammals is especially interesting, since these are the other group of highly evolved modern vertebrates. Their mode of evolution was very different. Instead of breaking out from the base into perfected types, they developed progressively with many failures and evolutionary dead-ends. Most modern types of mammal appeared only in the Pliocene or even the Pleistocene, separated by long lines of descent from the primitive forms which were contemporary with the 'modern' birds of the Eocene. To compare the two evolutionary trees, that of the mammals appears as a common stem dividing into many branches which separate at various heights, subdividing endlessly right up into the crown. In contrast the avian tree appears as if trained against a trellis, with the short common stem breaking up into an array of straight parallel branches which give off only minor twigs. As a whole the mainly terrestrial mammals were able to 'hesitate' and to pass through intermediate stages which would not have been viable in the long term, so that they adapted diversely by solving the same problems in different ways. In contrast the birds had to solve a single problem – flight and the conquest of the air – involving many rigorous conditions. As soon as these were fulfilled the perfected bird appeared, which did not need to alter its basic plan of organization, so that all later modifications were secondary. Although the passerines later took precedence over the non-passerines, and some lines disappeared, the basic type remains with a highly evolved structure which has been fixed since the Eocene.

Continuing avian evolution

Since their ancient origin, and especially their major differentiation in the Tertiary, birds have never ceased to evolve in detail. Avian evolution has been dealt with in innumerable studies and remarkable syntheses, especially those by Ernst Mayr (1942, 1963). Despite their relatively slow replacement of generations, birds are ideal material for the study of evolution – both from their own characteristics and because their systematics and biology are very well known.

Evolution is above all a phenomenon of *populations*, since selection acts not at the level of individuals but of large groups. The individuals of a sexually reproducing population are of course by no means identical, but since they can potentially interbreed in every possible way the population is a genetic community, possessing a common *gene-pool* which is the sole evolving entity. The differentiation of a new form (a neutral term for any systematic category, avoiding the need to specify here whether it is a species, subspecies or local race) requires well-defined conditions, since it is a process involving isolation, variation within the isolate, and selection between the variants under environmental pressures.

An originally united population may be broken up by very various types of barrier of which the most immediate and best known are geographical, such as mountain ranges, sea straits, or climatic boundaries. The effectiveness of the latter, some very narrow, can be gauged by studying the insular fauna of an archipelago. Thus in Polynesia and the Papuan region the multiplicity of islands allows various passerines (such as white-eyes *Zosterops*, monarch flycatchers *Monarcha* and thrushes of the genus *Turdus*) to differentiate, with a local race on each island. Climatic fluctuations play an equally important part, as is shown by the effects of Quaternary glaciations. Many species – for example several of the Turdidae such as nightingales, and several warblers – differentiated in Europe and northern Asia, when their original stocks were split by advancing glaciers. In North America the differentiation of the wood-warblers (Parulidae) probably dates from the glaciations at the end of the Tertiary and during the Pleistocene, which markedly influenced these birds – originally of deciduous woodlands – to colonize conifers (Mengel).

Isolation may also be ethological. Nuptial displays make use of very complicated behaviour patterns which act as specific releasers. The appearance of new behaviour in part of a population, as a result of mutation, can result in sexual incompatibility and thus in isolation. Such modifications often follow geographical or ecological isolation, quickly producing more effective barriers to interbreeding between the separated fractions of the original population, which can then evolve side by side without blending and progressively become more distinct (Hinde 1959).

Barriers may equally be ecological, since habitats are sometimes broken up by clearly-defined zones of very different ecology, as happens where tropical forest and savannah interdigitate. Forested patches may lie like islands in oceans of open habitat, each supporting an avifauna which does not venture outside. Ecological isolation may also be due to the acquisition of distinct needs through mutations which affect habitat preference, so that two or more fractions of the original population are confined to as many distinct environments. This contributes effectively to the origin of ecological races, which can then evolve more profoundly in isolation.

In fact the various isolating factors usually act in combination. This is especially true of ecological barriers, which can scarcely be separated from geographical ones. For example, mountain ranges present a succession of vegetation zones, each with definite ecological conditions. Ecological barriers are more effective against small birds with reduced home ranges, since those which are sedentary are restricted to clearly defined areas – as is clearly shown by comparing rate of speciation between small and large birds, and between sedentary ones and migrants.

The second condition for evolution is that the populations should show a high degree of variability, as recent studies on population genetics confirm that they do. While the probability of a particular mutation is low, the very large number of genes which makes up a genotype ensures that the total number of mutations is considerable. As a result all populations show a certain amount of heritable variability, arising both from gene mutations and from genotypic recombination.

The third condition is natural *selection*. The various genotypes which appear in each generation are never equally well adapted to the environment. Selection acts by favouring certain individuals (because of their superior vigour, fecundity, etc.) and thus producing *differential reproduction*, while penalizing others less well adapted to environmental conditions and producing *differential elimination*. This implies, among other things, that populations of the same species will take divergent evolutionary paths, according to their environments. The resulting geographical variation, reflecting adaptations of the species to the diversity of environments which it meets throughout its range, may be more or less continuous, following a geographical gradient and producing a *cline* in the characters concerned. A good example is the colour variation of the Great Tit *Parus major* across its enormous Palaearctic range. If on the other hand the range is broken up by impassable barriers the variation is correspondingly discontinuous, producing local *subspecies*. This pattern is seen especially well in species distributed across archipelagos like those of Oceania.

In an environment where conditions are steady, selection standardizes the population by favouring individuals with the best-adapted genotypes and eliminating the rest (*stabilizing selection*). However, few environments are actually stable even when they seem to be in equilibrium; and where conditions are changing selection ceaselessly acts to adapt the population to them (*dynamic selection*), resulting in evolution which is in a sense directed. Much greater changes, of which climatic fluctuations are only the most conspicuous, also occasionally take place. Birds are mobile, and so liable to invade new areas. Such emigrations open new evolutionary possibilities, by confronting populations with new environmental conditions. The evolutionary history of species is punctuated by crises of change, often resulting from profound alterations in the environment, during which evolution accelerates and differentiation takes on unaccustomed breadth.

Absence of competition encourages profound divergence from a common stock, as is seen on oceanic islands with few faunal elements. Darwin's finches in the Galapagos (Geospizinae), the Hawaiian honeycreepers (Drepanididae) and the vangas in Madagascar (Vangidae) are classic examples of this *adaptive radiation*. This very far-reaching type of evolution certainly allows us to conceive how the expansion of birds took place during the Tertiary, when even on the huge continents innumerable empty niches were available to the new type of vertebrate. Thus the differentiation of new forms is to some extent proportional to the number of ecological niches accessible to the type of animal concerned. This explains the faunistic richness of some environments, especially tropical rainforest, and the poverty of others such as deserts and arctic zones.

Racial differentiation may be much more rapid than was generally believed. Thus House Sparrows introduced to North America in 1852 and to Hawaii in 1870 have already evolved sufficiently to form clearly recognizable local populations, after a lapse of time much shorter than the 4,000 years which some authors had thought necessary for racial differentiation (Johnson & Selander).

What is known as *macroevolution*, because it has given rise to higher categories, relies on the same factors as this *microevolution*. Time plays a most important part, which must be taken into account in any explanation of evolution. Furthermore, the appearance of birds corresponded to a period of important geological and climatic changes; and their rapid diversification to the innumerable niches which were suddenly available to this new kind of vertebrate. Since the Pleistocene, these niches have apparently all been filled and since the environments have remained constant the species are now stable in the major features of their organization. This does not in the least indicate that evolution is not still taking place, nor that species are not now arising by the differentiation of local populations. Avian communities are certainly being continuously perfected, all of their elements evolving by simultaneous interactions. Only the reduced time-scale at which we are able to study these phenomena prevents us from actually perceiving these changes.

Chapter 17

The Classification of Birds

THE problem of classifying birds has long held the attention of ornithologists, and many systems have been proposed in order to establish their major lines. Until the last century the problem was solved in a rather over-simplified way by using only superficial characters – especially external form, way of life and diet – which are in fact in the highest degree adaptive. Thus all birds whose toes are joined by webs were grouped in the Natatores; those with long legs in the Grallatores; those distinguished by the arrangement of their toes and their ability to move up and down tree-trunks in the Scansores; and those with hooked bills and carnivorous diets in the Raptores. These classifications were convenient, but did not rest on any fundamental character. It is to Charles Darwin and the evolutionists that we owe a new approach to systematics, placing birds in a sequence which reflects their phylogeny and grouping them according to their actual affinities. The first attempts at such 'natural' classifications date from the end of the last century, and they were progressively improved in step with increasing knowledge. However, it must be admitted that they are still highly imperfect.

This unsatisfactory state of the higher classification results especially from two features of birds themselves and the course of their evolution. The first is the almost complete absence of fossil forms which are significant and usable in the phylogenetic study of recent species. Only a few rare exceptions help us out of the difficulty on particular points, such as *Elopteryx* of the European Cretaceous which was intermediate between pelicans and cormorants, and *Odontopteryx* which in the Tertiary formed a transition between the Procellariiformes and Pelicaniformes. This absence of fossils is especially regrettable where it concerns the passerines, a group which appeared gradually and acquired more and more importance in the course of evolution. The phylogenetic tree of birds is largely hypothetical. The second difficulty is inherent in the speed of evolution of birds themselves (though this itself may be nothing more than an illusion due to the lack of fossil evidence). Most types seem to have appeared already fully developed and very much as they are now.

Thus ornithologists have been obliged to take account only of contemporary forms, and to set up a classification by trying to make clear their relationships through the study of their morphological and anatomical characters. The false sunbirds (*Neodrepanis*) of Madagascar used to be considered closely related to the sunbirds (Nectariniidae), because of similarities in the forms of their bills and in their nectarivorous diets. It was not until 1951 that they were unquestionably associated with the asitys (*Philepitta*) in a family related to the American primitive passerines (the tyrant flycatchers, cotingas, etc.) by the structure of their syringes (Amadon). Many of the American ovenbirds (Furnariidae) were grouped with the larks by earlier systematists, because of similarities in their proportions, coloration and even ways of life. It is only necessary to leaf through any nineteenth-century work on avian systematics to come across many examples of erroneous classifications due to convergence.

In order to place the classification on a natural basis, modern ornithologists invoke criteria from many disciplines. Under the influence of evolutionists such as Julian Huxley and Ernst Mayr, they have long since replaced the morphological definition of species by a biological one. This movement, known as the 'New Systematics' (Huxley), allows us to bring all our conceptions up to date, and take a decisive step in the understanding of avian phylogeny. While morphological and anatomical characters remain valid, others are drawn from the biological study of birds – especially of their behaviour patterns, some of which are linked to physiological functioning and even to physico-chemical construction. All available characters have to be used, simultaneously and critically; any classification based on a single character is bound to be misleading.

Morphological and anatomical characters

This class of characters has been in use longest and still remains the most valuable to the systematist who understands how to eliminate the effects of the environment or the way of life (especially the diet). From this point of view the form of the bill in particular is highly adaptive, and most bill characters are useless, as is shown by the variation in this organ within otherwise homogeneous families. So too with the shape of the wing, which is closely linked to way of life and habitat.

In contrast morphological characters of the plumage provide much better indications. Some groups are distinguished by the fine structure of their feathers: such as the hummingbirds which in their display plumage have unique barbules provided with two superposed basal lamellae, or the birds of paradise whose feathers show a series of unparalleled specializations. Characters of coloration are similarly useful, some families showing special structures for the production of colour effects, or well defined pigments such as turacin which is confined to the turacos, while negative characters resulting from the absence of a type of coloration also have to be considered. The distribution of coloured areas – the colour *pattern* – is equally valuable, and often leads to sounder conclusions than coloration itself. Thus all thrushes (Turdidae) are characterized by spotted juvenile plumages, which distinguishes them from the neighbouring groups, the warblers (Sylviidae) and flycatchers (Muscicapidae), whose young are uniformly patterned. Study of the juvenile, and sometimes of the female plumages is especially interesting, since these have retained more primitive characters than that of the adult male which has often been secondarily acquired during evolution.

The sequence in which the moult of the remiges takes place has been used to determine differences and resemblances between the various groups studied (Stresemann). The legs also provide some systematic criteria for consideration, despite many secondary adaptations. Thus the scutellation of the tarsus – the arrangement of the scales which cover it are fused into a continuous plate in the most highly evolved birds – allows certain conclusions to be drawn.

In order to supplement external characters avian systematists have called upon internal anatomy, which provides many useful indicators. Thus old studies have recently been taken up again by authors who have widely increased our knowledge of the subject (including Beecher, Bock, Glenny, Tordoff, Verheyen and others). It is especially interesting to investigate the characters of parts of the body which are little affected by the environment, and on which selection has therefore scarcely acted. Such anatomical structures have been preserved unaltered during evolution, and therefore show primitive characters of the greatest interest. This is especially true of the lower part of the skull and particularly the palate, whose structural variety has held the attention of anatomists. Although early authors (such as T. H. Huxley and Pycraft) exaggerated the importance of palatal characters (expressing the relationships between the vomer, the palatine process of the maxillary, the palatine, and the pterygoid) these structures have preserved ancestral traits of some phylogenetic validity. This is also true of the tendons which move the toes, which are grouped or anastomosed differently in different groups, and of certain muscular insertions especially in the pelvic region and thigh. Their slight adaptive importance certainly allows them to be considered, at least in part, as survivals from a distant past, so that in the absence of more obvious relationships their variations reflect deep-seated affinities. Other characters reveal the progressive specialization of an organ such as the passerine syrinx, which has evolved more or less in accordance with the specialization of song from primitive forms like those of non-passerines towards highly developed types.

On the whole anatomical characters have the greatest interest for the systematist. Internal organs and structures, like others, may be affected by external influences, while certain anatomical criteria have been shown to be deceptive. This is especially true of the muscles of the lower jaw. Certain ornithologists have set up tables with columns showing the characters which link a given bird to one group or another. They have deduced affinities by summing the characters in each category, relating those groups which share the greatest number. This *numerical* evaluation is a source of errors, for it takes no account of the fact that many anatomical characters are *linked*. A group of correlated characters should be counted as only one unit, and not as the total number of elements.

Ecological and ethological characters

The biology of birds also provides many characters which can be used in their classification, provided of course that these are critically interpreted. Many of these characters are clearly ecological, such as the manner of nesting which often characterizes an entire family and distinguishes it from neighbouring ones. For example all weavers (Ploceidae) have closed nests in contrast to finches (Fringillidae) which build open ones, a character which leads the sparrows (*Passer*) to be placed with the Ploceidae. The stage attained by the young at hatching (which characterizes them as nidicolous or nidifugous), and to some extent other reproductive characters such as nest site, can be similarly used. In contrast the colour of eggs are on the whole of no systematic value.

The ethological characters shown in various types of behaviour are much more significant than ecological ones. During the differentiation of birds they have been subject to an evolution whose major lines can be followed through careful study. Certain behaviour patterns connected with vegetative life, notably feeding behaviour, have no systematic validity since they are eminently adaptive. In contrast sexual behaviour, especially nuptial displays, are of great importance as many recent studies have shown. No other type of activity has greater significance for the survival of a species, and most of these behaviour patterns are innate so as to ensure maximal reproductive efficiency. Their obvious stability undoubtedly implies that the characters derived from breeding biology are conservative.

However, the characters drawn from sexual behaviour must be treated selectively, since some of them appear only at the species level. This is because nuptial displays present a ritual which is intelligible only to conspecific mates, so as to avoid hybridization, and therefore include postures and movements which act as releasers proper to and characteristic of the species. The displays of drakes are strongly specific, sometimes differing widely between closely related species, and are thus of very limited use in the investigation of relationships. In contrast the displays of ducks have a different meaning, having evolved differently from those of the drakes, and are more conservative and hence more useful in phylogenetic research, as are precopulatory postures. It is thus possible to set up a hierarchy of systematic criteria drawn from the study of behaviour. Precopulatory behaviour is the most conservative, being scarcely affected by divergent selection, and can thus serve as a guide in phylogenetic analysis at higher levels. Ornithologists (notably Konrad Lorenz, and Johnsgard 1965) have carefully studied the breeding biology of ducks, and have thus been able to follow the phylogeny of these behaviour patterns and to distinguish certain groups. The distinctive male plumage of those species with marked sexual dimorphism has been secondarily evolved, and differentiates later in the course of individual development (*ontogeny*), whereas female and juvenile plumages are much more conservative – a fact which has not escaped the morphologists who have used them in their classifications. Considerations of this sort could be repeated for many other groups of birds.

Physico-chemical characters

It is known that a given species shows a combination of characteristics resulting from the very constitution of its substance, and the presence of chemical compounds arranged in a particular way, and the new methods involved in discovering these differences have been applied especially to the study of birds. Serology was one of the first of such techniques to be used, but analysis of birds sera (especially through precipitation reactions) gave only ambiguous results, or ones which merely confirmed points so obvious that they had long been agreed.

Analysis of the proteins in egg-white, begun by McCabe & Deutsch and considerably developed by Sibley, is much more promising. Egg albumen is made up of proteins, which are very large molecules formed of chains of amino acids. There are about twenty different amino acids which combine in very diverse arrangements – the number of theoretically possible combinations being astronomical – to make up the proteins which characterize a

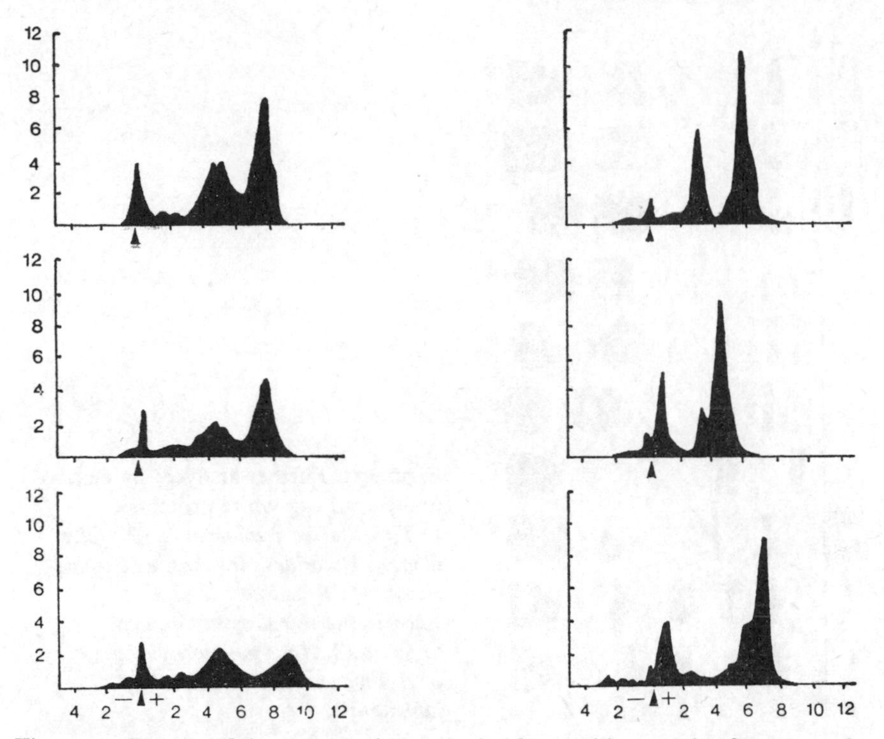

Figure 70. Results of the comparative analysis of egg-white proteins from several birds. Left, top to bottom: *Ardea cinerea*, *Egretta thula* and *Butorides virescens* (herons, family Ardeidae). Right, top to bottom: *Lophura leucomelana*, *Phasianus versicolor* and *Gallus sonnerati* (game-birds, family Phasianidae).

species individually and by their relative proportions. Thus it is tempting to analyse the proteins, so as to investigate the chemical nature of the species, and hence to study its ultimate genetic relationships. In a sense the white of an egg is at the same time living matter and an assemblage of inert chemical substances, stored ready for use by the developing embryo. The fact that its composition is characteristic of the species is shown by its constancy during embryogenesis, the embryo using the various components at proportional rates. Biochemists have further shown that the linear sequence of amino acids in a protein chain reflects the sequence of units making up the genetic code in a segment of DNA (desoxyribonucleic acid).

After various attempts at analysis based on serological and other techniques, the one adopted is paper electrophoresis which allows the separation of components diffusing by capillarity within paper subjected to an electromagnetic field. These substances emigrate and distribute themselves selectively, at differential rates depending on their ionic charges and other

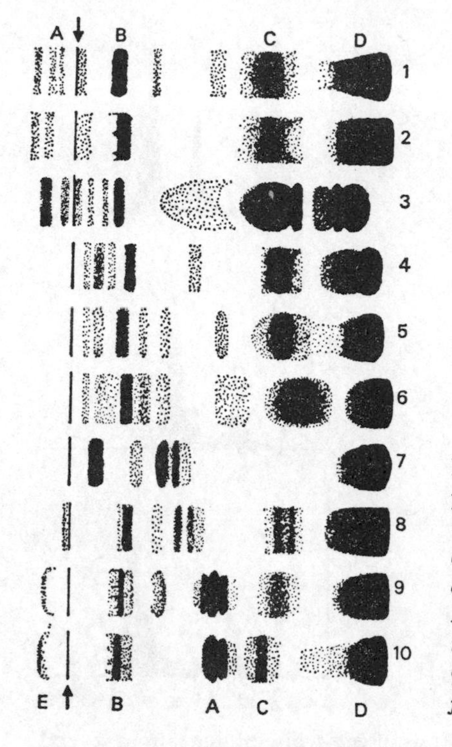

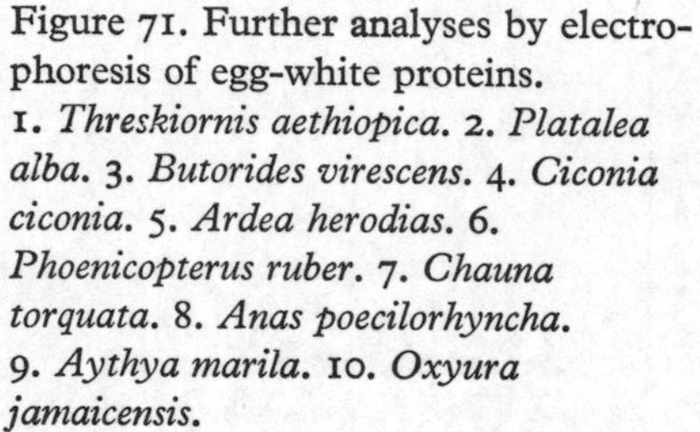

Figure 71. Further analyses by electrophoresis of egg-white proteins. 1. *Threskiornis aethiopica*. 2. *Platalea alba*. 3. *Butorides virescens*. 4. *Ciconia ciconia*. 5. *Ardea herodias*. 6. *Phoenicopterus ruber*. 7. *Chauna torquata*. 8. *Anas poecilorhyncha*. 9. *Aythya marila*. 10. *Oxyura jamaicensis*.

characteristics. Thus proteins separate from one another and can be studied after staining so as to obtain a characteristic density profile, according to classical chromatographic techniques. Though of course this technique does not allow every component to be analysed and its chemical nature understood, it does yield curves characteristic of particular species. These can then be compared in attempting to read a taxonomic message based on the chemical natures of the species concerned. In ten years Sibley has undertaken multiple analyses of 2,000 species of birds, belonging to twenty-seven recognized orders and 146 families. Ducks show close biochemical relationships among themselves as do game-birds, confirming what had been learned from the study of other characters and from their abilities to hybridize. The Charadriformes and the New World nine-primaried oscines both show up as homogeneous natural groups. In contrast the Pelecaniformes and the Coraciadiformes are remarkably heterogeneous, and probably polyphyletic. The flamingos are nearer to the herons than to the ducks. Among the passerines the larks (Alaudidae), swallows (Hirundinidae) and crows (Corvidae) form well-marked homogeneous families, while others such as the wood-warblers, icterids and tanagers (Parulidae, Icteridae and Thraupi-

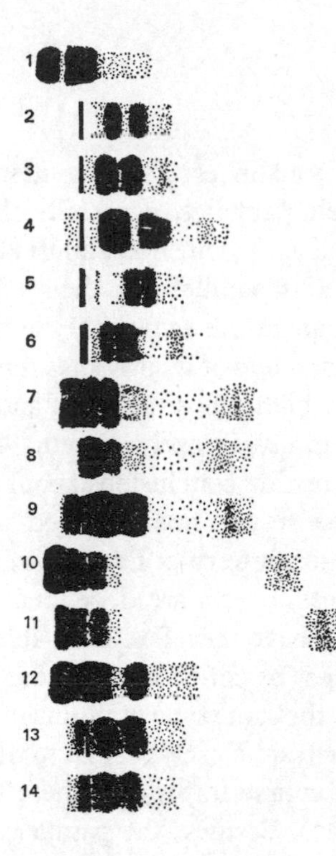

Figure 72. Analyses by electrophoresis of the blood haemoglobins of some birds. 1. *Guara alba*. 2. *Florida caerulea*. 3. *Butorides striatus*. 4. *Mycteris americana*. 5. *Ardea goliath*. 6. *Ardea herodias*. 7. *Phoenicopterus ruber*. 8. *Phoeniconaias minor*. 9. *Cereopsis novaehollandiae*. 10. *Anhima cornuta*. 11. *Chauna chavaria*. 12. *Anas acuta*. 13. *Anser fabalis*. 14. *Branta canadensis*.

dae) cannot be distinguished from one another electrophoretically. This method is not infallible, since it sometimes leads to associations which are clearly wrong on the basis of other criteria. According to its egg-white proteins the Turkey (*Meleagris*) should be intermediate between the common Pheasant and the domestic fowl, while the three gannets *Sula dactylatra*, *S. sula* and *S. bassana* show greater differences between themselves than are usually found between different genera or even families. Sibley has also used the same techniques on many haemoglobins, which are very homogeneous among the higher passerines (Oscines), while the proteins of the eye lens can also be used.

Attempts have been made to make use of chromosomes. However, of all vertebrates the birds are the least well known in this respect, because their chromosomes are very numerous (sixty to eighty-four) and very small. The facts available at present are notoriously insufficient to allow any conclusion whatever, and merely show that from this point of view, birds are remarkably homogeneous.

Other characters

Many other classes of character have been used in establishing natural classifications of birds. Among these are their parasites, especially their biting lice or Mallophaga, an insect order related to the true lice, almost all of which are parasitic on birds. These have evolved in parallel with their hosts, but more conservatively since they do not adapt in the same way; so that even when a host is secondarily adapted the characters of its parasites reflect the origins of both. Mallophaga in general are highly host-specific, and in most cases the parasites of related hosts are themselves closely related. Thus the study of these insects has yielded very interesting conclusions about the phylogeny of their avian hosts. However, these results must be used with great caution, since host-specificity is not an absolute rule. Because of the nature of their arrangements for clinging, these insects are dependent on hosts having a particular feather structure, and have therefore been able to transfer secondarily to very different birds which (by convergence under the action of the same environmental factors) have the same type of plumage. In this way the tropic-birds have Mallophaga which are closely related to those of the terns, though this transfer does not of course imply that these two groups of birds are really closely related (Clay). Besides, the parallel and comparative study of endoparasites such as cestodes and of ectoparasites such as Mallophaga yields directly opposing conclusions in certain disputable cases. While this does not indicate that the study of internal or external parasites cannot contribute to the solution of phylogenetic problems in the birds, this class of characters is no more absolute than others and equally needs interpretation.

Biogeography can bring information of some value to bear on conclusions drawn from the study of other characters. Often the general distribution of a systematic group is well-characterized biogeographically, and any abnormal distribution draws the systematist's attention. Although some species do have unexpected distributions completely unlike those of the other members of their groups, this is not the general rule.

These examples confirm that systematic arrangements should never be based on single criteria, since each character considered may be the result of environmental adaptations and hence of convergence.

The classification of birds is at a very advanced stage in comparison with that of many other animals. The attraction which birds have for man certainly plays a part in this, since they have always received more attention than less spectacular zoological groups. The reliability of ornithological classification at the species level and their relationships is explained primarily by the fact that the characters it uses have been selected in the same context. This classification has long been based on plumage pattern which plays a most important part in nuptial displays and specific behaviour, and is thus of great evolutionary importance.

Systematics reconstructs the phylogenetic tree and is intimately linked to and reflects the major problem of biology, the evolution of living things.

Modern classifications

By using these diverse criteria with discrimination, systematists have succeeded in defining the *species* of birds very satisfactorily. Apart from a few difficult groups in which species limits are obscure, the concept of a species is remarkably clear in ornithology. Species have been divided into subspecies or geographical races, whose definition is much more subjective. Our conceptions at this level need to be completely reconsidered, in the light of the findings of population genetics. The true biological unit is the local population or *deme*, and except in special cases it may soon be pointless to invoke the category and nomenclature of subspecies, which have already been shown by recent studies to be mere abstractions.

In contrast to the species, the genus is much less well defined in ornithology. Classification of the high categories, the families and orders, is a still more complex and largely unsolved problem, involved in the actual phylogeny of birds.

Several classifications have been proposed during the last forty years by those ornithologists most qualified to attempt the synthesis, the best known of which are those of Stresemman (1927–34), Berlioz (1950), Mayr & Amadon (1951) and Wetmore (1954 & 1960). Whether or not they are the best those of Wetmore and of Mayr & Amadon have been the most widely adopted, while they have recently been somewhat modified by Storer, Vaurie and various other systematists. A new classification is in course of development, as successive volumes are published of the *Checklist of Birds of the World* begun by J. L. Peters. These classifications have been subjected to interminable discussion, some highly stimulating and some utterly sterile.

What general agreement there is among ornithologists concerns the non-passerines, whereas agreement on the passerines is much further off. This group, which includes the most highly evolved birds and some 60 per cent of existing species, is a close-grained, dense combination of closely related forms, in which the family concept – well defined in most of the other orders as a result of 'pruning' during evolution – is much more fluid. One passes easily from one family to another, for example from the thrushes to the warblers and flycatchers, which are therefore now often included in a single family. Some birds have been placed in one family or another, like the kinglets (*Regulus*) which some considered as tits (Paridae) and some as warblers.

The most highly evolved passerines divide rather naturally into three major groups: the crows and related families; the other ten-primaried oscines; and those with nine primaries (in which the first or outermost primary has regressed, becoming vestigial or disappeared). Of course, many species can be attached to one or other of these groups only with difficulty. Disagreement has centred on the order in which the three assemblages should be arranged, on which Old World and New World ornithologists differ. The first place the crows and their relatives at the summit because they are more 'intelligent' and 'beautiful' (sexual dimorphism tending to be exaggerated in groups such as the birds of paradise). The second consider that the nine-primaried oscines should be placed in this position, because they show the most varied adaptations.

Our present knowledge allows us to establish a probable systematic sequence which is almost definitive for the lower categories, and more or less arbitary for the relative positions of passerine families.

The table below summarizes the major lines of avian classification as far as they can be expressed at present, adopting on the whole the classification proposed by Wetmore, while taking account of some changes suggested by Storer and the results of various recent studies. This classification gives an idea of the ornithological sequence, but with the above reservations and without implying any preference for one side or other in the dispute.

The Ratites are here divided into several orders, since the resemblances between these birds are convergences due to loss of the power of flight. They have to be classed in a particular group of orders, like the penguins (Sphenisciformes), which are highly specialized derivatives of a group from which the Procellariiformes too are probably descended. The Pelecaniformes (Steganopodes) may be a heterogeneous group, like the Accipitriformes (from which the falcons especially might properly be separated). The Anseriformes are a natural order, although the screamers (Anhimidae) show some affinity with the curassows (Cracidae) among the Galliformes. The Lariformes (gulls and terns), are certainly close to the waders (Charadriiformes), so that many authors include them in a single order, whereas the stone-curlews and jacanas are related to the Gruiformes. Pigeons (Columbiformes) and parrots (Psittaciformes) are well-defined groups, while the hummingbirds and swifts (Trochilidae and Apodidae) are related to one another.

The passerines are much more difficult to classify. It is fairly easy to separate in two suborders the less highly evolved forms, distinguished by the structure of the syrinx. The remainder, grouped in the suborder Oscines, comprise an assemblage of forms which cannot be classified naturally. Of the first two families, the larks (Alaudidae) are no doubt a relatively primitive type, notably on account of their tarsal scutellation, while the swallows are highly specialized and without close relatives.

Classification of birds

(modified from Wetmore, 1960)

Class AVES

Subclass ARCHAEORNITHES

+ ARCHAEOPTERYGIFORMES[1]
    + *Archaeopterygidae* – *Archaeopteryx*

Subclass NEORNITHES

ODONTOGNATHAE

+ HESPERORNITHIFORMES
    + *Hesperornithidae* – *Hesperornis*

ICHTHYORNITHES

+ ICHTHYORNITHIFORMES
    + *Ichthyornithidae* – *Ichthyornis*

IMPENNES

SPHENISCIFORMES
    Spheniscidae – penguins

NEOGNATHAE

STRUTHIONIFORMES
    Struthionidae – ostriches
RHEIFORMES
    Rheidae – rheas
CASUARIIFORMES
    Dromiceidae – emus
    Casuariidae – cassowaries
+ AEPYORNITHIFORMES
    + *Aepyornithidae* – *Aepyornis* (elephant birds)
+ DINORNITHIFORMES
    + *Dinornithidae* – moas
APTERYGIFORMES
    Apterygidae – kiwis
TINAMIFORMES
    Tinamidae – tinamous
GAVIIFORMES
    Gaviidae – divers

[1] Only the more important fossil groups are noted, and those consisting solely of extinct forms are preceded by a cross (+).

PODICIPITIFORMES
Podicipitidae – grebes
PROCELLARIIFORMES
Diomedeidae – albatrosses
Procellariidae – petrels, shearwaters
Hydrobatidae – storm-petrels
Pelecanoididae – diving-petrels
PELECANIFORMES (Steganopodes)
+Odontopteryges
+ *Odontopterygidae* – *Odontopteryx*
Phaethontes
Phaethontidae – tropic-birds
Pelecani
Pelecanidae – pelicans
Sulidae – gannets and boobies
Phalacrocoracidae – cormorants, darters
Fregatae
Fregatidae – frigate-birds
CICONIIFORMES (Ardeiformes)
Ardeae
Ardeidae – herons
Cochleariidae – Boatbill
Balaenicipites
Balaenicipitidae – Shoebill
Ciconiae
Scopidae – Hammerhead
Ciconiidae – storks
Threskiornithidae – ibises, spoonbills
PHOENICOPTERIFORMES
Phoenicopteridae – flamingos
ANSERIFORMES
Anhimae
Anhimidae – screamers
Anseres
Anatidae – waterfowl
FALCONIFORMES
Cathartae
Cathartidae – condors, American vultures
Falcones
Sagittariidae – Secretary Bird
Accipitridae – hawks, eagles, Old World vultures
Pandionidae – Osprey
Falconidae – falcons
GALLIFORMES
Cracoidea
Megapodiidae – megapodes (incubator birds)
Cracidae – curassows
Phasianoidea
Tetraonidae – grouse
Phasianidae – pheasants, fowl, quail, partridges
Numididae – guineafowl
Meleagridae – turkeys

Opisthocomi
Opisthocomidae – Hoazin
GRUIFORMES
Mesitornithides
Mesitornithidae – mesites
Turnices
Turnicidae – buttonquails (hemipodes)
Pedionomidae – Plains Wanderer
Grues
Gruidae – cranes
Aramidae – Limpkin
Psophiidae – trumpeters
Rallidae – rails
Heliornithes
Heliornithidae – finfoots (sungrebes)
Rhynocheti
Rhynochetidae – Kagu
Eurypygae
Eurypygidae – Sunbittern
Cariamae
Cariamidae – seriemas
Otides
Otidae – bustards
CHARADRIIFORMES
Jacanidae – jacanas (lilly-trotters)
Rostratulidae – painted snipe
Haematopodidae – oystertcatchers
Charadriidae – plovers
Scolopacidae – snipe, woodcocks
Recurvirostridae – avocets, stilts
Phalaropodidae – phalaropes
Dromadidae – Crab-Plover
Burhinidae – thickknees (stone-curlews)
Glareolidae – pratincoles
Thinocoridae – seed-snipe
Chionididae – sheathbills
LARIFORMES
Lari
Stercorariidae – skuas (jaegers)
Laridae – gulls, terns
Rhynchopidae – skimmers
Alcae
Alcidae – auks
COLUMBIFORMES
Pterocletes
Pteroclididae – sandgrouse
Columbac
+ *Raphidae* – Dodo, Solitaire
Columbidae – pigeons doves
PSITTACIFORMES
Psittacidae – parrots

CUCULIFORMES
- Musophagi
  - *Musophagidae* – turacos
- Cuculi
  - *Cuculidae* – cuckoos

STRIGIFORMES
  - *Tytonidae* – barn-owls
  - *Strigidae* – owls

CAPRIMULGIFORMES
- Steatornithes
  - *Steatornithidae* – Oilbird
- Caprimulgi
  - *Podargidae* – frogmouths
  - *Nyctibiidae* – potoos
  - *Aegothelidae* – owlet-nightjars
  - *Caprimulgidae* – nightjars

APODIFORMES
- Apodi
  - *Apodidae* – swifts
  - *Hemiprocnidae* – crested swifts
- Trochili
  - *Trochilidae* – hummingbirds

COLIIFORMES
  - *Coliidae* – mousebirds

TROGONIFORMES
  - *Trogonidae* – trogons

CORACIADIFORMES
- Alcedines
  - *Alcedinidae* – kingfishers
  - *Todidae* – todies
  - *Momotidae* – motmots
- Meropes
  - *Meropidae* – bee-eaters
- Coracii
  - *Coraciadidae* – rollers
  - *Brachypteraciidae* – ground rollers
  - *Leptosomatidae* – Cuckoo-Roller
  - *Upupidae* – Hoopoe, wood-hoopoes
- Bucerotes
  - *Bucerotidae* – hornbills

PICIFORMES
- Galbulae
  - *Galbulidae* – jacamars
  - *Bucconidae* – puffbirds
  - *Capitonidae* – barbets
  - *Indicatoridae* – honeyguides
  - *Ramphastidae* – toucans
- Pici
  - *Picidae* – woodpeckers

PASSERIFORMES
- Eurylaimi
  - *Eurylaimidae* – broadbills

Tyranni
Dendrocolaptidae – woodhewers
Furnariidae – ovenbirds
Formicariidae – antbirds
Conopophagidae – antpipits
Rhinocryptidae – tapaculos
Cotingidae – cotingas
Pipridae – manakins
Tyranidae – tyrant flycatchers
Oxyruncidae – Sharpbills
Phytotomidae – plantcutters
Pittidae – pittas
Acanthisittidae (*Xenicidae*) – New Zealand wrens
Philepittidae – asitys
Menurae
Menuridae – lyrebirds
Atrichornithidae – scrub-birds
Passeres (Oscines)
Alaudidae – larks
Hirundinidae – swallows
Dicruridae – drongos
Oriolidae – Old World orioles
Corvidae – crows, magpies, jays
Cracticidae – Australian butcherbirds, etc.
Grallinidae – Australian mudnest-builders
Ptilonorhynchidae – bowerbirds
Paradisaeidae – birds of paradise
Paridae – tits
Sittidae – nuthatches
Certhiidae – treecreepers
Paradoxornithidae – parrotbills
Chamaeidae – Wren-Tit, etc.
Timaliidae – babblers
Campephagidae – cuckoo-shrikes, minivets
Pycnonotidae – bulbuls
Chloropsidae – (*Irenidae*) – leafbirds, etc.
Cinclidae – dippers
Troglodytidae – wrens
Mimidae – mockingbirds
Regulidae – kinglets (goldcrests)
Muscicapidae – Old World flycatchers
Turdidae – thrushes, chats
Sylviidae – Old World warblers
Prunellidae – accentors
Motacillidae – pipits, wagtails
Bombycillidae – waxwings
Ptilogonatidae – silky flycatchers (phainopeplas)
Dulidae – palm-chats
Artamidae – woodswallows
Vangidae – vangas
Laniidae – shrikes
Cyclarhidae – pepper-shrikes

Prionopidae – helmet-shrikes
Vireolaniidae – shrike-vireos
Callaeidae – New Zealand wattlebirds
Sturnidae – starlings
Meliphagidae – honeyeaters
Nectariniidae – sunbirds
Dicaeidae – flowerpeckers
Zosteropidae – white-eyes
Vireonidae – vireos
Coerebidae – American honeycreepers
Drepanididae – Hawaiian honeycreepers
Parulidae – wood-warblers (American warblers)
Ploceidae – weavers
Icteridae – icterids (American orioles, etc.)
Tersinidae – Swallow-Tanager
Thraupidae – tanagers
Catamblyrhynchidae – Plushcapped Finch
Fringillidae – finches, buntings

Thus the birds are placed in a more or less arbitary sequence, leading in principle from the most primitive to the most highly evolved. Almost all the orders and families of existing birds are in fact evolved to similar levels of specialization in their own particular directions. Within the birds as a whole, as in many other animal groups, some characters may remain primitive while others are highly specialized. Since in different groups it has not been the same characters which evolution has most deeply affected, it is of course impossible to classify them in a rigorous order.

The total number of recognized bird species is about 8,600. However, this number varies widely according to the authors consulted. The figure of 8,600 quoted from Mayr may perhaps be regarded as a minimum, which we can use as a base in comparing the number of existing bird species with those of other vertebrates. Adopting uniform systematic concepts, we get the following figures:

Fishes	20,000
Reptiles and amphibians	6,000
Birds	8,600
Mammals	3,200

Except for the more numerous fishes, birds have the widest present distribution across the world, bird species being almost ubiquitous. Only about six new species a year are described, and nothing has recently overturned the classification, though this does not mean that very interesting discoveries do not remain to be made.

The predominance of the passerines over other groups is noteworthy. Of the 8,600 species the passerines amount to 5,100, of which 4,000 are oscines.

Speciation has been exceptionally active during the recent history of this group of vertebrates, while the others have slowed down at the same time as they have lost many species by extinction. The speed of speciation is well shown by the ratio of number of species to number of genera, which is much higher among the passerines. Thus the difficulty of classifying them is not at all surprising, since their evolution has been recent and is far from ended.

Chapter 18

Bird Distribution

BIRDS show great diversity across the world, forming well defined faunistic assemblages which are characteristic of the regions in which they occur. Their distribution certainly results from their evolution during the distant geological past. Palaeontological research has shown that there were birds in many parts of the world during the Cretaceous, but also that they did not have the same distribution as at present. Thus trogons and parrots, birds now confined to tropical regions, were living during the Tertiary in France, England and Germany (which also shows that at that time the climate of western Europe was very different).

The present distribution of birds depends above all on recent geological events. These phenomena continue up to the present, as is shown by the natural contractions or expansions in the ranges of many species.

While bird distribution follows from the history of dry land and the distribution of centres of differentiation and dispersal, it depends above all on ecological needs and the distribution of habitats suitable for each species. Biogeography must lean at least as much on ecology as on palaeogeography. While the distribution of birds can be studied at a given moment, its dynamic aspect must never be forgotten: distribution is constantly changing as a result of environmental fluctuations. Thus biogeography is essentially the study of a phenomenon in course of evolution.

Ubiquity of the avian class

Birds are distributed through almost every part of the globe from pole to pole. While the great majority occupy dry land the oceans have their share, since petrels and albatrosses are effectively pelagic for parts of their annual cycle. Of all vertebrates, birds reach farthest towards the north. Gulls and even passerines such as the Snow Bunting *Plectrophenax nivalis*, have been reported about 240km from the North Pole at latitude 88°. Equally, they are found nearest the South Pole. Sir Edmund Hilary met Giant Petrels *Macronectes giganteus* only 80km from the Pole while crossing the Antarctic continent. Their altitudinal distribution is also wide. They even nest to

5,000m in the Andes, almost at the limit between vegetation and eternal snow, and climb even higher during migratory flights. Geese have been observed travelling at more than 9,000m, while ducks and waders fly over the Himalayan ranges.

Not only birds as a whole but many much more restricted groups among them have worldwide distributions, such as some plovers (*Charadrius*), ducks (*Anas*), herons (*Nycticorax*) and pigeons (*Columba*). Gulls of the genus *Larus* are found along the coasts of the whole world, many inland waters, and even on lakes at high altitude in Tibet and the Andes to a height of about 5,000m. Similar cosmopolitan distributions are shown even by single species, such as the Osprey *Pandion haliaetus*, the Barn Owl *Tyto alba* and the Turnstone *Arenaria interpres* – a wader which is met with throughout the world, either as a breeding bird or as a wintering migrant. Such birds belong mainly to relatively primitive orders and families, and are either highly adaptable or dependent on environments which are themselves widely distributed across the globe.

However, such worldwide distributions are rare, even in groups which are themselves of extended range – they are particularly rare among passerines. In contrast to them are birds with very local distributions. Many families are confined to a single continent or even part of one, such as the Pipridae, Cotingidae, Dendrocolaptidae, Furnaridae and Trochilidae of Tropical America and the African Musophagidae. Others are dependent on a particular environment, such as the trogons which are confined to tropical rainforests in the Old as well as the New World. Some species, and even certain genera, have remarkably restricted distributions. Such endemism, while it is common enough on islands where isolation provides a sufficient explanation, does occur occasionally even on the great continental masses. The most surprising examples are those of Kirtland's Warbler *Dendroica kirtlandi*, which nests only in a small area of northern Michigan; and the Ipswich Sparrow *Passerculus princeps*, which nests only on Sable Island off Nova Scotia which has an area of about 50km^2. Mountain birds – such as certain Andean species and in particular the hummingbirds – are often confined to one mountain range, or sometimes to one valley. Such distributions are no doubt explicable in terms of a habitat with well-defined conditions, closely hemmed in with 'uncrossable' barriers where ecological conditions change sharply.

Limited ranges are much more difficult to explain where the species concerned occupies an area without sharp relief, in the midst of a habitat which is apparently uniform over wide expanses. Possibly this results from a limited potential for expansion, the species' population dynamics involving no numerical increase such as would oblige it to expand its range; or perhaps these birds are confined to their well-defined ranges by specific ecological needs which have so far defied analysis. Competition with another species explains only a limited number of these cases, and must itself be related to environmental factors so that one of the species dominates in one habitat and one in another.

Range changes during geological history

Some avian groups once had much wider distributions than they have now. These total disappearances from parts of former ranges can be explained by climatic fluctuations. Some birds, such as the parrots and especially the trogons, have probably always been dependent on a particular type of habitat and consequently on a very special climate, as a result of their ecological and especially their dietary needs. Other birds in contrast seem to have evolved during their whole histories within well-defined areas from which they have never escaped. This is true of the wingless moas of New Zealand and *Aepyornis* of Madagascar, no wider traces of which have been found even in the most distant geological history, and of the penguins which have always had a strictly southern distribution.

Centres of bird differentiation were probably in the northern hemisphere, and possibly even in Eurasia. From there they would have spread, probably by rapid emigrations, to invade the southern continental masses of Africa, South America and Australia as well as the Pacific islands. This explains the relative homogeneity of the northern faunas (the more so since the climatic zones are evenly distributed around the pole), while the southern faunas differ greatly from one continent to the next. These peripheral regions have themselves become the centres of distribution of highly specialized avifaunas. This is particularly true of South America, where a remarkable number of types of bird have differentiated, but also of New Guinea, Australia and New Zealand. In these last regions, insular isolation has prevented the groups which had thus evolved there from undertaking reverse colonizations. Australo-Papuan birds, which are certainly descended from very ancient asiatic stocks, have scarcely recolonized Malaysia – because the land areas are very discontinuous and also to some extent because of the resistance to such an invasion set up by the birds of South-east Asia, which form a thriving avifauna of closely-knit communities. In contrast there is clear evidence for a reverse migration in the New World, since the North American avifauna contains a high proportion of neotropical elements originating in South America. The tanagers, hummingbirds and icterids are only the most conspicuous of these characteristically tropical birds to have settled in a temperate country. However, not a single neotropical bird has penetrated the Old World, in exchange for the stocks which the latter has contributed to tropical America.

The tropical Old World avifauna has scarcely invaded Europe, except for a very few elements (the Roller, Bee-eater and Golden Oriole, and perhaps the swifts). This is explicable by geographical circumstances and the profound differences in structure between the continents. Temperate Europe is separated from tropical Africa both by the Mediterranean Sea and by the Sahara Desert, which do not encourage species to spread. Farther east Asia is cut in two by a barrier of deserts reaching from the Near East to Mongolia, and by great mountain ranges. Thus the tropical and the temperate and cold faunas are very clearly separated by extensive geographical and ecological barriers. Faunal mixtures are rare, the Ethiopian and Oriental faunas being clearly distinct from that of the Palaearctic and the change taking place abruptly on either side of definite barriers. Only in Japan, as a result of a favourable climate and gradual climatic gradient, is there a more intimate mixing of the faunas.

The configuration of the New World is entirely different. The mountain chains run generally north and south, and there are no such extensive desert or marine barriers. Even the Gulf of Mexico is easily skirted by way of Central America or the West Indies. Thus one can pass gradually from cold through temperate to tropical environments, while in each zone wet and dry habitats intermingle. As a result many neotropical birds have returned as far as the United States or even Canada. Thus, through ecological and geographical causes, the mixture of nearctic and neotropical birds is very intimate and gradual, in marked contrast to the situation elsewhere. Correspondingly, certain northern elements have penetrated as far as Patagonia, thanks to the high Andean chain, which provides an almost uninterrupted corridor of open, cold or temperate habitats between Colombia and Chile, across the hot and humid intertropical zone. This explains for example the distribution of pipits (*Anthus*), a group of arctic origin found as far as Tiera del Fuego.

The avian lines which have colonized oceanic islands have evolved still more profoundly than those on the peripheral continental masses. As is true among other animals, insular avifaunas essentially comprise two elements: firstly the primitive forms which have disappeared elsewhere, but have survived in the shelter of evolutionary backwaters; and secondly forms which are over-specialized in particular directions, and sometimes resemble members of very different lines. As a result these islands always have a very high proportion of endemic forms.

At certain periods the extension of glaciers has driven the northern faunas southwards. After the retreat of the glaciers the avifauna again moved northwards as fast as the land became habitable; but part of the northern fauna remains in more southerly regions, where high altitudes allow. This explains those discontinuous distributions which biogeographers call boreo-alpine, where certain species which are widely distributed in the cold zones of high latitudes have left isolated *relicts* in the mountain chains farther south. The ranges of these populations, which have often evolved into distinct races, are ecologically like continental islands, separated from the main range of the species by wide intrusive zones which for obvious ecological reasons it cannot occupy. Thus Ptarmigan are found in the Alps; and crossbills, crows and juncos in the mountains of Central America, surrounded by the wide-spread tropical avifauna of the lowlands.

The Quaternary glaciations have likewise had profound effects on ocean levels. During these great withdrawals of water in the form of ice, Malaysia and the bordering islands of Java, Borneo and Sumatra formed a single continent, later broken up by straits which remain shallow. This explains why these islands on the same continental shelf are occupied by a rather uniform fauna, in strong contrast to those of Celebes and the Lesser Sunda Islands. The Macassar Strait between Celebes and Borneo is much deeper, and marks the western edge of a much more unstable area between the Malaysian and the Australo-Papuan or Sahul shelves. The Oriental and Australasian faunas have never been in free communication across this zone: hence the faunistic differences on either side of this biogeographic frontier, Wallace's Line.

The present distribution of birds

It is a surprising fact that the birds of England and Japan are very similar (in avifaunal facies, even if the species are different) despite the intervening distance of 11,000km. In contrast it is only necessary to cross the Sahara to find oneself among birds which are entirely different from those of Europe and North Africa. Similar differences are sometimes obvious over quite short distances, as between Africa and Madagascar, or Borneo and Celebes. These very clear faunistic differences between one region and another allow faunas to be defined, and the surface of the globe to be divided into a number of biogeographical regions. Birds above all other animal groups are used to characterize these regions. In 1858, Sclater traced the limits of natural regions which we still recognize; and in 1876, Wallace pushed the analysis further and virtually founded the new science, based largely on ornithological considerations. The classical biogeographical regions are indicated below, together with the avian groups which characterize them. Sometimes they have clear boundaries, especially where they are bounded by extensive seas, but more often their limits are gradual, passing through transitional zones in which elements belonging to different faunistic assemblages blend.

Palaearctic region: comprising Europe, North Africa and northern Asia (as far as Iran, Afghanistan, the Himalayas and northern China). As an area this region is poor in species, and very few birds are confined to it. Of the sixty-three families found there, only one is endemic, the Prunellidae (accentors). Several other families (the Tetraonidae, Bombycillidae and Certhiidae) are found elsewhere than in the Palaearctic region, but are more highly diversified in Eurasia and thus to some extent characteristic, as are the many members of the circumpolar families, the Alcidae (auks) and Gaviidae (divers). More avian genera are characteristic of the Palaearctic, while the region can equally well be characterized negatively, since a large number of tropical birds typical of the Ethiopian and Oriental regions stop at its boundaries.

About 1,100 bird species of sixty-three families, of which 579 species are passerines, occupy the Palaearctic region. The dominant types are raptors, owls, woodpeckers, thrushes, tits, crows, wagtails and pipits, waders, and ducks. Many of these birds are migrants because of the rigorous winter climate, some reaching the tropical regions of Africa and Asia.

Nearctic region: comprising North America, and Greenland, as far south as Mexico where the limit is often placed at the Isthmus of Tehuantapec. This region shows indisputable affinities with the Palaearctic, so that some biogeographers unite them as the Holarctic region. A considerable number of birds of the most northerly zones are common to both parts of the world, because of very similar ecological conditions, but these common circumboreal elements are of less and less importance in the more southerly zones. During geological history North America has received a substantial influx from the eastern Palaearctic, across the Bering Strait. Even the ancestors of some groups which later evolved in the Neotropical region had arrived by this route a very long time ago. The most recent immigrants, such as the Tetraonidae and Phasianidae and various passerines which are better represented in the Old than the New World, remain in the coldest parts of the northern hemisphere thus betraying their origin. In contrast few typically American birds have invaded eastern Siberia, and none have reached Europe except the wrens – which originated in the New World where they are well represented, whereas they have but a single representative in the Old World.

The Nearctic region is also characterized by an important Neotropical contribution. These elements, which are more and more numerous towards the south, form the dominant part of the *Sonorian fauna*, which preponderates from the central USA to Mexico and is characterized by many species and fewer genera which are peculiar to it, though at higher taxonomic levels it is only moderately distinct. It provides a gradual transition with the Neotropical fauna.

Because of their composite origin the birds of the Nearctic region do not include a high proportion of endemic forms. Despite their variety, in sixty-two families of nesting birds, only the Meleagridae (turkeys) are characteristic of this part of the world, and even they are not confined to it. The avifauna of the Nearctic is relatively poor, containing only about 750 species, in which the dominant groups are the mockingbirds, tyrant-flycatchers, vireos, wood-warblers, icterids, and the graminivorous passerines of which the buntings or American sparrows are the most important. Many of these birds, especially within the Neotropical element, are migrants.

Neotropical region: extending from the edge of the Mexican plateau across Central America and the West Indies to the tip of South America. The Amazon basin and surrounding areas form the heart of the tropical part, but wide areas have very different climates culminating in those which are extremely cold. The Andes bring areas of high altitude right into the tropical region, while South America reaches farther southwards than any other continent into the subpolar zone. Thus the Neotropical region presents a greater variety of habitats than any other part of the world.

Hornbills, tits, and true shrikes, and such almost ubiquitous birds as crows of the genus *Corvus* are absent here due to early isolation. Pantropical families are here mostly represented by endemic genera, especially among the trogons, barbets and parrots (of which only a single subfamily occurs in America).

Despite the notable absences, this region has been the centre of origin of a particularly rich avifauna, unequalled anywhere in the world. The title 'Continent of birds' bestowed on South America is well deserved, for the neotropical region is occupied by eighty-nine avian families represented by about 2,500 species. Here speciation, and even differentiation at higher taxonomic levels, have reached a very advanced state. Several groups with world-wide distributions have undergone truly explosive radiations on this continent. Thus thrushes of the genus *Turdus* are represented here by as many species as in the whole of the rest of the world, where they are found everywhere except in Australia. In North America there is only one species. The proportion of forms endemic to the Neotropical region is very high, in accordance with the relative isolation of this part of the world. Twenty-five bird families are endemic or scarcely transgress the boundaries of this region, in which they certainly originated. Among these are notably the rheas (Rheidae), tinamous (Tinamidae), curassows (Cracidae), the Hoazin (Opisthocomidae), the Oilbird (Steatornithidae), todies (Todidae), motmots (Momotidae), jacamars (Galbulidae), puffbirds (Bucconidae), toucans (Rhamphastidae), manakins (Pipridae), cotingas (Cotingidae), and a long series of oscine passerines. The hummingbirds (Trochilidae) also originated in this region.

It is noteworthy that among this group of birds the non-passerines occupy a relatively more important place than elsewhere in the world. South America is also the only region where primitive passerines of the Mesomyodes (the Furnariidae, Formicariidae, Dendrocolaptidae, Cotingidae, etc.) doimnate the oscines, which elsewhere have replaced them during the course of evolution. It has thus provided a refuge for a relatively primitive fauna, and even for a few 'living fossils' such as the Hoazin. However, this has not prevented the Neotropical region from giving rise to highly evolved forms among the higher passerines, as the tanagers, wrens and finches prove.

Apart from a great number of water birds (ducks, herons and egrets, etc.), the dominant Neotropical birds are hummingbirds, parrots, toucans, barbets, pigeons, cuckoos, manakins, cotingas, tyrant-flycatchers, woodhewers, ovenbirds, icterids, tanagers and finches.

The West Indies are especially notable. Apart from a very few endemics including the todies, this area is characterized by a marked relative impoverishment due to its insular isolation. Many characteristic Neotropical groups are totally lacking – especially the tinamous, curassows, the Piciformes apart from a few woodpeckers, the motmots and tracheophone passerines – while others are represented only by a very few species. However, a sizeable Nearctic component has reached the Greater Antilles. Thus the West Indies, despite their faunal impoverishment, provide a meeting area for the faunas of the two regions, in which because of their climate the Neotropical contribution is dominant.

The Galapagos, a volcanic archipelago in the Pacific just below the Equator and 1,000km from the continent, is much poorer still in species. Its very unusual climatic and oceanographic conditions, bathed by the cold

Humboldt Current, have allowed Antarctic species including a penguin to reach these low latitudes. Furthermore, it has been the place of origin of remarkable endemic forms – among them the Geospizinae or Darwin's finches, a long independent branch of the Fringillidae, which play a key part in our understanding of the evolutionary process.

Ethiopian region: extending over the whole of Africa south of the Sahara, and Madagascar, plus south-western Arabia. It forms a comparatively homogeneous block; especially since its southern tip, which does not extend as far southwards as South America, does not constitute a truly temperate zone. All biotopes which can occur within the tropics are represented here, from rain-forest to desert by way of various types of savannah.

The Ethiopian avifauna as a whole is rich, containing about 1,750 species in sixty-seven families, seven of which are endemic: the Shoebill (Balaenicipitidae), the Hammerhead (Scopidae), guineafowl (Numididae), the Secretary-Bird (Sagittariidae), turacos (Musophagidae), mousebirds (Coliidae) and helmet-shrikes (Prionopidae). Several well characterized groups of birds are confined to the African continent, such as the ostriches, wood-hoopoes (*Phoeniculus*), viduine whydahs, oxpeckers and sugarbirds (*Promerops*), while the majority of honeyguides, bee-eaters, rollers, starlings, shrikes and sunbirds are also found there. The Ethiopian region is clearly differentiated from the Palaearctic (although it receives from there many migrants during the northern winter), but shows close affinities to the Oriental region (Moreau). Not less than 30 per cent of genera, and even 2 per cent of species, are common to the two regions, although such a small proportion of common species indicates a long history of diverging evolution. Apart from those already mentioned, the most characteristic birds are pigeons, bustards, cuckoos, nightjars, swallows, hornbills and orioles, and many other passerines among which the Turdidae and Ploceidae predominate.

Madagascar is indisputably a biogeographic entity, although its inclusion in the Ethiopian region cannot be seriously doubted. The island has been isolated for a very long time from the neighbouring continent, so that its fauna is considerably impoverished and contains a very high proportion of endemic forms, mostly related to African birds but with a few elements of clearly Asiatic affinity. Madagascar has also served as a refuge area for certain primitive types which have disappeared elsewhere, and has been the site of specialized differentiation. For example the vangas (Vangidae), related to the shrikes, have undergone adaptive radiation so as to occupy a series of ecological niches left empty by the insular impoverishment.

Oriental region: extending over the tropical part of Asia to Malaysia and the Philippines. In the west the great Himalayan ranges form a sharp natural frontier separating it from the Palaearctic region. In the east its frontiers are vaguer, notably in China where one passes gradually from one fauna to the other, while even in Japan the tropical element is still clearly perceptible.

The Oriental region is occupied by a rich fauna comprising about 1,500 species in eighty-three families. Endemism at the family level is very slight, since only the Chloropsidae are confined to the region. However, a few others – such as the game-birds, certain groups of pigeons, the woodpeckers and the pittas – though they are represented elsewhere, have their principal centres of dispersal in South-East Asia. The most characteristic Oriental birds are pheasants, pigeons, parrots, trogons, woodpeckers, babblers and sunbirds, as well as graminivorous passerines. Many Palaearctic migrants winter in the warmer parts of Asia.

Australian (or Oceanian) region: To the east of the preceding area, this region extends over New Guinea, Australia, New Zealand and Oceania to the most distant Pacific islands. It is of profoundly insular character, being broken up into a great many widely-separated islands, of which the largest is the Australian continent itself. These islands have had no continental connections during the evolution of modern birds, hence the great faunal peculiarity of the area. Although poorer in species than the Neotropical region, Australasia shares with South America the distinction of having the most peculiar avifauna. Birds have only been able to reach it by 'island-hopping'; and the increasing separation and diminishing sizes of the islands, and their decreasing diversity of habitats, in passing towards eastern Polynesia explains the progressive faunal impoverishment and higher degree of endemism encountered towards the east of the region.

No less than eighty-three avian families make up this fauna, of which seventeen are endemic, though some of these contain only a very few or even a single species. The most striking endemic families are the cassowaries (Casuaridae), emus (Dromiceidae) kiwis (Apterygidae), megapodes (Megapodidae), the Kagu (Rhinochetidae), lyre-birds (Menuridae), butcherbirds (Cracticidae), mudnest-builders (Grallinidae), birds-of-paradise (Paradiseidae), bowerbirds (Ptilonorhynchidae) and honeyeaters (Meliphagidae). The parrots (especially cockatoos and nectarivorous brush-tongued lories), kingfishers, pigeons, and several passerine groups such as the flycatchers, are also very numerous. Pheasants, trogons, barbets, woodpeckers and finches are conspicuously absent. The islands are wintering grounds for a certain number of migrants from northern Asia and Australia, and a few waders from North America.

In mid-Pacific the volcanic Hawaiian islands form a rather peculiar district – the meeting place for birds with American affinities, such as the Hawaiian Goose or Ne-ne *Nesochen sandvicensis*, and others which are clearly Australasian such as honeyeaters. Very much impoverished in fauna because of their isolation, these islands have been the scene of the most interesting differentiations. Thus the Hawaiian honeycreepers (Drepanididae), a passerine group possibly related to the Neotropical honeycreepers (Coerebidae), have undergone a very characteristic adaptive radiation.

The oceans. Biogeographers distinguish two polar and subpolar cold zones, rich in marine organisms serving as food for birds, separated by a poorer tropical zone. The oceans can be further divided into provinces and districts according to the physical properties of their surface waters, which control the faunal composition of their characteristic communities. Despite their mobility, many seabirds are very strictly tied to one or other of these communities.

Dynamic biogeography

Birds' ranges are continually changing, if only in detail on our time-scale, since they are the result of dynamic equilibria between species and their environments.

As a result of the progressive warming of the arctic zones, many birds are expanding northwards. The climatic amelioration opens the successive zones to colonization, as each begins to meet the needs of the species in question. Twenty-seven species in Sweden (Curry-Lindahl), and eleven in Finland including the Lapwing, Blackbird and *Phylloscopus* warblers (Kakela), have extended their ranges towards the north. A score of Canadian species are reaching higher and higher latitudes. Certain migrants are becoming sedentary, such as the American Robin in many northern states of the USA. In other cases the favouring factor remains mysterious. Thus the Collared Dove *Streptopelia decaocto*, confined forty years ago to parts of south-eastern Europe, began a slow but continuous colonization of the rest of the continent shortly before 1930. Its nesting range now extends over West Germany, France, the Netherlands and England, where it arrived in 1952.

In contrast to such slow and gradual colonizations are examples of sudden irruptions, such as that of the Cattle Egret in America. A particular species seems all of a sudden to be fired with an expansive power which drives it to extend its range. It is sufficient for one among the various factors which check the growth potential of species to relax, for the stocks to increase. In that case the excess population tends to spread out towards the species' boundaries, since it is usually made up of young birds which show only a very limited attachment to their home grounds. As soon as they are capable of flight these wander in a very characteristic way, notably during the phenomenon known as *post-juvenile dispersion* which is a regular part of the life-cycle of many species.

The population itself may change, by an actual ecological mutation. This will lead to an expansion of the species' range if the mutation makes it more adaptable, or to contraction if its ecological demands become more stringent. The effects of changes in the external environment may thus be reinforced by those in the animal itself.

The better travellers birds are, the better they are fitted to increase their ranges, and the annual migrations which many of them undertake may be particularly significant in dispersal. During these voyages birds fly over extensive territories, where normally they merely pass through, but in some of which they would be able to nest. A few populations, such as the Hooded Crows of the Near East, have thus become sedentary along their migration routes. The same may be true of certain populations of wood-warblers (Parulidae) in Central America. It has been noted that migratory species have been quicker than sedentary ones to take advantage of the present warming of the northern zones.

The colonization of new territories may originate by passive transport. If a group of individuals is carried away by the wind, the majority may die, but a few may survive to colonize a new area. This founding population may die out without descendants, if a close-knit and well structured resident avian community sets up sufficient resistance to the invaders. This happens especially when the ecological niche which might be occupied by the newcomer is already firmly 'held' by a native species. In January 1937 a flock of Fieldfares *Turdus pilaris*, on migration from Scandinavia towards the British Isles, was taken unawares by a south-easterly storm. Blown to Greenland, these birds settled there thanks to the warming climate, and have established a population in equilibrium with its new environment. A still more spectacular example is provided by the Cattle Egret *Bubulcus ibis*, which until 1930 was confined to the warm regions of the Old World. About that time a few appeared in Guyana, which although attributed to human intervention were by all appearances natural colonists: a stock of these egrets could have been carried by the great winds, which blow across the

Atlantic from east to west between Western Africa and the Caribbean. It is known from the recapture of ringed birds that such unaccustomed crossings are not exceptional, especially among the Ardeidae. Several Herons ringed in France and a Little Egret ringed in Spain have been retaken in the West Indies. Since then, the Cattle Egret has set about colonizing the New World, which it now occupies as far as Canada and Argentina, establishing itself in very varied habitats. The presence of certain birds on islands which are regularly swept by winds, blowing from countries where the birds may have originated, is no doubt explicable in this way. Thus the presence of a pipit (*Anthus antarcticus*) and a duck (*Anas georgica*) in South Georgia is probably due to strong winds blowing from Patagonia, where very closely related species occur.

However, it must be stressed that colonization of a new area by a species is not dependent only on its arrival, by immigration or passive transport. It is equally indispensable that the founding stock should have been able to install itself in a favourable habitat, and to resist the effects of predation and competition from the well established native community.

Thus the expansion of species is controlled by many factors. Some species have all the potentialities needed to occupy new territories, but are stopped by geographical barriers. Other species are limited by environmental factors, and especially by the nature of the habitat. Competition with elements of native communities is an equally important factor in the barriers to expansion by birds, and is directly linked to ecological environmental differences, since the same species may be successful in one environment and not in another. Such bird-habitat complexes are without any doubt the explanation for the clear-cut boundaries observed by many birds, and for their diversity around the world.

Whenever one of the parameters of the ecological equation is changed, the equilibrium is displaced, which results biogeographically in an extension or regression by the species or the community as a whole. Such changes become much larger during geographical or climatic upsets. At such times distributions and the composition of local avifaunas may be profoundly modified, as has repeatedly happened during geological history.

Avifaunal diversity

Thus the faunistic complexes distributed round the globe are remarkably diverse, their composition varying largely in accordance with local geological history and with various geographical and ecological factors. The isolation of certain land masses for long periods, especially during the major evolution of birds, has resulted in their obvious avifaunal impoverishment. There is no question, for example, that Madagascar was ever connected to the neighbouring continent, at any rate since the Triassic, and it is inhabited by only 184 nesting species, in strong contrast to the avifaunal richness of Africa. In contrast Borneo, which has been connected as recently as the Pleistocene with the land areas of the huge Malaysian shelf, is occupied by 554 species of birds. Oceanic islands, of volcanic origin and remote from any continental mass, are poorer still. The Hawaiian islands are occupied by only twenty-five passerine species, while the Galapagos total about eighty species including seabirds. Even Australia has only 651 species, of which 531 are nesting land and freshwater birds.

As long as the area of an isolated island is sufficient and provides a variety of habitats, the degree of endemism on it is always very high. Thus the Malagasy avifauna includes 125 endemic species out of 184 and forty-six endemic genera out of 133, with no less than seventeen endemic families and subfamilies.

Geographical and ecological conditions also profoundly influence faunal richness in other ways. Within all animal groups, and birds in particular, the number of species increases as one goes from high latitudes towards the equator (apart from clearly unfavourable zones such as the upper levels of mountains and especially deserts). Thus Greenland has only fifty-six species of birds, Europe 473, eastern North America 440 and western 583. In contrast Borneo has 554 species, New Guinea 650 (of which 568 nest there, seabirds excluded), and southern Africa 875. The avifauna of Colombia includes 1,556 species, and that of Ecuador as many. These comparisons should of course take account of the areas of the regions considered, which would make the figures still more meaningful.

Certain niches, such as all those occupied by nectarivorous or frugivorous birds in the tropics, are totally absent from the temperate zones. Presumably these birds are unable to establish themselves outside the warm parts of the globe, where alone the resources they need are abundantly available throughout the year. On the other hand the populations of each species seem to vary inversely with the number of species. In tropical environments, and especially in rainforest, the number of species is high but their stocks are only moderate; whereas in arctic environments there are few species, but most of these are represented in considerable numbers.

The foregoing conclusions apply only to terrestrial habitats. As far as the seas are concerned, the situation is partly reversed. Tropical seas are poor in marine birds: their stocks are reduced and there are relatively few species, in contrast to the situation in cold seas. This difference agrees with the amounts of food available at the various latitudes, and with the complexity of marine ecosystems in cold seas.

We have seen that faunistic complexes sometimes have elements in common. A single phyletic line is distributed around the world, and has evolved into local races, species or genera as the case may be, with each representative form occupying the same ecological place in its own community. But we have also seen that the geographical isolation of the various parts of the world led their faunas to evolve independently of one another, which explains their profound faunistic divergences and the endemism of certain groups. However, in comparing certain similar habitats in different parts of the world, one cannot fail to be struck by their physiognomic resemblances. In different regions the same niche is occupied by birds of very different origins, belonging to unrelated taxonomic groups; yet through convergent evolution they show common characters which, though superficial and adaptive, are none the less real. It seems as though parallel biological communities may be formed from very different ancestral stocks, in which species which are obviously unrelated occupy corresponding niches in identical ways. Thus parallel or convergent evolution has been brought about by the environment. Avifaunas in similar environments throughout the world show the same physiognomy, thus leading many early systematists into error when they based their classifications on characters which reflected parallel adaptations. Different ecosystems may thus function in the same general way. It is as though there were a kind of set of pigeon-holes characteristic of each *biome*, each compartment of which was occupied by a particular element, often of very diverse origins in different areas, but similar in its adaptations and habits. These correspondences are sometimes only relative. Some of the pigeon-holes may remain empty even in the fauna of a huge continent, or a single species may occupy several which are shared out among as many species in an equivalent biome. None the less faunas in different parts of the world do correspond in general, although insular faunas are so impoverished that it is often difficult to trace the equivalences.

Correspondences appear when the faunas of different temperate biomes are compared. Thus in the New World the wood-warblers (Parulidae) replace our warblers (Sylviidae), certain tyrant-flycatchers (Tyranidae) our true flycatchers (Muscicapidae), American vultures (Cathartidae) the true vultures (of the Accipitridae), and American quails (Odontophorinae) our partridges (of the Phasianinae). Similarly between tropical forest biomes, New World toucans replace Old World hornbills, hummingbirds replace sunbirds, and antpittas replace pittas. Such resemblances are sometimes detailed, with striking parallels between individual species which resemble one another closely in general morphology, biology and behaviour. This is true between the American icterid meadowlarks (*Sturnella*) and African motacillid longclaws (*Macronyx*), whose ways of life and even coloration arc exceedingly alike.

Adaptive radiation may bring one family to resemble several families elsewhere, so that the New World Tyranidae have evolved in very diverse ways to occupy niches which in the Old World belong to various Sylviidae, Muscicapidae and even Turdidae. Unrelated families may show parallel adaptive radiations, as do the American Icteridae and African Ploceidae, some of which are remarkably similar because they occupy corresponding niches in equivalent biomes of the New and Old Worlds (Lack). Thus the troupials (*Icterus*) take an insectivorous diet supplemented with fruit and are monogamous solitary nesters, like certain forest weavers such as *Malimbus* (and also like the true orioles, Oriolidae). Both the American blackbirds (*Agelaius*) and the African bishops (*Euplectes*, etc.) are insectivorous and polygynous, nesting in tall grass, while some (such as *A. tricolor* and *Quelea quelea*) form nesting colonies. The oropendolas (*Zarhynchus*, *Ostinops*, etc.) build hanging nests just like the weavers (*Ploceus*). Both families have even developed brood parasites, with the cowbirds (*Molothrus*) more or less equivalent to the Parasitic Weaver *Anomalospiza imberbis*.

Bibliography

This includes only a few of the several thousand works on which this book is based, far too many to list here. Those cited are only the principal works, to which the reader may usefully turn for further reading on the various topics.

All the cited references are indicated in the text by the name of the author followed by the date of publication. Those names given without dates merely underline the parts the authors played in obtaining the results. Bibliographical indexes such as the *Zoological Record*, published by the Zoological Society of London, provide an outline to the ornithological literature, which grows more rapidly every year.

General Works

Austin, O. L. (1961) *Birds of the World.* Golden Press, New York.

Grassé, P. P. (Ed.) (1950) *Traité de Zoologie*, XV. *Oiseaux.* Masson, Paris

Lack, D. (1968) *Ecological Adaptations for Breeding in Birds*, Methuen, London

Marshall, A. J. (Ed.) (1960–1) *Biology and Comparative Physiology of Birds.* 2 vols Academic Press, New York and London

Odum, E. P. (1959) *Fundamentals of Ecology.* Saunders, Philadelphia

Stresemann, E. (1927–34) *Aves.* In: W. Kukenthal and T. Krumbach (Eds) *Handbuch der Zoologie*, Berlin and Leipzig

Thomson, A. L. (Ed.) (1964) *A New Dictionary of Birds.* Nelson, London

Van Tyne, J., and Berger, A. J. (1959) *Fundamentals of Ornithology.* J. Wiley, New York

Welty, J. C. (1962) *The Life of Birds.* Saunders, Philadelphia and London

Chapter 2

Assenmacher, I. (1958) 'La mue des Oiseaux et son déterminisme endocrinien' *Alauda*, **26**: 241–89.

Brown, R. H. J. (1963) 'The flight of birds'. *Biol. Rev.* **38**: 460–89.

George, J. C., and Berger, A. J. (1966): *Avian Myology.* Academic Press, New York and London

Greenewalt, C. H. (1960) *Hummingbirds.* Amer. Mus Nat. Hist., New York

Greenewalt, C. H. (1962) 'Dimensional relationships for flying animals'. *Smithson. Misc. Coll.*, **144 (2)** :1–46

Lissaman, P. B. S., and Shollenberger, C. A. (1970) 'Formation flight of birds'. *Science*, **168**: 1003–5

Salomonsen, F. (1968) 'The moult migration'. *Wildfowl*, **19**: 5–24

Savile, D. B. O. (1957) 'Adaptative evolution in the avian wing'. *Evolution*, **11**: 212–24.

Schmidt, H. (Ed.) (1960): *Der Flug der Tiere*. W. Kramer, Frankfurt
Storer, J. H. (1948) *The Flight of Birds*. Cranbrook Inst. of Science: Bloomfield Hills, Michigan
Wetmore, A. (1936) 'The number of contour feathers in passeriform and related birds'. *Auk*. 53: 159–69

Chapter 3

Richardson, F. (1942) 'Adaptative modifications for treetrunk foraging in birds'. *Univ. Calif. Publ. Zool.*, **46**: 317–68.
Ruggeberg, T. (1960) Zur funktionellen Anatomie der hinteren Extremiät einiger mitteleuropäischer Singvogelarten' *Zeitschr. Wiss. Zool.*, **164**: 1–106
Rutschke, E. (1960) Untersuchüngen über Wasserfestigkeit und Struktur des Gefieders von Schwimmvögeln. *Zool. Jb.* (Abt. 7), **87**: 441–506

Chapter 4

Berthold, P. (1966) 'Über Haftfarben bei Vögeln: Rostförbung durch Eisenoxid beim Bartgeier (*Gypaetus barbatus*) und bei anderen Arten. *Zool. Jb. Syst.*, **93**: 507–95
Dementiev, G. P. (1958) 'La faune désertique du Turkestan'. *Terre et Vie*, **105**: 3–44
Desselberger, H. (1930) 'Über das Lipochrom der Vogelfcdern'. *J. Orn.*, **78**: 328–76
Dorst, J. (1951) 'Recherches sur la structure des plumes des Trochilidés. *Mém. Mus. nat. Hist. nat.*, A. Zool. **1**: 125–260
Durrer, H., and Villiger, W. (1966) 'Schillerfarben der Trogoniden'. *J. Orn.*, **107**: 1–26
Fox, D. J., (1953) *Animal biochromes and structural colours*. Cambridge Univ. Press, Cambridge
Frank, F. (1939) 'Die Färbung der Vogelgefieder durch Pigment und Struktur'. *J. Orn.*, **87**: 426–523
Greenewalt, C. H., Brandt, W., and Friel, D. D. (1960 a) 'Iridescent colors of hummingbird feathers'. *J. Optical Soc. Amer.*, **50**: 1005–13
Greenewalt, C. H., Brandt, W., and Friel, D. D. (1960 b) 'The iridescent colors of hummingbird feathers'. *Proc. Amer. Phil. Soc.*, **104**: 249–53
Schmidt, W. J. (1952) 'Wie entstehen die Schillerfarben der Federn?' *Naturwiss.*, **39**: 313–18.
Schmidt, W. J., and Ruska, H. (1962) 'Tyndallblau-Struktur von Federn im Elektronenmikroskop'. *Zeits. Zellforsch.*, **56**: 693–708.
Schmidt, W. J., and Ruska, H. (1962) 'Über das schillernde Federmelanin bei *Heliangelus* und *Lophophorus*'. *Zeitschr. Zellforsch.*, **57**: 1–36
Volker, O. (1934) 'Die Abhängigkeit der Lipochrombildung bei Vögel von pflanzlichen Karotinoiden'. *J. Orn.*, **82**: 439–50
Volker, O. (1938) 'Porphyrin in Vogelfedern'. *J. Orn*, **86**: 436–56

Chapter 5

Attwell, R. I. G. (1966) 'Oxpeckers, and their association with mammals in Zambia'. *Puku*, **4**: 17–48

Beecher, W. J. (1953) 'Feeding adaptations and systematics in the avian order Piciformes'. *J. Wash. Acad. Sci.*, **43**: 293–9

Beecher, W. J. (1962) 'The bio-mechanics of the bird skull'. *Bull. Chicago Acad. Sci.*, **11**: 10–33

Bock, W. J. (1964) 'Kinetics of the avian skull'. *J. Morph.*, **114**: 1–42

Bock, W. J. (1966) 'An approach to the functional analysis of bill shape'. *Auk*, **83**: 10–51

Bock, W. J., and Miller, W. D. (1959) 'The scansorial foot of the woodpeckers with comments on the evolution of perching and climbing feet in birds'. *Amer. Mus. Nov.*, **1931**: 1–44

Dorst, J. (1947) 'Le rôle disséminateur des Oiseaux dans la vie des plantes' *Terre et Vie*, **94**: 106–119

Hespenheide, H. A. (1966) 'The selection of seed size by finches'. *Wilson Bull.*, **78**: 191–7

Huxley, J. (1960) 'The Openbill's open bill: a teleonomic enquiry'. *Zool. Jb. Syst.*, **88**: 10–30

Leandri, J. (1950) 'Points de vue sur le problème de l'ornithophilie'. *Terre et Vie* **97**: 86–100

Rand, A. L. (1954) 'Social feeding behaviour of birds'. *Fieldiana, Zool.*, 36, **1**: 1–71

Spring, L. W (1965) 'Climbing and pecking adaptations in some North American woodpeckers' *Condor*, **67**: 457–88

Weymouth, R. D., Lasiewski, R. C., and Berger, A. J. (1964) 'The tongue apparatus in hummingbirds'. *Acta Anat.*, **58**: 252–70

Ziswiler, V. (1965) 'Zur Kenntnis des Samenöffnens und der Struktur des hörnernen Gaumens bei körnerfressenden Oscines'. *J. Orn.*, **106**: 1–48

Ziswiler, V. (1967) 'Vergleichend morphologische Untersuchungen am Verdauungstrakt körnerfressender Singvögel zur Abklärung ihrer systematischen Stellung'. *Zool. Jb. Syst.*, **94**: 427–520

Chapter 6

Baldwin, S. P., and Kendeigh, S. C. (1932) 'Physiology of the Temperature of Birds'. *Sci. Publ. Cleveland Mus. Nat. Hist.*, 3

Brenner, F. J. (1965) 'Metabolism and survival time of grouped Starlings at various temperatures'. *Wilson Bull.*, 77: 388–95

Hartman, F. A. (1955) 'Heart weight in birds'. *Condor*, **57**: 221–38

Hartman, F. A., and Lessler, M. A. (1963) 'Erythrocyte measurements in birds'. *Auk*, **80**: 467–73

Howell, T. R., and Bartholomew, G. A. (1962) 'Temperature regulation in the Red-tailed Tropic Bird and the Red-footed Booby'. *Condor*, **64**: 6–18

Irving, L., and Krog, J. (1955) 'Skin temperature in the arctic as a regulator of heat'. *J. appl. Physiol.*, **7**: 354–63

Johnston, D. (1963) 'Heart weights of some Alaskan birds'. *Wilson Bull.*, **75**: 435–46

Kahl, M. P. (1963) 'Thermoregulation in the Wood Stork, with special reference to the role of the legs'. *Physiol. Zool.*, **36**: 141–51

King J. R., and Farner, D. S. (1961) 'Energy Metabolism, Thermoregulation and

Body Temperature'. In: *Biology and Comparative Physiology of Birds*, 215–88. Academic Press, New York and London.
Koskimies, J. (1950) 'The life of the swift, *Micropus apus* (L.), in relation to the weather'. *Ann. Acad. Scient. Fenn.*, A, 4, **15**: 1–151
Macdonald, J. D. (1960) 'Secondary external nares of the gannet'. *Proc. Zool. Soc. London*, **135**: 357–63
Pearson, O. P. (1950) 'The metabolism of hummingbirds'. *Condor*, **52**: 145–52
Peiponen, V. A. (1966) 'The diurnal heterothermy of the nightjar (*Caprimulgus europaeus* L.)'. *Ann. Acad. Scient. Fenn.*, A, 4, **101**: 1–35
Sturkie, P. D. (1954) *Avian Physiology*. Comstock Publ. Associates, Ithaca

Chapter 7

Bang, B. G. (1966) 'The olfactory apparatus of tubenosed birds'. *Acta Anat.*, **65**: 391–415
Donner, (1951) 'The visual acuity of some passerine birds'. *Acta Zool. Fenn.*, **66**: 1–40
Neuhaus, W. (1963) 'On the olfactory sense of birds'. In: *Olfaction and Taste*, 111–23. London (Pergamon Press)
Schwartzkopf, J. (1955) 'Schallsinnesorgane, ihre Funktion und biologische Bedeutung bei Vögeln'. *Acta XI Congr. Int. Orn.* (Basel), 189–208
Schwartzkopf, J. (1963) 'Morphological and physiological properties of the auditory system in birds'. *Proc. XIII Int. Orn. Congr.* (Ithaca), 1059–68
Stager, K. E. (1964) 'The role of olfaction in food location in the Turkey Vulture (*Cathartes aura*)'. *Contributions in Science*, **81**: 1–63 (Los Angeles County Museum)
Walls, G. L. (1942) *The Vertebrate Eye and its Adaptative Radiation*. Cranbrook Institute of Science: Bloomfield Hills, Michigan

Chapter 8

Armstrong, E. A. (1963) *A Study of Bird Song*. Oxford Univ. Press, London
Brémond, J.-C. (1968): 'Recherches sur la sémantique et les éléments vecteurs d'information dans les signaux acoustiques du Rouge-gorge (*Erithacus rubecula* L.). *Terre et Vie*, **114**: 109–220.
Busnel, R. G. (Ed.) (1963) *Acoustic Behaviour of Animals*. Elsevier, Amsterdam
Griffin, D. R. (1958) *Listening in the Dark*. Yale Univ. Press, New Haven
Lanyon, W. E., and Tavolga, W. N. (Eds) (1960) *Animal sounds and communication*. Amer. Inst. Biol Sciences, Washington
Thorpe, W. H. (1958) 'The learning of song patterns by birds, with especial reference to the song of the Chaffinch *Fringilla coelebs*'. *Ibis*, **100**: 535–70
Thorpe, W. H. (1961) *Bird Song*. Univ. Press, Cambridge

Chapter 9

Ashmole, N. P. (1963) 'Biology of the Wideawake or Sooty Tern *Sterna fuscata* on Ascension Island.' *Ibis*, **103 b**: 297–364
Chapin, J. P., and Wing, L. W. (1959) 'The Wide-awake calendar', 1953 and 1958. *Auk*, **76**: 153–58
Coulson, J. C., and White, E. (1960) 'The effect of age and density of breeding birds on the time of breeding of the Kittiwake *Rissa tridactyla*.' *Ibis*, **102**: 71–92

Farner, D. S. (1964) 'The photoperiodic control of reproductive cycles in birds.' *Amer. Scientist*, **52**: 137–56

Farner, D. S. (1967) 'The control of avian reproductive cycles.' *Proc. XIV Int. Orn. Congr:* (Oxford), 107–33

Hamilton, T. H., and R. H., Barth Jr (1962) 'The biological significance of season change in male plumage appearance in some new world migratory bird species.' *Amer. Nat.*, **46**: 129–44

Lehrman, D. S. (1965) 'Interaction between internal and external environments in the regulation of the reproductive cycle of the Ring Dove'. In: *Sex and Behaviour*, F. A. Beach (Ed), 355–80. Wiley, New York

Lofts, B., and Murton, R. K. (1966) 'The role of weather, food and biological factors in timing the sexual cycle of Woodpigeons.' *Br. Birds*, **59**: 261–80

Lofts, B., and Westwood, N. J. (1966) 'Gonadal cycles and the evolution of breeding season in British Columbidae.' *J. Zool. Lond.*, **150**: 249–72

Miller, A. H. (1959) 'Reproductive cycles in an equatorial sparrow.' *Proc. Nat. Acad. Sci.*, **45**: 105–10

Rowan, W. (1939) 'Light and seasonal reproduction in animals.' *Biol. Rev.*, **13**: 374–402

Chapter 10

Carpenter, C. R. (1958), 'Territoriality. A review of concepts and problems.' In: *Behaviour and Evolution*, Roe and Simpson (Eds), 224–50. Yale Univ. Press, New Haven.

Hatch, J. J. (1966) 'Collective territories in Galapagos mocking-birds, with notes on other behaviour.' *Wilson Bull.*, **78**: 198–207

Hinde, R. A. (1956) 'The biological significance of the territories of birds.' *Ibis*, **98**: 340–69

Howard, E. H. (1920) *Territory in Bird Life*. New edition, 1964. Collins, London

Kalela, O. (1958) 'Über ausserbrutzeitliches Territorialverhalten bei Vögeln.' *Ann. Acad. Sci. Fenn*

Kilham, L. (1958) 'Territorial behaviour of wintering Red-headed Woodpeckers.' *Wilson Bull.*, **70**: 347–58.

Lack, D. (1953): *The Life of the Robin*. Penguin Books, London

Nice, M. M. (1941) 'The Role of Territory in Bird Life.' *Amer. Midl. Nat.*, **26**: 441–87

Tinbergen, N. (1957) 'The functions of territory'. *Bird Study*, **4**: 14–27

Chapter 11

Armstrong, E. A. (1947) *Bird Display and Behaviour*. Lindsay Drummond, London

Gilliard, E. T. (1959) 'The Courtship Behaviour of Sandford's Bowerbird (*Archboldia sanfordi*).' *Amer. Mus. Nov.*, **1935**: 1–18

Lorenz, K. (1951–3) 'Comparative Studies on the Behaviour of the *Anatidae*'. *Avic. Mag.*, **57**: 157–82; **58**: 8–17, 61–72, 86–94, 172–84; **59**: 24–34, 80–91

Marshall, A. J. (1954) *Bower-birds*. Oxford Univ. Press. London.

Snow, D. W. (1963) 'The Evolution of Manakin Displays.' *Proc. XIII Int. Orn. Congr.* (Ithaca). 553–61

Chapter 12

Collias, N. E. (1964) 'The evolution of nests and nestbuilding in birds.' *Amer. Zoologist*, **4**: 175–90

Collias, N. E., and Collias, E. C. (1964) 'The development of nest-building behaviour in a Weaver-bird.' *Auk*, **81**: 42–52

Haartman, L. von (1957) 'Adaptation in hole-nesting birds.' *Evolution*, **11**: 339–47

Hindwood, V. A. (1959) 'The nesting of birds in the nests of social insects.' *Emu*, **59**: 1–36

Chapter 13

Bailey, R. E. (1952) 'The incubation patch of passerine birds.' *Condor*, **54**: 121–36

Fischer, H. I. (1966) 'Hatching and the hatching muscle in some North American ducks.' *Trans. III. St Acad. Sci.*, **59**: 305–25

Frith, H. J. (1956) 'Breeding habits in the family Megapodiidae.' *Ibis*, 98: 620–40

Kendeigh, S. C. (1952) 'Parental Care and its Evolution in Birds.' *Illinois Biol. Monographs.*, **22**: 1–356

Lack, D. (1954) *The Natural Regulation of Animal Numbers*. Oxford Univ. Press, London

Lehrman, D. S. (1961) 'Hormonal regulation of parental behaviour in birds and infrahuman mammals.' In: *Sex and internal secretion*, W. C. Young (Ed.), 1268–382

Nice, M. M. (1962) 'Development of Behaviour in Precocial Birds.' *Trans. Linnaean Soc. New York*, 8

Romanoff, A. L., and Romanoff, A. J. (1949) *The Avian Egg*. Y. J. Wiley, New York

Skutch, A. F. (1957) 'The incubation patterns of birds.' *Ibis*, **99**: 69–93

Skutch, A. F. (1961) 'Helpers among birds.' *Condor*, **63**: 198–226

Sutter, E. (1951) 'Growth and differentiation of the brain in nidifugous and nidicolous birds.' *Proc. Xth Int. Orn. Congr.* (Uppsala), 636–43

Tinbergen, N. (1953) *The Herring Gull's World*. Collins, London

Chapter 14

Friedmann, H. (1948) *The Parasitic Cuckoos of Africa*. Wash. Acad. Sci., Washington

Friedmann, H. (1955) 'The Honey-guides.' *U.S. Nat. Mus. Bull.*, 208

Friedmann, H. (1960) 'The Parasitic Weaver-birds.' *U.S. Nat. Mus. Bull.*, **223**

Friedmann, H. (1963) 'Host Relations of the Parasitic Cowbirds.' *U.S. Nat. Mus. Bull.*, 233

Friedmann, H. (1968) 'The Evolutionary History of the Avian Genus *Chrysococcyx*.' *U.S. Nat. Mus. Bull.*, 265

Neunzig, R. (1929) 'Zum Brutparasitismus der Viduinen'. *J. Orn.*, **77**: 1–21

Nicolai, J. (1964) 'Der Brutparasitismus der Viduinae als ethologisches Problem: Prägungsphänomene als Faktoren der Rassen und Artbildung'. *Z. Tierpsychol.* **21**: 129–204

Nicolai, J. (1967) 'Rassen- une Artbildung in der Viduinen-Gattung *Hypochera*'. *J. Orn.*, **108**: 309–19

Perrins, C. M. (1967) 'The short apparent incubation period of the Cuckoo.' *Br. Birds*, **60**: 51–2.

Rothschild, M., and Clay, T. (1952): *Fleas, flukes and cuckoos*. Collins, London
Selander, R. K., and Yang, S. Y. (1966): 'Behavioural responses of Brown-headed Cowbirds to nests and eggs'. *Auk*, **83**: 207–32

Chapter 15

Austin, O. L., and Austin, O. L. Jr (1956) 'Some demographic aspects of the Cape Cod population of Common Terns (*Sterna hirundo*)'. *Bird Banding.*, **27**: 55–66
Einarsen, A. S. (1942) 'Specific results from Ringnecked Pheasant studies in the Pacific Northwest'. *Trans. North Amer. Wildlife Conf.*, **7**: 130–45
Emlen, J. T. (1940) 'Sex and age ratios in survival of the California Quail'. *Journ. Wildlife Management*, **4**: 92–9
Farner, D. S. (1955) 'Bird Banding in the study of population dynamics'. In: *Recent Studies in Avian Biology*. A. Wolfson (Ed.), Univ. Illinois Press: Urbana, 397–449
Flower, S. S. (1938) 'Further notes on the duration of life in animals'. IV Birds. *Proc. Zool. Soc. London*, **108 (A)**: 195–235
Hickey, J. J. (1952) 'Survival studies of Banded Birds'. *U.S. Dept. Int. Fish and Wildlife Ser. Spec. Sci. Rep Wildlife*, 15
Johnston, R F. (1956) 'Population structure in Salt marsh Song Sparrow'. *Condor* **58**: 24–44, 254–72
Lack, D. (1954) *The Natural Regulation of Animal Numbers*. Clarendon Press, Oxford
Lack, D. (1966) *Population Studies of Birds*. Clarendon Press, Oxford
Morel, M. Y. (1964) 'Natalité et mortalité dans une population naturelle d'un Passereau tropical, le *Lagonosticta senegala*'. *Terre et Vie*, **111**: 436–51
Nice, M. M. (1957) 'Nesting success in altricial birds'. *Auk*, **74**: 305–21
Paynter, R. A. (1966) 'A new attempt to construct life tables for Kent Island Herring Gulls'. *Bull. Mus. comp. Zool.*, **133**: 491–521
Ribaut, J. P. (1964) 'Dynamique d'une population de Merles noirs, *Turdus merula* L.' *Rev. Suisse Zool.*, **71**: 816–901
Rydzewski, W. (1962) 'Longevity of ringed birds.' *Ring*, **3 (33)**: 147–52
Snow, D. (1962) 'A field study of the Black and White Manakin *Manacus manacus* in Trinidad.' *Zoologica*, **47**: 65–104
Wynne-Edwards, V. C. (1962) *Animal Dispersion in relation to Social Behaviour*. Oliver and Boyd, Edinburgh and London

Chapter 16

Heilmann, G. (1927) *The Origin of Birds*. Appleton and Co., New York
Hinde, R. A. (1959) 'Behaviour and speciation in birds and lower vertebrates'. *Biol. Rev.*, **34**: 85–128
Howard, H. (1950) 'Fossil evidence of avian-evolution'. *Ibis*, **92**: 1–21
Howard, H. (1955) 'Fossil birds with especial reference to the birds of rancho La Brea'. *Los Angeles County Museum. Science Ser.*, 17
Martin, P. E., and H. E. Wright Jr (Eds) (1947) *Pleistocene extinctions. The Search for a Cause*. Yale Univ. Press, New Haven and London
Mayr, E. (1942) *Systematics and the Origin of species*. Columbia Univ. Press, New York

Mayr, E. (1963) *Animal Species and Evolution*. Harvard Univ. Press, Cambridge

Piveteau, J. (1955) *Oiseaux*. In: *Traité de Paléontologie*, J. Piveteau (Ed.), 15. Masson, Paris

Selander, R. K. (1965) 'Avian speciation in the Quaternary'. In: H. E. Wright Jr and D. C. Fry (Eds.) *The Quaternary of the United States*, 527–42

Chapter 17

Johnsgard, P. A. (1965) *Handbook of Waterfowl Behaviour*. Constable and Co., London

Mayr, E. (1959) 'Trends in Avian Systematics.' *Ibis*, **101**: 293–302

Mayr, E., and Amadon, D. (1951) 'A Classification of Recent Birds.' *Amer. Mus. Nov.*, 1496

Mayr, E., Linsley, E. G., and Usinger, R. L. (1953) *Methods and principles of Systematic zoology*. McGraw-Hill, New York

Sibley, C. G. (1960). 'The electrophoretic patterns of avian egg-white proteins as taxonomic characters'. *Ibis*, **102**: 215–84

Sibley, C. G. (1965) 'Molecular systematics: new techniques applied to old problems'. *Oiseau R. F. O.*, **35**, suppl.: 112–24

Wetmore, A. (1960) 'A Classification for the Birds of the World'. *Smithsonian Misc. Coll.*, 139, No. 11

Wickler, W. (1961) Über die Stammesgeschichte und den taxonomischen Wert einiger Verhaltensweisen der Vögel'. *Z. Tierpsych.*, **18**: 320–42

Chapter 18

Darlington, P J Jr (1957) *Zoogeography: The geographical distribution of animals* J. Wiley, New York

Mayr, E. (1946) 'History of North American bird fauna'. *Wilson Bull.*, **58**: 3–41

Mayr, E. (1965) 'What is a fauna ?' *Zool. Jb. Syst.*, **92**: 473–86

Moreau, R. E. (1966) *The Bird Faunas of Africa and its Islands*. Academic Press, London